Smart Innovation, Systems and Technologies

Volume 287

The Smart Innovation, Systems and Technologies book series encompasses the topics of knowledge, intelligence, innovation and sustainability. The aim of the series is to make available a platform for the publication of books on all aspects of single and multi-disciplinary research on these themes in order to make the latest results available in a readily-accessible form. Volumes on interdisciplinary research combining two or more of these areas is particularly sought.

The series covers systems and paradigms that employ knowledge and intelligence in a broad sense. Its scope is systems having embedded knowledge and intelligence, which may be applied to the solution of world problems in industry, the environment and the community. It also focusses on the knowledge-transfer methodologies and innovation strategies employed to make this happen effectively. The combination of intelligent systems tools and a broad range of applications introduces a need for a synergy of disciplines from science, technology, business and the humanities. The series will include conference proceedings, edited collections, monographs, handbooks, reference books, and other relevant types of book in areas of science and technology where smart systems and technologies can offer innovative solutions.

High quality content is an essential feature for all book proposals accepted for the series. It is expected that editors of all accepted volumes will ensure that contributions are subjected to an appropriate level of reviewing process and adhere to KES quality principles.

Indexed by SCOPUS, EI Compendex, INSPEC, WTI Frankfurt eG, zbMATH, Japanese Science and Technology Agency (JST), SCImago, DBLP.

All books published in the series are submitted for consideration in Web of Science.

More information about this series at https://link.springer.com/bookseries/8767

Elena I. Inshakova · Agnessa O. Inshakova
Editors

New Technology for Inclusive and Sustainable Growth

Perception, Challenges and Opportunities

Springer

Editors
Elena I. Inshakova
Institute of Economy and Management
Volgograd State University
Volgograd, Russia

Agnessa O. Inshakova
Institute of Law
Volgograd State University
Volgograd, Russia

ISSN 2190-3018 ISSN 2190-3026 (electronic)
Smart Innovation, Systems and Technologies
ISBN 978-981-16-9803-3 ISBN 978-981-16-9804-0 (eBook)
https://doi.org/10.1007/978-981-16-9804-0

This Springer imprint is published by the registered company Springer Nature Singapore Pte Ltd.
The registered company address is: 152 Beach Road, #21-01/04 Gateway East, Singapore 189721, Singapore

This book is dedicated to Oleg V. Inshakov (1952–2018) with love and gratitude.

Preface

The book is dedicated to Doctor of Economics, Prof. Oleg Vasilievich Inshakov (1952–2018), on the 70th anniversary of his birth. At Volgograd State University (Russia), Prof. Inshakov held the positions of Rector (1995–2014), President (2014–2016), and Director of Research Institute of Regional Social and Economic Development (2016–2017).

Oleg Vasilievich Inshakov is a widely renowned economist in the Russian Federation, best known as the founder of the production factors evolutionary theory and economic genetics theory. Professor Inshakov laid the foundation for theoretical studies in the field of nanoeconomics. He contributed to evolutionary economics by proposing his model of an inherent 'development nucleus' in economic systems and revealed the dynamics of global socio-economic modes based on the meta-production function. Oleg Inshakov is recognized for building a multi-criteria classification of institutional mechanisms, shaping a general model of the cyclic functioning of the economic mechanism, and developing his theory of global economic system based on the evolution of its hierarchical structure that extends the concepts of the object and subject of the economic theory. His scientific output included more than 500 publications in Russian and foreign peer-reviewed journals and books. In the last decade of his life, Prof. Inshakov devoted himself to examining socio-economic problems arising from the sixth technological mode formation, investigating macro-generations emerging from its base, and finding a way to overcome the crisis in domestic economic theory and practice based on the evolutionary theory.

Professor Inshakov is the founder of a powerful school of institutional and evolutionary economics, strategic management, and planning of regional economic development. He supervised 26 doctoral and 51 candidate dissertations that examined various themes related to the functions, structure, and evolution in economic systems of different types, levels, and scales.

Oleg Inshakov was awarded state and departmental awards and given honorary titles of the Russian Federation in recognition of his notable contribution to science and higher education in the Russian Federation and his promotion of international relations in education. He earned accolades from Russian and foreign research organizations and higher educational institutions.

The book bears witness to the fact that theoretical grounds and methodological tools developed by Prof. Inshakov provide a fruitful basis for the multi-dimensional interdisciplinary research that draws on concepts from various branches of social and natural sciences within the framework of the proposed themes. Drawing on the systematic evolutionary approach, the contributions to the book suggest guidelines to ensure inclusive and sustainable growth for national economies and regions in the context of the controversial impacts of the emerging Industry 4.0, as well as new challenges and threats, including the COVID-19 pandemic.

We wish the findings and insights gained from the theoretical and empirical studies will contribute to the development of conceptual foundations and practical recommendations, which in turn will refine economic, environmental, and social mechanisms and policies that regulate the use of Industry 4.0 technologies to achieve inclusive and sustainable regional and country growth development.

Volgograd, Russia

Elena I. Inshakova
Agnessa O. Inshakova

Introduction

The book focuses on the effects and consequences of the emerging Industry 4.0, which offers unprecedented opportunities for the acceleration of economic growth and enables its inclusive characteristics. Given this fact, the world community shares a common interest in the rapid development and dissemination of Industry 4.0 technology geared toward a fairer distribution of growth across society and creating opportunities for the advancement in the economic, technological, social, and environmental spheres.

The other side of the coin is that new technology generates challenges and bears potential risks, thus making the case that national economies can take different growth paths. As a result, divergence processes arising from digitalization in the world economy contravene the UN Sustainable Development Goals (SDGs). To tackle this challenge, regulatory mechanisms capable of identifying, overcoming, and preventing these risks in a timely fashion are required to enhance the competitiveness of the national economic systems in pursuit of inclusive and sustainable growth.

With the view to investigate the problems related to the provision of inclusive growth under emerging Industry 4.0 issues, the book adopts a far broader perspective, which goes beyond the constraints of industry-specific and territory-related approaches, conceptual and methodological framework, and research techniques restricted to a particular branch of science.

The vast majority of studies predominantly are bound by common and related themes that focus on problems and sources of achieving inclusive growth, including new technological developments, in different sectors of the economy or specific regions or countries. This book, therefore, contributes to redressing the current imbalance in the field by offering a multidisciplinary investigation into the economic, technological, environmental, and social impacts of Industry 4.0 technology that ensures inclusive and sustainable growth development of regions and countries. Along with identifying new opportunities that new technology provides for inclusive growth, the book aims to propose theoretical substantiation and develop economic, institutional, organizational, and information mechanisms that aid to reduce and eliminate the potential economic, social, and environmental risks. Navigation of these complex demands is becoming a pressing scientific and practical task.

The five parts of the book comprise twenty-seven chapters, addressing the following major themes. Part I discusses theoretical assumptions underpinning issues of inclusive growth achievement in the age of new technology within the framework of the ecosystem approach, in particular, in the context of the COVID-19 pandemic. Part II proposes original statistical and econometric approaches to assess the current state of the economy, based on the values of the key indicators of inclusive and sustainable development of the environmental, production, and social spheres of the national economic system in the context of digitalization. Part III reveals the contribution of key enabling technologies such as artificial intelligence, neural networks, and nanotechnology to the inclusive development of modern society concerning the provision of business continuity and health protection in the context of the COVID-19 pandemic. Part IV gives special attention to financial and investment technologies boosting inclusive growth of the economy and human capital quality. Part V shows that an inclusive economy has many dimensions, linguistic and pedagogical dimensions, in particular. It discusses cognition technology capable of providing inclusive human development in the digitalizing economy.

The book responds to challenges facing new Industry 4.0 technology application in the Russian Federation by focusing on (i) economic and institutional changes relevant to the appropriate sectoral structural shifts; (ii) modeling and forecasting the impact of new technology on the growth in urban and rural areas; (iii) investment policies contributing to the sustained development of economy and society of the regions in the digital age; (iv) state regulation of technology-based transition to a resource-efficient and green, carbon-free economy; (v) measures and efforts to tackle global climate change and eliminate its adverse impact on the environment and society; (vi) reproduction of human capital.

The methodological basis underpinning the analysis of the opportunities and challenges of Industry 4.0. is the systemic evolutionary approach. It employs a combination of methods such as structural and functional, temporal and spatial, comparative and documentary, statistical and econometric analysis, accompanied by theoretical modeling, and, in some cases, computer simulation of economic systems.

Thus, a broad multidisciplinary approach integrating research capabilities of economic and administrative sciences, artificial intelligence and computer sciences, pedagogy and linguistics, latest findings in the above-mentioned scientific areas, as well as empirical evidence and pilot innovative research projects conducted by the contributors allowed them to draw conclusions and develop recommendations for achieving inclusive growth in various spheres such as industrial and agricultural production, innovation and investment activities, management and environment protection, health care, and education associated with the use of new technology.

Drawing on the methodological approach described above, the contributors performed a comprehensive study of the development of the economic system considering the different structural levels of the global economic system in the context of the impacts of the emerging Industry 4.0, digital transformation of the economy, and society, and adverse effects of the COVID-19 pandemic.

With the focus on the proposed themes, the study reports the following results. In the context of challenges for implementing a multi-faceted and multi-level systemic transformation of the economy in Russia, the contributors conclude that the embraced on all the levels of governance project-oriented approach, which is often responsible for the heterogeneity of the economic system, needs to be replaced by an ecosystem approach. The development of ecosystems is bound to strengthen the interdependence between the various sectors of the economy and enhance the degree of continuity and homogeneity of the economic landscape of the country.

The study substantiates the expediency of designing the economic doctrine of Russia, a scientifically grounded and practically oriented program document that will provide a solid foundation for comprehensive long-term inclusive growth. The development of ecosystems, independent socio-economic entities retaining relative stability over time that carry out efficient economic activities, is considered the focus of economic governance.

In addition, dynamics of the socio-economic growth and environment in the urban and rural ecosystems along with the economic and institutional changes resulting from the sectoral structural shifts determined by the Fourth Industrial Revolution are to receive the focus of attention of the government development institutions, researchers, businesses, and society.

Within the framework of the ecosystem approach to achieving inclusive growth, the study contributes to identifying the factors that generate different conditions for inclusive economic growth in urban and rural ecosystems. The concept of inclusive growth was analyzed with a focus on the technological integration of urban and rural areas. The contributors conducted a comparative analysis of the innovation activity in the urban and rural regions of Russia, and its impact on economic development, along with the comparison of dynamics of innovation activity and changes in the indicators of living standards in the regions and territories under consideration. The results allowed them to define the principles for building a unified ecosystem, integrating large urban agglomerations, middle-sized cities, small cities, and rural settlements.

The contributors argue the existence of an inextricable link between the provision of inclusive growth of regions and the resilience of regional economies to external shocks. The asymmetric impact of the COVID-19 pandemic on the development of some territories necessitated the identification of factors that are distinctive for regional adaptation capabilities to the adverse effect of the COVID-19 pandemic. It is concluded that the assessment of the adaptation potential of the regions to the coronavirus crisis will predict the regional development path as a result of the potential realization.

Focusing on temporal, spatial, and comparative analysis, the contributors study the level of entrepreneurship development as a determinant of economic growth in the regions of the Russian Federation.

The book suggests digital transformation directions in agricultural production management in the turbulent business environment, including the effects of the COVID-19 pandemic by disclosing the industries that have received a powerful impetus for further development and determining the problematic areas by identifying challenges and risks for the business structures' activities. The contributors demonstrate that Industry 4.0 technologies create new opportunities for a digital management system in agricultural production based on the virtualization of contractual relationships and transactions undertaken using smart contracts, blockchain technology, etc.

Re-thinking the impact of external factors on inclusive and sustainable economic growth in the regions of the world countries, the book discusses the problem of cross-border cooperation (CBC). Based on the experience of the European Union (EU) and the Eurasian Economic Union countries (EAEU), and the One Belt One Road (OBOR) initiative, the implication is that CBC is essential for creating an environment for entrepreneurship and investment, infrastructure construction planning, and implementation. In this regard, the study considers the identification of success factors in the management of cross-border projects critical for achieving sustainable development due to consistent improvement in the living standards in the border regions.

As emerging Industry 4.0 has a profound impact on the transformation of the global economic system, the contributors underline the need for improved methods for assessing and forecasting economic and social processes to provide the timely correction of the mechanisms for economic and social policy implementation. Thus, there is an increasing need for the efficient provision of the relevant data on the current state of the economy and society for effective management and decision-making.

The contributors prove that designing and developing productive approaches and instrumental software tools to preserve reliability and business continuity support have become vital in Industry 4.0 progress. Current experience in Industry 4.0 solutions design and exploitation demonstrates that to carry out Big Data analysis in a limited time (computer Big Data analysis and decision-making support) requires operating mathematical models and algorithms based on the application of artificial intelligence (AI). The study discusses the AI capabilities to satisfy the urgent need in Industry 4.0—the personalization of products, services, and decisions.

The book demonstrates functional and application potential for new Industry 4.0 technologies and systems: AI technologies—on the examples from high-technology evidence-based medicine and management audit procedures, and nanotechnology—on the cases from health care, energy sector, road infrastructure construction, etc., as significant factors in ensuring inclusive growth in Russia. The book also reveals large-scale opportunities of applying new digital technologies as drivers of inclusive economic and social development for enhancing business and social communication in the world economy during the COVID-19 pandemic lockdowns, global and large national data centers, and data center networks as the example.

Inclusive and sustainable development, the contributors argue, is mainly associated with the preservation of the biosphere and natural capital, together with the technological and social spheres. The role of new Industry 4.0 technology application for greening the economy warrants a more careful study. The book reflects on the experience of the largest post-Soviet countries in ensuring transition to a resource-efficient and green economy. The contributors identify challenges arising from the inclusion of Kazakhstan, Russia, and Ukraine on the global climate agenda to support managerial decisions for ensuring the sustainable development of these largest post-Soviet countries.

The successful identification of institutional drivers and removal of barriers to inclusive development of the financial services market in modern Russia will contribute to the development of financial self-activity and inclusive involvement of people in investment processes. The successful solution to this problem is likely to reduce the financial burden on the governments, provide passive income self-sufficiency for the population, and increase the level of prosperity.

Inclusive and sustainable development has multiple dimensions, of which the human capital is an indispensable part. Within the framework of studying the problems of achieving growth based on the adoption of new Industry 4.0 technologies, evaluation of the effectiveness of investment in human capital is one of the crucial issues. The book focuses on the problem of ineffective investment in the main components of the human capital, such as children, adolescents, and youth in the post-Soviet countries. The proposed method to assess investment effectiveness in human capital in the Russian Federation is entitled to improve the mechanism of reproducing human capital on the regional scale or the country level.

The book discusses the theory and practice of human intellectual abilities development as a vital factor in the inclusive growth of the quality of human capital in the context of Industry 4.0 and the pandemic lockdowns, based on the results of theoretical and empirical research, as well as linguistic projects implemented in higher education.

In this respect, multilingualism is viewed as one of the ways to ensure the inclusive and sustainable development of multi-ethnic and multi-cultural societies. It is shown that digital transformation is beneficial for the maintenance of language diversity and preservation of ethnic identity as it provides several technological opportunities for social and economic inclusion for small-numbered and indigenous peoples. The study discusses the need to implement specific enhanced language policies ensured by state institutions and public organizations.

To provide an inclusive educational environment for non-native speaker learners in universities of host countries, the contributors propose a universal methodology for organizing foreign language teaching based on modular training using mobile technologies of Industry 4.0. The study showed that the proposed training module enhances learning satisfaction that is crucial in non-native learners' adaptation to a foreign-language learning environment.

The analytical studies relied on the instrumental basis of the statistical and econometric analysis, along with theoretical modeling and computer simulation of economic systems, which enabled drawing conclusions and developing recommendations.

Statistical analysis underpinned, for instance, a comparative analytical study of the best practical experience of the largest post-Soviet countries, Kazakhstan, Russia, and Ukraine, in ensuring their participation in the global climate agenda. The contributors analyzed the aggregated indicators of the decarbonization levels in Kazakhstan, Russia, and Ukraine that characterize the relevance of the current state of their economies to the UN's SDGs. These revealed a significant gap between even the best values of the considered countries' indicators reviewed and the corresponding world values, and entailed implications for the state regulation in the energy sector development, which will enhance the post-Soviet countries' transition to sustainable and inclusive growth.

To forecast regional (territorial) economic growth associated with technological developments in the context of Industry 4.0, the contributors analyzed the statistical outputs from national statistical systems of France, Germany, Italy, the United Kingdom, and the Russian Federation. Based on a correlation and regression analysis, the study assessed the impact of innovative technologies on the formation of the domestic economic growth points to substantiate the priorities of state support for industries, technologies, and territories that ensure the transition to an inclusive growth scenario and meet national interests.

Using econometric analysis tools, the contributors managed to identify the factors determining the digital transformation of economic systems, which can accelerate their adaptation to the impacts arising from the emergence of Industry 4.0. The findings can enable national programs for the digital transformation of the economy to develop according to the most promising scenarios, taking into account the divergent development conditions of the world countries.

The econometric analysis assessed the impact of factors that arise from digital technologies implementation in economic practice and social interactions on inclusive economic development. The constructed econometric model allowed ranking the indicators of the use of digital technologies according to the degree of their influence on the values of inclusive economy development indicators.

The statistical and econometric analysis also yielded sound results to solve the problem of developing methods for evaluating the effectiveness of investments in human capital. The proposed statistical assessment of return on investment correlates costs in the healthcare system with the dynamics of morbidity and several other indicators in the studied area, based on the inter-generational analysis over long periods to ensure that the results are statistically significant.

The research employed theoretical modeling and computer simulation of economic systems as efficient tools to achieve the goals. Some of them are listed as follows:

- the ecosystem model represents a tetrad that includes organizational, infrastructural, communication, logistics, and innovation components and serves as an effective tool for studying the behavior and structure of ecosystems, determining the most efficient ways of distribution and consumption of the main types of ecosystem resources;
- a model of an integrated urban-rural ecosystem which allows enhancing sustainability and inclusiveness of economic growth due to the use of a diversified portfolio of resources, and availability of growth results to a wide range of economic agents, final consumers, regardless of the territorial location of their residence;
- the predictive model, built on a neural network, to forecast the economic growth of a territory related to the structural changes caused by the technological developments, which is associated primarily with the formation of Industry 4.0;

- a model of the intelligence system for automated auditing of management systems that can significantly reduce the costs and risks associated with face-to-face audits implemented as an integrated software product;
- the project management success model in cross-border cooperation that helps to identify success factors in managing cross-border projects, therefore, is critical for achieving sustainable development through CBC strategies;
- a model for transforming the regulations for financial services and products design and consumption, which provides the inclusive growth of Russia's market for financial services based on enhancing integration and social partnership of financial institutions, society, and state;
- a tax administration system model developed on the best tax administration digitalization practices employed worldwide and in the Russian Federation. It relies on the principles of interaction transparency between subjects of tax relations, the conjuncture of the fulfillment of tax obligations, and the quality of digital services, including some others.

The book addresses a professional audience, such as academicians exploring themes related to the impacts of the use of Industry 4.0 new technologies on the economy, society, and environment, and the development of economic, organizational, information, and institutional mechanisms to achieve inclusive and sustainable growth under the transition to a new technological mode. It can be of interest to broad and varied audiences such as officials and practitioners engaged in the economic, environmental, and social regulation of new technology implementation in the industrial and agricultural sectors, public and corporate management, environmental protection and investment activities, health care, and education. The findings might be helpful for postgraduate students as they will serve as a valuable information source. The book contributes to training support while teaching and learning the curricula disciplines for master's degree students who major in economic policy, economic theory, management, innovative entrepreneurship, and information and communication technologies at higher educational institutions.

The contributors hope that empirical materials, innovative developments, and suggestions inspire scientific research, encourage applied studies, and supplement training programs in economic, administrative, social, and computer sciences at the advanced universities and research institutions, in the post-Soviet territory, in particular.

Elena I. Inshakova
Agnessa O. Inshakova

Contents

Editors and Contributors

About the Editors

Elena I. Inshakova Professor, Doctor of Economics is Honoured Worker of Higher School of the Russian Federation; Professor of the Department of Economic Theory, Regional Economy and Entrepreneurship at Volgograd State University, Volgograd, Russian Federation; Professor of Basic Department of the Southern Scientific Center of the Russian Academy of Sciences (RAS) at Volgograd State University; Chief Research Fellow at the Institute of Social and Economic Development of Region at Volgograd State University. She is the author of more then 200 scientific and educational publications: about 20 monographs, including foreign publications (in English) indexed in the largest international abstract and citation databases Scopus and Web of Science Core Collection as a leading author and editor; 15 textbooks and training manuals for Master's degree, graduate and postgraduate programmes; more than 30 articles in the journals reviewed by the Higher Attestation Commission of the Russian Federation and included in the database Russian Science Citation Index (RSCI) on the Web of Science platform. She is the author of more than 20 articles in publications indexed in Scopus; 10 articles in publications indexed in Web of Science Core Collection. Scientific and metric indicators: citations in RSCI—1094, Scopus—118, Web of Science Core Collection—63; h-index: RSCI—16, Scopus—4, Web of Science Core Collection—2.

Agnessa O. Inshakova Professor, Doctor of Law is Honorary Worker of the sphere of education of the Russian Federation, head of the master's degree program "Civil law, business law, private international law," head of the scientific school "Law of economic entities in the context of technological and socio-economic transformations," head of the scientific and educational center "Modernization of the legal system of modern Russia" at Volgograd State University, Institute of Law, and head of the Basic Department of the Southern Scientific Center of the Russian Academy of Sciences (RAS). She is Author of more than 400 scientific, educational and methodological publications, notably: more than 30 monographs, including in leading foreign publications indexed in the international scientific-metric databases SCOPUS / Web of Science Core Collection in English (as a leading author and editor); 36 textbooks, workshops and training manuals (including more than 10 publications with the stamp recommendations of the Ministry of Science and Higher Education, as well as leading Federal Publishers, including Yurayt, Zertsalo, etc.). She is Author of more than 200 articles in journals in the List Higher Attestation Commission of the Ministry of Education and Science of the Russian Federation; 82 articles in journals indexed in the international database Scopus, including Q1 and Q2; 43 articles in journals indexed in the international database Web of Science Core Collection, including Q1 and Q2. It ranks first in the country in the number of publications in SCOPUS and Web of Science Core Collection in the specialty—Law. She has high scientific-metric indicators: citations in RSCI—1702, Scopus—260, Web of Science Core Collection—54; h-index in RSCI—17; Scopus—8; Web of Science Core Collection—4.

Contributors

Dina Alshimbayeva Satbayev University, Almaty, Kazakhstan

Farida G. Alzhanova Institute of Economics SC MES RK, Almaty, Kazakhstan

Veronika V. Antonenko Volgograd State University, Volgograd, Russia

Anastasia V. Arzhanovskaya Volgograd State University, Volgograd, Russia

Irina S. Averina Volgograd State University, Volgograd, Russia

Lyudmila Yu. Bogachkova Volgograd State University, Volgograd, Russia

Petr V. Bondarenko Volgograd State University, Volgograd, Russia

Natalya P. Boroznina Volgograd State University, Volgograd, Russia

Sergey V. Boroznin Volgograd State University, Volgograd, Russia

Hans-Christian Brauweiler WHZ Zwickau University of Applied Sciences, Zwickau, Germany

Marina E. Buyanova Volgograd State University, Volgograd, Russia

Olga A. Chernova Southern Federal University, Rostov-on-Don, Russia

Valentin Dzedik Volgograd State University, Volgograd, Russia

Elena A. Eltanskaya Volgograd State University, Volgograd, Russia

Veronica S. Epinina Volgograd State University, Volgograd, Russia

Alex Ezrakhovich AEConformity Pty Ltd, Sydney, NSW, Australia

Olga V. Fetisova Volgograd State University, Volgograd, Russia

Lubov I. Galamyan The Russian Presidential Academy of National Economy and Public Administration, Moscow, Russia

Larisa M. Generalova Volgograd State University, Volgograd, Russia

Natalia V. Gorshkova Volgograd State University, Volgograd, Russia

Irina V. Grigorenko Volgograd State University, Volgograd, Russia

Lyubov V. Grigoryeva Volgograd State University, Volgograd, Russia

Lidiya S. Guryanova Kharkiv National University of Economics, Kharkiv, Ukraine

Mikhail M. Guzev Volgograd State University, Volzhsky, Russia

Dmitriy Y. Ilyin Volgograd State University, Volgograd, Russia

Elena I. Inshakova Volgograd State University, Volgograd, Russia

Elena A. Ivantsova Volgograd State University, Volgograd, Russia

Roman M. Kachalov Central Economics and Mathematics Institute, Russian Academy of Sciences, Moscow, Russia

Alla E. Kalinina Volgograd State University, Volgograd, Russia

Venera A. Karpinskaya Department of Modeling of Industrial Objects and Complexes, Central Economics and Mathematics Institute of the Russian Academy of Sciences (CEMI RAS), Moscow, Russia

Anna V. Kholodenko Volgograd State University, Volgograd, Russia

George B. Kleiner Department of Modeling of Industrial Objects and Complexes, Central Economics and Mathematics Institute of the Russian Academy of Sciences (CEMI RAS), Moscow, Russia;
Department of System Analysis in Economics, Financial University Under the Government of the Russian Federation, Moscow, Russia

Larisa A. Kochetova Volgograd State University, Volgograd, Russia

Daria A. Koneva Volgograd State University, Volgograd, Russia

Sergey A. Korobov Volgograd State University, Volgograd, Russia

Viktoria M. Ksenda Volgograd State University, Volgograd, Russia

Vladimir V. Kurchenkov Volgograd State University, Volgograd, Russia

Elena A. Kurchenkova Volgograd State University, Volgograd, Russia

Elena V. Loginova Volgograd State University, Volzhsky, Russia

Alexander G. Losev Volgograd State University, Volgograd, Russia

Leyla A. Markevich Volgograd State University, Volgograd, Russia

Anna A. Matveeva Volgograd State University, Volgograd, Russia

Denis E. Matytsin Volgograd State University, Volgograd, Russia;
Volga Branch of the International Law Institute, Volzhsky, Russia

Marina V. Milovanova Volgograd State University, Volgograd, Russia

Inna V. Mitrofanova Federal Research Centre the Southern Scientific Centre of the Russian Academy of Sciences (SSC RAS), Rostov-on-Don, Russia;
Volgograd State University, Volgograd, Russia

Vera A. Mityagina Volgograd State University, Volgograd, Russia

Valentina Moiseeva Volgograd State University, Volgograd, Russia

Aida Myrzakhmetova al-Farabi Kazakh National University, Almaty, Kazakhstan

Henrietta Nagy Kodolányi János University, Budapest, Hungary

Olga S. Oleinik Branch of Russian Presidential Academy of National Economy and Public Administration, Volgograd Institute of Management, Volgograd, Russia

Elena A. Petrova Volgograd State University, Volgograd, Russia

Marina V. Pleshakova Moscow City University, Moscow, Russia

Aleksander A. Polkovnikov Volgograd State University, Volzhsky, Russia

Larisa V. Ponomareva Volgograd State University, Volgograd, Russia

Elena G. Popkova Moscow State Institute of International Relations (MGIMO), Moscow, Russia

Ilya V. Pshenichnikov Volgograd State University, Volgograd, Russia

Elena G. Russkova Volgograd State University, Volgograd, Russia

Maxim A. Rybachuk Department of Modeling of Industrial Objects and Complexes, Central Economics and Mathematics Institute of the Russian Academy of Sciences (CEMI RAS), Moscow, Russia;
Department of System Analysis in Economics, Financial University Under the Government of the Russian Federation, Moscow, Russia

Bruno S. Sergi Harvard University, USA and University of Messina, Messina, Italy

Nikolay L. Shamne Volgograd State University, Volgograd, Russia

Dildarakhon A. Shelestova Volgograd State University, Volgograd, Russia

Ekaterina A. Shkarupa Volgograd State University, Volgograd, Russia

Inna M. Shor Volgograd State University, Volgograd, Russia

Elena G. Sidorova Volgograd State University, Volgograd, Russia

Sergey N. Sokolov Volgograd State University, Volgograd, Russia

Ekaterina V. Stepanova Branch of Russian Presidential Academy of National Economy and Public Administration, Volgograd Institute of Management, Volgograd, Russia

Andrey V. Svetlov Volgograd State University, Volgograd, Russia

Elena V. Terentyeva Volgograd State University, Volgograd, Russia

Yu. Yu. Trunin Department of Radiology and Radiosurgery, N.N. Burdenko National Medical Research Center of Neurosurgery, Ministry of Health of the Russian Federation, Moscow, Russia

Nadezhda Yu. Usacheva Volgograd State University, Volgograd, Russia

Irina D. Volkova Volgograd State University, Volgograd, Russia

Stephan Walter Johannes Gutenberg University, Mainz, Germany

Vasily A. Yakhtin Volgograd State University, Volgograd, Russia

Aida Yerimpasheva al-Farabi Kazakh National University, Almaty, Kazakhstan

Michael I. Zabezhailo Federal Research Center "Computer Science and Control", Russian Academy of Sciences (RAS), Moscow Institute of Physics and Technology (MIPT—National Research University), Moscow, Russia

Elena A. Zaliznyak Volgograd State University, Volgograd, Russia

Irina V. Zaporotskova Volgograd State University, Volgograd, Russia

Part I
Achieving Growth Across Economy and Society in the Age of New Technology: Theoretical Modeling

Chapter 1
Systemic Factors and Prerequisites for the Inclusive Growth of the Russian Economy

George B. Kleiner, Maxim A. Rybachuk, and Venera A. Karpinskaya

Abstract The chapter provides the conditions for the transition of the Russian economy and society to inclusive systemic growth. The essential systemic components of the socio-economic space of Russia are determined. The transformation of these components is necessary for the transition to systemic inclusive growth. Two points of view on the socio-economic space of Russia are considered, which are of particular importance for the change to inclusive growth. The first is a system-wide point of view wherein the Russian economy is represented in the form of four systemic sectors interacting with each other in a chain: "the object sector—the environment sector—the process sector—the project sector." The second point of view represents society in the form of four relatively independent macro-entities. The interaction between these entities is also carried out along the chain: "state—society—economy—business." Taken together, the study of the two above-mentioned structures makes it possible to determine the directions of the systemic transformation of society, ensuring movement toward inclusive growth.

1.1 Introduction

The concept of inclusive growth as a target function of socio-economic development entered public discourse in the early 2010s. Usually, inclusive growth is understood as "long-term sustainable growth in productivity and employment, opportunities for a wide range of firms and households" [1]. The concept of inclusive growth is often reduced to the even distribution of income among different population segments [2].

G. B. Kleiner (✉) · M. A. Rybachuk · V. A. Karpinskaya
Department of Modeling of Industrial Objects and Complexes, Central Economics and Mathematics Institute of the Russian Academy of Sciences (CEMI RAS), Moscow, Russia
e-mail: george.kleiner@inbox.ru

G. B. Kleiner · M. A. Rybachuk
Department of System Analysis in Economics, Financial University Under the Government of the Russian Federation, Moscow, Russia

E. I. Inshakova and A. O. Inshakova (eds.), *New Technology for Inclusive and Sustainable Growth*, Smart Innovation, Systems and Technologies 287,
https://doi.org/10.1007/978-981-16-9804-0_1

In our opinion, the increasing complexity of the socio-economic and administrative-political space structure, which is rapidly covering Russia and most of the world's developed countries, requires a transition to a more accurate perception of inclusive growth. It should be considered as a coherent and synchronous development of all significant society components, not just firms and households. To determine the main activities that support inclusive economic growth, one should analyze the structure of society, highlighting the elements and connections in it that are essential for social growth, and determine the conditions for their inclusive development. If the integral concept of economic growth does not require such structuring in the general case, then inclusive growth is bound to rely on such structuring.

The problem of inclusive growth is of particular importance for the Russian economy and society. Over the past decades, the Russian economy has alternately transitioned from crisis to stagnation. At the same time, both the crisis and stagnation affected various segments of the population, various industries, and territories of the country in different ways. The economic downturn affected the population's poorest segments to the greatest extent. In contrast, economic growth led to additional enrichment for the upper class, which is already characterized by high incomes. According to BCG, less than 0.0001% of the adult population in Russia (about 500 "super-rich" citizens) own 40% of all financial assets of Russians, or more than $ 600 billion [3]. Therefore, it is not just growth that is important for Russia but also the inclusive and systemic growth of the economy and society. But, unfortunately, a similar phenomenon of uneven development occurs in almost all economic components: the economic theory, economic policy, system and methods of economic management, and national economy. In each area and the relationship between them, systemic problems have emerged that require systemic solutions. Moreover, serious tensions permeate Russian society as well. Relations between people, organizations, and social groups are increasingly acquiring the character of mutual bitterness. Meanwhile, the principal attention of the leaders of both government agencies and organizations and business structures at all economic levels is concentrated mainly on formulating and solving immediate local tasks. Many of these tasks are purely bureaucratic and are associated with the achievement of secondary formal indicators. Therefore, it is necessary to move from traditional methods and one-sided approaches to managing social production to meaningful conceptual criteria and systemic means of ensuring socio-economic development.

In this chapter, we identify the essential systemic components of the socio-economic space of Russia, the transformation of which is necessary for the transition to systemic inclusive growth. The main strategic goals, which should become guidelines on the chosen path, are substantiated. Two points of view on the socio-economic space of Russia are considered, which are of particular importance for the change to inclusive growth. The first is a system-wide point of view wherein the Russian economy is represented in the form of four systemic sectors interacting with each other in a chain: "the object sector—the environment sector—the process sector—the project sector." The second point of view represents society in the form of four relatively independent macro-entities. The interaction between these entities is also carried out along the chain: "state—society—economy—business." Taken together, the study of the two above-mentioned structures makes it possible to determine the directions of the systemic transformation of society, ensuring movement toward inclusive growth.

1.2 Methodology

We proceed from the assumption that the concept of inclusive growth should be based on one or another theoretical paradigm that determines the vision of the structure of social development. This chapter adopts a systemic paradigm as a scientific basis for the study, which considers society as a set of interacting and transforming socio-economic systems [4, 5]. Analysis of the structure of such systems' internal content and external environment allows one to determine the components, the development of which is critical for inclusive growth [6]. Thus, it is systemic inclusive growth that should act as the general goal of social development. Note that only in the simplest cases, the system's growth is the sum of the growth of its components. In more complex cases, the result is a complex function of the constituents. In this case, it is crucial to distinguish between integral growth and systemic growth. In the latter case, we should talk about the growth of components ("summands") and the transformation of connections between these components. When this is the case, genuinely systemic, inclusive growth emerges.

1.3 Inclusive Growth: A Systemic View

The systems paradigm underlying this study presents each country's economy as a "system of systems" [7, 8].

Here is the information necessary to further present the features and classification of socio-economic systems [9]. A system is understood as a relatively stable in space and time part of the economic space–time, accessible to observation. Systems differ in their location in space and time. For the primary identification of these differences, attributes of the presence/absence of subsystem's clear boundaries in space and the presence/absence of such limitations in time are used. The corresponding grouping gives four classes (types) of subsystems of the economy:

1. *Objects* are systems with known boundaries in space and indefinite boundaries in time (example: an enterprise).
2. *Processes* are systems with indefinite boundaries in space and definite boundaries in time (example: dissemination of important information through television or radio transmission).
3. *Projects* (events) are systems with certain boundaries in space and time (example: building construction).

4. *Environments* are systems with indefinite boundaries in space and time (example: the Internet).

Thus, there are object, process, project, and environment sectors of the economy. All combined, they can be considered as components of a four-sector division of the economy.

Object-type systems (objects) have access to unlimited resources of time (T). The use of space resources thus demonstrates the inherent ability to use the available space (an intensive ability I) efficiently. Process-type systems (processes) provide unrestricted access to spatial resources S, limited reserves of time T of their functioning "without rebooting," and have an inherent ability to use it efficiently (activity abilities A). Environment-type systems (environments) have certain unambiguous spatial time boundaries, provided with unlimited access to the resources of space S and time T, but are not endowed with sufficient capabilities I, A for the efficient use of these resources. Project-type systems (projects) have limited reserves of time T and space S and sufficient capabilities A, I for the efficient use of these resources.

For the stable functioning of each subsystem, all four types of resources and abilities (S, T, A, I) are required. Uneven distribution of resources by system types leads to groupings in which systems exchange resources/abilities in abundance. Thus, the so-called tetrads appear. Tetrads are complexes consisting of four types of systems (object, process, environment, and project) that interact with each other stably [9]. In this case, the basis for the stable functioning of the tetrad is the interaction of its subsystems for the joint use of resources/abilities S, T, A, I (Fig. 1.1).

The structure of interaction of systemic sectors of the economy can be presented similarly. Let us illustrate this with the example of system analysis of production and reproduction and planning and financial cycles, i.e., the circulation of the economy's planning, production, and sale of goods and services. As a set of object systems, the object sector transfers the potential for intensive economic space use to the environment. The real bearer of this potential is the flow of goods and services that

Fig. 1.1 The structure and function of the tetrad. *Source* [9]

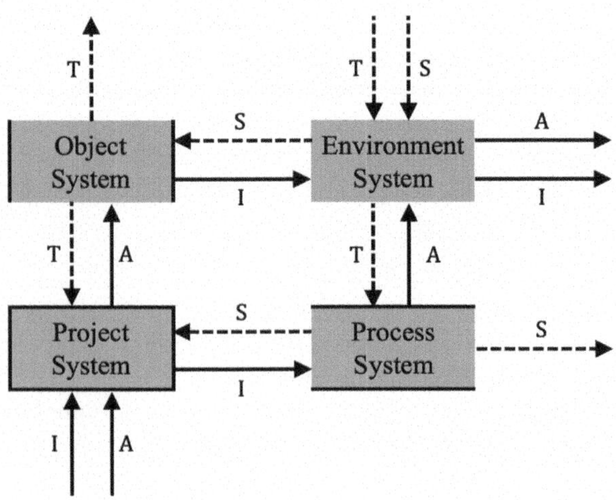

are produced by autonomous objects and enter the environment sector for sale. The environment sector, in turn, carries out the transfer of the part of the space to the object sector. This part of space is necessary to place the results of the objects' activities outside their spatial boundaries.

Each of the projects that make up the project sector of the economy, within the framework of the corresponding tetrad, carries out activities to change the state of the object included in this tetrad. In particular, equipment is being modernized or replaced, measures are being taken to improve production technologies, and new units of equipment, raw materials, components, and labor resources are involved in production. As part of this activity, the transition to the release of new products is being carried out. Thanks to this, the object is able to use every period effectively and continue to operate indefinitely. In turn, the object provides the project with the ability to function during its life established for the project. It is due to the object's need for the project's activity results during its life and is shown in Fig. 1.1 in the form of a vertical arrow T from object to project.

The interaction of the project and process sectors is organized in the same way. Since the process does not have the inherent ability to use space efficiently, these functions are carried out through interaction with the project. On the other hand, since the functioning of the process-type system in relation to space is extensive, the processes need external sources of intensive use of each space unit. It is achieved through communication with the project sector (horizontal arrow I). In turn, the process sector provides the project space resource S for functioning. The interaction of the environment and the process sector is also carried out in two directions. The environment sector provides the process time resource T for the functioning of the processes included in the sector. The process sector also contributes an environment resource for activity A. In the context of the production and reproduction cycle, the sale of products produced by one of the elements of the object sector is included in the trading and intermediary environment within the environment sector, where the process of product sales takes place. As a result, conditions are created to complete this cycle by implementing a set of projects that support the object sector by acquiring new units of resources for simple or expanded reproduction.

The relationship between the sectors can be characterized as follows, based on the premise that the main product of the object and project sectors is goods, and the environment and process sectors is a service. These relationships are carried out using the following two chains. The first chain consists of elements: goods in the form of a material object (thing) (object sector—environment sector), a service in the form of provision of time's resource (environment sector—process sector), a service in the form of provision of space's resource (process sector—project sector), goods in the form of a material object (thing) (project sector—object sector). This chain reflects the formation of a commodity supply in the economy. In Fig. 1.1, the chain is represented by horizontal and vertical arrows, the sequence of which indicates clockwise movement. The second chain reflects the formation of demand for the goods of the object sector. The identification of individual needs for goods and services of the object sector is carried out within the framework of the project sector of the economy, after which the identified needs go through the processes of aggregation and financing, implemented by the process sector based on the environment sector. Here, the aggregate demand for the goods of the object sector is determined in the next production and reproduction cycle, which is reflected in the level of the spatial resource that is provided by the environment sector to the object sector under the generated demand. This chain consists of movement along with the horizontal and vertical arrows in Fig. 1.1 counterclockwise. In general, both chains reflect, in a systemic context, the formation, and interaction of supply and demand in the context of production and reproduction and planning and financial economic cycles.

In accordance with the above information about the features of economic systems and systemic sectors of the economy, we conclude that inclusive systemic growth is possible if the following conditions are met:

(1) coordinated development of four sectors of the economy: object, project, process, and environment;
(2) parallel development of intersectoral interactions in the form of transfer (delivery) of space and time resources, as well as the ability to intensively use the space resource and actively use the time resource;
(3) creating a system for effectively distributing these resources and abilities, including planning and financial mechanisms and mechanisms for the accumulation and transfer of information and intellectual resources (abilities).

Obviously, for the implementation of these conditions of inclusive growth, a significant restructuring of the organization of management of the economy, the creation of modern information and consultative complexes that monitor the development of systemic sectors of the economy and the links between them, and systems of indicative planning and decision-making are required.

1.4 Inclusive Growth: The Macro-Entity View

Traditionally, when structuring society's socio-economic and administrative-political space at the macro level, there are three main relatively independent macro-entities: state, society, and business. An in-depth analysis of the components of such a structure reveals a significant heterogeneity of business as a subsystem that carries out the vital activity of the economy. The concept of "business" (in a broad sense) includes such diverse activities and events as the creation and liquidation of economic entities; investment, support of financial relationships between them; work with securities; accumulation of investment funds and aggregation of investment sources; production of products using equipment, buildings, and structures; preparation of production programs; marketing and sales of manufactured products; implementation of innovations; attraction and placement of personnel and others. Studies show that this activity area consists of two qualitatively different, more or less internally homogeneous subsystems. First is the business itself as a sphere of accumulating financial resources and investment in various investment projects and securities. Second is the economy itself as a sphere of production using the means of production, labor resources, and production knowledge. Thus, the subject area of business (in the narrow sense) is various investment projects, while the subject area of the economy (in the narrow sense) is the production and economic processes. With this approach, the following relatively independent structural components of public space are found:

- the *state* as a political system organizing the economic, social, and political life of entities located or registered in the territory of a given country;
- *society* as a population of citizens, structured with the help of various kinds of social movements and organizations located on the territory of a given country;
- the *business* as a sphere of accumulating financial resources and investing them in socio-economic projects and processes;
- the *economy* as a sphere of implementation of production processes and circulation of goods that ensure the vital activity of other components.

These four components of public space have the properties of the four types of socio-economic systems specified earlier. Namely, the state is one of the object systems; society is the number of environmental systems; the economy is process systems; the business is project systems [10]. The most significant relationships between macro-entities are shown in Fig. 1.2 [11].

With such a socio-economic space configuration, inclusive growth is ensured, on the one hand, due to the development of macro-entities of society, on the other hand, due to the institution strengthening and developing of interaction between them (arrows 1–8, Fig. 1.2).

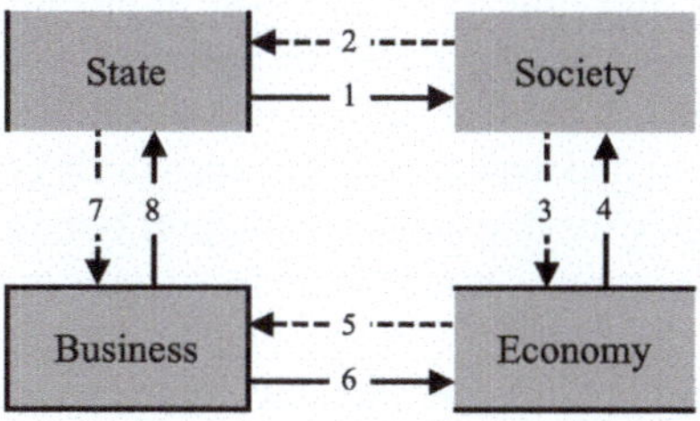

Legend:

1 – creation of conditions for safe life and development of society;

2 – delegation of power to the state by society;

3 – provision of labor resources to the economy by society;

4 – transfer to society of material benefits necessary for its life;

5 – providing businesses with opportunities (economic niches) to finance business projects;

6 – transfer of capital resources for economic development;

7 – creating conditions for a safe life and business development;

8 – payment of taxes based on the results of business activities.

Fig. 1.2 Macro-entity structure of society. *Source* [12]

In current conditions, the inclusive development of the state involves the implementation of the following main functions.

Integration function. It consists of the organization in space and time of the activities of social, economic, and administrative-political entities and systems to ensure the unlimited continuation of the country's functioning. It includes the function of producing public goods, the function of reproduction of the resources and conditions of activity used in this case, and other subfunctions that ensure the functioning of society, the economy, and the state itself. Note that economic modernization as a whole refers in this context to the reproductive function of the state. At the same time, the reproduction function belongs to the prerogatives of the state, while the production function is the prerogative of the economy (mainly).

Institutional function. We are talking about the role of the state in the creation (borrowing, transfer, transplantation, etc.) and consolidation of institutions—formal and informal laws, rules, and norms of public life. In this case, the norm is understood as a regulation, standard, and average characteristic of a phenomenon.

Benchmarking function. The state should promote the creation or fixation of the best models of behavior (functioning) of objects of material, intellectual or artistic culture, and determine the methodology and criteria for comparison, according to which certain artifacts are exemplary. This activity presupposes, in particular, the formation of various (formal and informal, explicit, and implicit) ratings in different areas of activity. In addition, government benchmarking should be carried out concerning in-country phenomena and cross-country comparisons.

Protective function. This function aims to ensure the protection and security of territories, maintaining law and order, control over strategic material, financial, energy, information resources, and the rights of social entities (individuals) and economic entities (enterprises), including the rights to livelihoods and others.

The state, society, economy, and business interact in a fairly complicated way [12, 13] (Fig. 1.2). Economic policy mechanisms that support outsiders and "pull" them up to the basic level of development established in society must be implemented to ensure the inclusive growth of macro-entities. Public authorities should have the power to choose a strategy for socio-economic development and adjust it to ensure society's life and preserve its integrity. Citizens as subjects of society can count on a certain subsistence minimum or unconditional income, equal access to education systems, health care, etc. Business entities should be able to find reliable partners to meet the socio-economic needs of the population. Business innovation and investment projects must be ensured with equal access to finance. At present, in Russia, the state occupies a commanding position; business is relegated to second place, the economy to the third, and society to the fourth.

Ideally, the interaction structure between the state, society, economy, and business should be balanced. Four components of this structure in the course of reconstruction should a) strengthen the statuses of relatively independent macro-subjects of society; b) institutionalize specific missions related to ensuring the crisis-free current and future development of society; c) build a system of the most important relationships between macro-entities (following Fig. 1.2), ensuring equal representation of the interests of each of the entities in the governing bodies and decision-making in the country [13].

In a normal socio-economic situation, these macro-entities are characterized by specific strategic goals or principles of behavior:

- goal of the state is sustainable unrestricted socio-economic development of society on the territory of the state;
- goal of society is ensuring well-being for us, our families, children, and grandchildren, regardless of where we live;
- goal of the economy is to combine production resources to meet the economic needs of social and economic entities;
- goal of the business is to obtain a financial result (profit) here and now.

Taken together, these attitudes ensure the implementation of consistent socio-economic development of the country within its territory on an unlimited time horizon. Disproportions in the scale and "negotiating power" of macro-entities violate this principle in spatial or temporal terms. Thus, the subordinate position of society concerning the state, the economy, and business leads to social inequality, a decrease in labor productivity, and the rejection of many creative, innovative solutions.

The inclusive systemic growth of society presupposes self-sustaining development, mutual coherence, and the absence of imbalances in the state, society, economy, and business development. Therefore, in the case of a lag in any macro-entities development, other macro-entities should contribute to their inclusion in the processes of mutual exchange of resources and support their performance of basic functions.

One of the directions of movement toward inclusive growth of the Russian economy and society should be the creation of a system that realizes the representation of the interests of these macro-entities both at the highest level of the management hierarchy and at the level of management of meso- and microeconomic systems [11, 13].

1.5 Conclusion

By now, conditions and imperatives have emerged for the Russian economy's multidimensional and multilevel systemic transformation. The fight against the consequences of the new coronavirus infection required the effective mobilization of economic resources and the ability of leaders to plan and implement socio-economic measures aimed at suppressing the pandemic and preserving essential components of the economy and business. As a result, a system of national projects has been developed, which determines the critical directions for developing the economy and society. At the same time, the content of these projects (1) does not cover all levels, and therefore, the entire volume of the Russian economy; (2) does not apply to the long term; (3) does not always answer questions about the possibility of joining some projects with each other.

In this regard, it seems that at this stage of development of the Russian economy, the priority should be not the project approach, which is now widely spread at all levels of government and often leads to an increase in the fragmentation of the economy, but the ecosystem approach focused on increasing the interdependence of various components of the economy and the degree of its integration.

In these conditions, it seems reasonable to prepare the economic doctrine of Russia as a scientific and applied document designed to form a reliable foundation for comprehensive long-term inclusive growth. Such a document should reflect the primary principles of economic policy and its implementation at basic levels of management (macro-, meso-, and microeconomic levels) to overcome the current socio-economic crisis, solve systemic problems of the economy, and enter the path of systemic inclusive growth. Therefore, the economic doctrine should reflect measures for the development of the structural components of the socio-economic space of Russia both in the general system context and in the macro-subject aspect. Such measures, as shown above, should allow, on the one hand, the provision of opportunities for the relatively independent development of the elements of these structures, provided they are balanced in terms of the volume and quality of attracted resources. On the other hand, the coordination of these structures functions to preserve the integrity of the economy and society as necessary conditions for inclusive growth.

The principle that can be called the principle of the ecosystem worldview should become fundamental in developing the economic doctrine [14]. By ecosystem, we mean "a spatially localized complex of organizations, business processes, innovative projects and infrastructure systems interacting with each other in the course of the creation and circulation of material and symbolic goods and values, capable of long-term independent functioning due to the circulation of these goods and systems and free from strictly centralized control" [15]. In accordance with the systems approach, a full-fledged ecosystem is a tetrad and includes organizational, infrastructural, communication, logistics, and innovation components. The cluster as an object subsystem represents the first component; the platform as an environment subsystem—the second component; the network as a process subsystem—the third component; and, finally, the business incubator as a project subsystem—the fourth component. Therefore, the ecosystem model in the form of a tetrad is an effective tool for studying the behavior and structure of ecosystems, determining the most efficient ways of distribution and consumption of the main types of ecosystem resources.

The economic doctrine should include sections devoted to forming and implementing policies for organizing and regulating the economy at the meso-level and, consequently, organizing and regulating socio-economic ecosystems. Despite the widespread development of the cluster, network, platform, and incubation approaches as independent areas of organization and regulation of the economy at the micro-level, experience in the formation of attitudes of coherent and co-evolutionary development of these microeconomic formations is still insufficient.

The center of gravity of economic management at all levels should be shifted to the development of ecosystems as independent socio-economic entities that demonstrate relative stability in space and time and accumulate the ability to conduct economic activities effectively [16]. As part of the management of ecosystems themselves and their relationship with the outside world, ensuring the S, T, A, I balance is critical for ensuring the resilience of ecosystems.

The connection of clusters, platforms, networks, and business incubators under the "umbrella" of ecosystems is the implementation of the principles of a circular economy since the initial components for each subsystem operation are the results of the activity of one of these subsystems. It can be assumed that such integration is in line with the development of the fourth industrial revolution, associated with an increase in the integration of socio-economic space–time. Therefore, expanding the population of ecosystems will help to increase the degree of integration of the economy, facilitate the process of creating innovations (innovation incubators), diffusion of innovations (network structures and information and logistics environments), implementation of innovations (clusters), and, as a result, inclusive growth [17].

Acknowledgements The chapter was accomplished with the support of the Russian Science Foundation (project No. 19-18-00335).

References

1. George, G., McGahan, A.M., Prabhu, J.: Innovation for inclusive growth: towards a theoretical framework and a research agenda. J. Manage. Stud. **49**(4), 661–683 (2012)
2. Albert, J.R.G., Dumagan, J.C., Martinez Jr, A.: Inequalities in income, labor, and education: the challenge of inclusive growth. PIDS Discussion Paper Series, No. 2015–01. https://www.econstor.eu/handle/10419/127037 (2015). Last accessed 23 Nov 2021
3. Tkachev, I., Kotchenko, K.: Experts have estimated the value of the assets of 500 "super-rich" Russians. RBC https://www.rbc.ru/economics/10/06/2021/60c0c14f9a79476c014a3263 (2021). Last accessed 22 Nov 2021
4. Kornai, J.: The system paradigm. William Davidson Institute Working Papers Series 278. William Davidson Institute at the University of Michigan, Ann Arbor (1998)
5. Kleiner, G.B.: System resource of economy. Vopr. Ekon. **1**, 89–100 (2011). https://doi.org/10.32609/0042-8736-2011-1-89-100
6. Kleiner, G.B.: The principles of duality in the light of the system economic theory. Vopr. Ekon. **11**, 127–149 (2019). https://doi.org/10.32609/0042-8736-2019-11-127-149
7. Allison, M.A., Batdorf, R., Chen, H., Generazio, H., Singh, H., Tucker, S., Bar-Yam, Y.: The characteristics and emerging behaviors of system of systems. New England Complex Systems Institute (NECSI), Cambridge (2004). http://www.necsi.edu/education/oneweek/winter05/NECSISoS.pdf Last accessed 21 Nov 2021
8. Carlock, P.G., Fenton, R.E.: System of systems (SoS) enterprise system engineering for information-intensive organizations. Syst. Eng. **4**, 242–261 (2001)
9. Kleiner, G.B.: New theory of economic systems and its applications. Her. Russ. Acad. Sci. **81**(9), 794–808 (2011)
10. Kleiner, G.B.: Socio-economic ecosystems in the light of system paradigm. In: System Analysis in Economics—2018, Proceedings of the V International Research Conference–biennale (21–23 November 2018), pp. 3–13. Prometheus publishing house, Moscow, 2018. https://doi.org/10.33278/SAE-2018.eng.004-012
11. Kleiner, G., Rybachuk, M.: System structure of the economy: qualitative time-space analysis. Front. J. Soc. Technol. Environ. Sci. **5**(2), 61–81 (2016)

12. Kleiner, G.: Sustainability of Russian economy in the mirror of the system economic theory (Part 1). Vopr. Ekon. **12**, 107–123 (2015). https://doi.org/10.32609/0042-8736-2015-12-107-123
13. Kleiner, G.: Sustainability of Russian economy in the mirror of the system economic theory (Part 2). Vopr. Ekon. **1**, 117–138 (2016). https://doi.org/10.32609/0042-8736-2016-1-117-138
14. Afshar, V., King, H., Wilt, S.: From controlling to empowering: Five practices for building business ecosystems. https://www.zdnet.com/article/from-controlling-to-connecting-five-practices-for-building-business-ecosystems/ (2018). Last accessed 23 Nov 2021
15. Kleiner, G.B., Rybachuk, M.A., Karpinskaya, V.A.: Development of ecosystems in the financial sector of Russia. Upravlenets—The Manager **11**(4), 2–15 (2020). https://doi.org/10.29141/2218-5003-2020-11-4-1
16. Braat, L.C., De Groot, R.: The ecosystem services agenda: bridging the worlds of natural science and economics, conservation and development, and public and private policy. Ecosyst. Serv. **1**(1), 4–15 (2012)
17. Acs, Z.J., Estrin, S., Mickiewicz, T., Szerb, L.: Entrepreneurship, institutional economics, and economic growth: an ecosystem perspective. Small Bus. Econ. **51**(2), 501–514 (2018)

Chapter 2
High-Tech Economic Growth from the Standpoint of the Theory of Economic Time: Modelling and Reducing Space–Time Inequality

Elena G. Popkova and Bruno S. Sergi

Abstract This chapter focuses on the following problem: the methodology used to examine the dynamics of the development of economic systems does not meet the new realities of our time. It does not consider the specifics of the Fourth Industrial Revolution. This chapter intends to solve the problem presented by rethinking high-tech economic growth from the standpoint of economic time. To accomplish this, based on reliable statistics for 2020–2021 from authoritative sources (IMD, World Bank, International Monetary Fund, etc.), economic and mathematical modelling of space–time inequality is carried out using the "underdevelopment whirlpools" method; also, prospects are determined and recommendations are developed to reduce this inequality and achieve a well-balanced world economic system. The chapter's contribution to literature consists in specifying the quality of economic growth, which has to be high-tech to contribute to the increase of living standards in the conditions of the Fourth Industrial Revolution. The considered experience of countries of BRICS could be helpful and could be extended to other countries of the world. Implementation of the authors' recommendations will reduce countries' inequality in the conditions of the Fourth Industrial Revolution.

2.1 Introduction

In the age of High Technologies, the concept of economic time is undoubtedly changing—cross-country comparisons are aimed at identifying not differences in the rate of socio-economic development of countries but at identifying differences in their technological development. The annual growth of some indicators (e.g., GDP—the rate of economic growth) does not adequately reflect the change in economic

E. G. Popkova (✉)
Moscow State Institute of International Relations (MGIMO), Moscow, Russia
e-mail: E.Popkova@inno.mgimo.ru

B. S. Sergi
Harvard University, USA and University of Messina, Messina, Italy
e-mail: bsergi@fas.harvard.edu

E. I. Inshakova and A. O. Inshakova (eds.), *New Technology for Inclusive and Sustainable Growth*, Smart Innovation, Systems and Technologies 287,
https://doi.org/10.1007/978-981-16-9804-0_2

systems over time. It is equally necessary to consider the technological order to obtain a comprehensive picture of economic time. We can only compare countries within the broad framework of a general Technological Order since only in this case, they compete on equal terms in global markets, and the annual growth of their indicators is comparable.

The complex problem is that the methodology used to examine the dynamics of the development of economic systems does not meet the new realities of our time and does not consider the specifics of the Fourth Industrial Revolution. As before, most country rankings consider only the annual growth of their indicators. However, the economic implication of such a comparison is contradictory. For example, a more rapid rate of economic growth in pre-digital economies leads to a crisis of overproduction due to the apparent inability to sell low-tech products in global markets—because of such economic growth, the quality of life in the country may decrease.

In contrast, digital economies' considerably slower economic growth rate contributes to increased foreign trade and improved quality of life. The traditional opposition of developed and developing countries does not overcome this problem since many developing countries have already moved to the new technological order, and the structure of developing countries is heterogeneous. This chapter aims to solve this problem by reconsidering high-tech economic growth from the theory of economic time.

2.2 Literature Review

This chapter uses the framework of the theory of economic growth and the theory of economic time. An essential role of high technologies for inclusive and sustainable development is emphasized in the works of [1, 2, 4, 9–15, 20]. Disproportions in the world economy and space–time inequality are studied in the articles of [3, 5, 6, 22].

The performed literature review shows that the issues of high-tech economic growth and space–time inequality are sufficiently studied and presented in the existing scientific literature. Still, space–time inequality of countries is measured primarily from the positions of GDP per capita and economic growth rate—without regard to technological mode, which is a research gap. This distorts the created models of space–time inequality. This gap is filled in this chapter with the help of taking the technological mode into account during the modelling of space–time inequality of countries in the modern world economy.

2.3 Materials and Method

This research is conducted based on the proprietary methodology of calculation and analysis of "underdevelopment whirlpools" of Prof. Dr. Elena G. Popkova. This methodology forms the foundations of the theory of economic time and is described and tested in multiple works, among which are [16–19]. For the additional regard to the factor of technological mode in this chapter, space–time inequality is assessed not only from the positions of differences in GDP per capita (EG) of countries but also from the positions of their differences in the volume of high-technology exports (THE), which implies multidimensional modelling of "underdevelopment whirlpools". According to the methodology of "underdevelopment whirlpools", the order of their calculation is as follows:

1. Determining the value of the studied indicator—GDP per capita (EG) or volume of high-technology exports (THE), accordingly—in the set year (HTEco): in this chapter, the assessment is performed for 2013 (t_0) and 2021 (t_1);
2. Determining the year in which the reference country (model) had the same (or as close as possible) value of the studied indicator (HTEet);
3. Calculating the depth of the "underdevelopment whirlpool" (Duw) as a difference between the set year (2013 or 2021—Y) and the year in which the reference country had a similar (or as close as possible) value of the studied indicator by the formula Duw = Y- HTEet (measured in years);
4. Calculating the speed of sucking in the "underdevelopment whirlpool" (Suw) according to the following formula: Suw = (Duwt$_1$-Duwt$_0$)/(t_1-t_0) = (Duwt$_1$-Duwt$_0$)/(2021–2013) = (Duwt$_1$-Duwt$_0$)/9 (measured in years per one year).

For the complete coverage of the international experience of high-tech economic growth in the conditions of the Fourth Industrial Revolution, this research is conducted by the example of countries of BRICS, which correspond to different technological modes and demonstrate a large scatter in the rate of economic growth. The reference country is the USA, as the global leader of the Fourth Industrial Revolution (ranked 1st in the World Digital Competitiveness Ranking in 2021, according to [7]).

As shown in Fig. 2.1, in the studied period (2013–2021), digital competitiveness—position in the ranking of [7]—grew the most in China (from 38 to 15th position). It increased for all the considered countries as well. The economic growth rate also increased in all countries; the most significant growth was observed in the USA (from 1.84% in 2013 to 53.97% in 2021). In China, the economic growth rate changed to a lesser extent than in other countries (from 7.77% in 2013 to 8.02% in 2021).

Dynamics of high-technology exports and GDP per capita in the reference country (USA), required for the following calculation of "underdevelopment whirlpools", are given in Tables 2.1 and 2.2, accordingly.

2.4 Results

To critically reconsider high-tech economic growth from the theory of economic time, modelling space–time inequality (underrun from the USA) by the example of countries of BRICS in 2013 and 2021 is performed in Table 2.3.

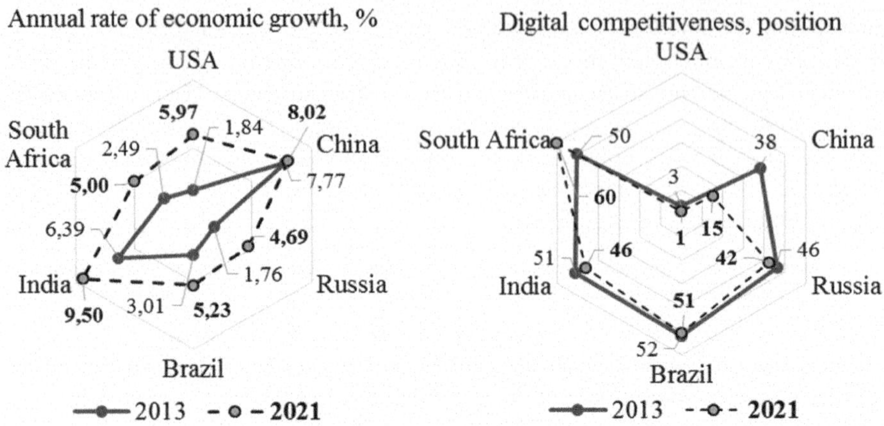

Fig. 2.1 Dynamics of digital competitiveness and economic growth rate in the USA and countries of BRICS during 2013–2021. *Source* Compiled by the authors based on IMD [7] and International Monetary Fund [8]

Table 2.1 Dynamics of high-technology exports in the USA during 2007–2020, proprietary assessment for 1985–2006, and forecast for 2021–2032, to determine Duw_{HTE}, % of industrial exports

Year	High-technology exports	Year	High-technology exports	Year	High-technology exports	Year	High-technology exports
1985	1.27	1997	7.56	2009	24.87	2021	20.61
1986	1.47	1998	8.78	2010	23.00	2022	21.50
1987	1.71	1999	10.18	2011	20.98	2023	22.43
1988	1.98	2000	11.82	2012	20.56	2024	23.40
1989	2.30	2001	13.71	2013	20.53	2025	24.41
1990	2.67	2002	15.91	2014	20.86	2026	25.47
1991	3.10	2003	18.47	2015	21.76	2027	26.57
1992	3.59	2004	21.43	2016	22.72	2028	27.72
1993	4.17	2005	24.87	2017	19.52	2029	28.91
1994	4.84	2006	28.86	2018	18.74	2030	30.16
1995	5.62	2007	30.37	2019	18.93	2031	31.47
1996	6.52	2008	28.86	2020	19.75	2032	32.83

Source Calculated by the authors based on World Bank [21]

As shown in Table 2.3, all countries of BRICS are peculiar for an increase of underrun from the USA in 2021 compared to 2013. The technological leader among countries of BRICS is China—it is in the fourth (Industry 4.0) technological mode (the level of its economy's robotization is very high, similar to developed countries). By the share of high-technology exports in the structure of industrial exports in 2013, China was ahead of the USA by 18 years, but the difference reduced to 9 years by

Table 2.2 Dynamics of GDP per capita in the USA during 1980–2026 (Gross domestic product per capita, current prices) to calculate EGet, USD

Year	GDP per capita	Year	GDP per capita	Year	GDP per capita	Year	GDP per capita	Year	GDP per capita
1968**	3,542.31	1980	12,552.94	1992	25,392.93	2004	41,641.62	2016	57,839.99
1969**	3,936.18	1981	13,948.70	1993	26,364.19	2005	44,034.26	2017	59,885.71
1970**	4,373.85	1982	14,404.99	1994	27,674.02	2006	46,216.85	2018	62,769.66
1971**	4,860.17	1983	15,513.68	1995	28,671.48	2007	47,943.35	2019	65,051.88
1972**	5,400.57	1984	17,086.44	1996	29,946.97	2008	48,470.55	2020	63,358.49
1973**	6,001.06	1985	18,199.32	1997	31,440.09	2009	47,102.43	2021*	69,375.38
1974**	6,668.32	1986	19,034.77	1998	32,833.67	2010	48,586.29	2022*	74,725.00
1975**	7,409.77	1987	20,000.97	1999	34,496.24	2011	50,008.11	2023*	77,881.31
1976**	8,233.66	1988	21,376.00	2000	36,312.78	2012	51,736.74	2024*	80,714.78
1977**	9,149.15	1989	22,814.08	2001	37,101.45	2013	53,245.52	2025*	83,563.82
1978**	10,166.45	1990	23,847.98	2002	37,945.76	2014	55,083.51	2026*	86,428.68
1979**	11,296.85	1991	24,302.78	2003	39,405.35	2015	56,729.68	–	–

* Forecast by International Monetary Fund [8] based on the factual data for 2020
**Proprietary assessment for 1968–1979
Source Compiled by the authors based on International Monetary Fund [8]

Table 2.3 Modelling space–time inequality (underrun from the USA) by the example of countries of BRICS in 2013 and 2021

Country	High-technology exports				GDP per capita			
	2013							
	HTEco	HTEet	Duw_{HTE}	Suw_{HTE}	EGco	EGet	Duw_{EG}	Suw_{EG}
China	31.58	2031	-18	–	7,039.57	1975	38	–
Russia	10.73	2000	13	–	15,928.70	1983	30	–
Brazil	10.65	2000	13	–	12,358.34	1980	33	–
India	8.88	1998	15	–	1,449.60	1968	45	–
South Africa	6.52	1996	17	–	7,425.04	1975	38	–
Country	2021							
	HTEco	HTEet	Duw_{HTE}	Suw_{HTE}	EGco	EGet	Duw_{EG}	Suw_{EG}
China	30.70	2031	-9	1.00	11,891.20	1980	41	0.33
Russia	13.00	2001	20	0.78	11,273.24	1979	42	1.33
Brazil	13.27	2001	20	0.78	7,741.15	1975	46	1.44
India	10.30	1999	22	0.78	2,116.44	1968	53	0.89
South Africa	5.63	1995	26	1.00	6,861.17	1974	47	1.00

Source Authors

2021; the speed of formation of the "underdevelopment whirlpool": 1 year per 1 year. By GDP per capita, China was 38 years behind the USA in 2013, and this underrun reached 41 years by 2021; the speed of sucking in the "underdevelopment whirlpool" is 0.33 years per 1 year.

Other countries of BRICS are in the third (digital) technological mode. For Russia, the depth of the "underdevelopment whirlpool" (in terms of underrun from the USA) by high-technology exports was 13 years in 2013, and it reached 20 years in 2021; the speed of sucking in the "underdevelopment whirlpool" is 0.78 years per 1 year. By GDP per capita, Russia was 30 years behind the USA in 2013, and this underrun reached 42 years in 2021; the speed of sucking in the "underdevelopment whirlpool"—1.33 years per 1 year.

2.5 Conclusion

Thus, the performed modelling of space–time inequality (underrun from the USA) by the example of countries of BRICS in 2013 and 2021 allowed reconsidering the high-tech economic growth of these countries from the positions of the theory of economic time.

In China, high digital competitiveness and the fourth technological mode ensure a positive contribution of high-technology exports to increase living standards and quality of life (GDP per capita). However, a high economic growth rate is not enough for achieving high living standards (as of 2021, China is 41 years behind the USA by GDP per capita)—this requires an increase of high-technology exports. The recommendation for China is to increase the rate of high-tech economic growth.

In other countries of BRICS, which belong to the third (digital) technological mode, the advantages of high-technology exports are less noticeable. That's why it is expedient for them to move towards higher living standards in two consecutive stages. At the first stage, overcoming the "underdevelopment whirlpools" is recommended to increase economic growth and digital competitiveness. At the second state—after the transition to the fourth technological mode—it is offered to increase the volume of high-technology exports.

The contribution of this research to literature lies in specifying the quality of economic growth, which, as the obtained result showed, should be high-tech to contribute to the increase of living standards in the conditions of the Fourth Industrial Revolution. The studied experience of BRICS countries could be helpful and extended to other countries worldwide. Implementing the authors' recommendations will reduce countries' inequality amid the Fourth Industrial Revolution.

References

1. Abendin, S., Duan, P.: International trade and economic growth in Africa: the role of the digital economy. Cogent Econ. Finance **9**(1) (2021). https://doi.org/10.1080/23322039.2021.1911767
2. Ahmad, M., Majeed, A., Khan, M.A., Sohaib, M., Shehzad, K.: Digital financial inclusion and economic growth: provincial data analysis of China. China Econ. J. **14**(3), 291–310 (2021). https://doi.org/10.1080/17538963.2021.1882064
3. Asamoah, L.A., Figari, F., Vezzulli, A.: Spillover effects of innovation and entrepreneurial activity on income inequality in developing countries: a spatial panel approach. Reg. Sci. Policy Pract. **13**(5), 1661–1686 (2021). https://doi.org/10.1111/rsp3.12470
4. Astafeva, O.V., Astafyev, E.V., Khalikova, E.A., Leybert, T.B., Osipova, I.A.: XBRL reporting in the conditions of digital business transformation. Lect. Notes Netw. Syst. **84**, 373–381 (2020)
5. Ezcurra, R., Del Villar, A.: Globalization and spatial inequality: does economic integration affect regional disparities? Ann. Reg. Sci. **67**(2), 335–358 (2021). https://doi.org/10.1007/s00168-021-01050-5
6. Farrokhi, F.: Skill, agglomeration, and inequality in the spatial economy. Int. Econ. Rev. **62**(2), 671–721 (2021)
7. IMD: World Digital Competitiveness Ranking 2021. https://www.imd.org/centers/world-competitiveness-center/rankings/world-digital-competitiveness/ (2021). Last accessed 23 Nov 2021
8. International Monetary Fund: World Economic Outlook Database. https://www.imf.org/en/Publications/WEO/weo-database/2021/October (2021). Last accessed 23 Nov 2021
9. Jiao, S., Sun, Q.: Digital economic development and its impact on economic growth in China: research based on the perspective of sustainability. Sustainability (Switzerland). **13**(18), 10245 (2021). https://doi.org/10.3390/su131810245
10. Inshakov, O.V.: Strategy and tactics of nanoindustry development in the Russian Federation. Econ. Region **3**, 246–250 (2011). https://doi.org/10.17059/2011-3-29
11. Liu, Y., Luan, L., Wu, W., Zhang, Z., Hsu, Y.: Can digital financial inclusion promote China's economic growth? Int. Rev. Financ. Anal. 78 (2021) https://doi.org/10.1016/j.irfa.2021.101889

12. Masłoń-Oracz, A., Ojiambo, J., Kevin, O.: Digital economy as a driver of sustainable and inclusive growth in Africa—Case Study. In: Springer Proceedings in Complexity, pp. 605–61 (2021). https://doi.org/10.1007/978-3-030-62066-0_46
13. Moody, K.: High tech, low growth: robots and the future of work. Hist. Mater. **26**(4), 3–34 (2018). https://doi.org/10.1163/1569206X-00001745
14. Morozova, I.A., Kuzmina, E.V., Kuzmina, M.I., Shevchenko, S.A., Avdeyuk, O.A.: Economic development of the territory on the basis of formation of regional innovation system in conditions of digitalization. Lect. Notes Netw. Syst. **198**, 974–982 (2021). https://doi.org/10.1007/978-3-030-69415-9_108
15. O'Meara, S.: Taking the silk road to high-tech growth. Nature **563**(7729), S25–S27 (2018). https://doi.org/10.1038/d41586-018-07202-6
16. Popkova, E.: Problem analysis of "underdevelopment whirlpools" as an obstacle for economic growth in Asian countries. Int. J. Bus. Global. **19**(2), 167–174 (2017). https://doi.org/10.1504/IJBG.2017.085923
17. Popkova, E., Gornostaeva, Z., Tregulova, N.: Role of innovations in provision of competitiveness and innovational development of economy and overcoming of "underdevelopment whirlpools" in Russia and countries of Eastern Europe. J. Entrepren. Emerg. Econ. **10**(3), 511–523 (2018). https://doi.org/10.1108/JEEE-12-2017-0100
18. Popkova, E.G., Bogoviz, A.V., Lobova, S.V., Alekseev, A.N.: "Underdevelopment whirlpools" as manifestation of disproportions of economic growth in modern Russia. Stud. Syst. Decis. Control **135**, 155–162 (2018). https://doi.org/10.1007/978-3-319-72613-7_14
19. Popkova, E.G., Ekimova, K.V., Sergi, B.S.: Data set of balance of Russia's regional economy in 2005–2024 based on the methodology of calculation of "underdevelopment whirlpools". Data Brief 31 (2020). https://doi.org/10.1016/j.dib.2020.105821
20. Tian, D., Guo, X., Wang, P.: Innovation vouchers and the sustainable growth of high-tech SMEs: evidence from China. Sustainability (Switzerland) **13**(20), 11176 (2021). https://doi.org/10.3390/su132011176
21. World Bank: High-technology exports (% of manufactured exports). https://data.worldbank.org/indicator/TX.VAL.TECH.MF.ZS (2021). Last accessed 23 Nov 2021
22. Zhang, B., Nozawa, W., Managi, S.: Spatial inequality of inclusive wealth in China and Japan. Econ. Anal. Policy **71**, 164–179 (2021). https://doi.org/10.1016/j.eap.2021.04.014

Chapter 3
Innovative Technologies as a Factor in Ensuring Inclusive Growth in the Unified Ecosystem of Urban and Rural Areas

Vladimir V. Kurchenkov, Olga V. Fetisova, Daria A. Koneva, and Elena A. Kurchenkova

Abstract The chapter describes innovative technologies as a factor in the development of the urban environment and rural areas in modern Russia. It reveals the proportion of implemented technologies of various technological stages in Russia. Also, it compares the dynamics of innovation activity in the regions and territories under consideration, the movement in a standard of living criteria of the population in these territories, the level of migration, and natural population decline. The work reveals the reasons for different conditions for inclusive economic growth in urban and rural ecosystems. The concept of inclusive growth is considered in relation to the technological integration of urban and rural areas. The authors reveal principles of building a unified ecosystem, including large urban agglomerations, medium and small cities, as well as rural settlements. The work examines the peculiarities of the use of innovative technologies of the sixth technological stage for economic growth in rural areas of Russia, the growth of living standards of the population in them, as well as inclusive growth in the unified urban–rural ecosystem.

3.1 Introduction

The importance of innovative technologies in the development of society is now becoming more and more obvious. Despite the current situation associated with the increase of coronavirus infection in the world [1], they are considered as an integral

V. V. Kurchenkov (✉) · O. V. Fetisova · D. A. Koneva · E. A. Kurchenkova
Volgograd State University, Volgograd, Russia
e-mail: kurchenkov@volsu.ru

O. V. Fetisova
e-mail: fetissova@volsu.ru

D. A. Koneva
e-mail: dakoneva@volsu.ru

E. A. Kurchenkova
e-mail: kurlena@volsu.ru

E. I. Inshakova and A. O. Inshakova (eds.), *New Technology for Inclusive and Sustainable Growth*, Smart Innovation, Systems and Technologies 287,
https://doi.org/10.1007/978-981-16-9804-0_3

part of scientific and technological progress and modern science. Considering the complex transition of the national economy to the sixth technological stage, it is becoming obvious that there is a need for better support for innovations for the even and systematic development of urban and rural areas in our country.

The introduction of new technologies contributes to the integration of economic space and provides conditions for the sustainable development of cities and towns, as well as rural settlements. At the same time, an important aspect in the implementation of the concept of sustainable development on the territories of various types and different geographic locations is the achievement of inclusiveness of this development, which manifests itself in increasing the availability of economic and social benefits for the general population, regardless of their living conditions.

The difficulty of the solution to this problem in the near future for Russia is connected, on the one hand, with the complex geographic location of cities and rural settlements. At the same time, historically, there are disparities in the standard of living of the population in large cities and rural areas. On the other hand, today Russia still lags behind economically developed countries in terms of the rate of advanced technologies adoption in the growth of urban environment and rural areas. At the same time, the relatively stable position of Russia in the GII rating allows drawing conclusions about unused opportunities for the development of the innovation sphere, also in the context of the natural, cultural, industrial, and agricultural potential of its urban and rural areas.

3.2 Methods

Many domestic and foreign scientists pay attention to the problem of the development of innovative technologies in modern economy [2, 3]. Actual issues are related to innovations within the framework of the sixth technological stage, which is currently very problematic due to the uneven location of scientific and production potential in modern Russia and the dominance of the centrist configuration of the national innovation system. The main peculiarity of this model is the attenuation of innovation processes in the direction from the center to the periphery where rural areas suffer first of all, because they cannot compete with megacities in terms of scientific and industrial development [4–9].

We had investigated the specificity of urban and rural areas, their need for innovative development, and introduction of innovative technologies into production processes, and this was reflected in a number of scientific works [10–13]. Methods of statistic and comparative analyses, expert assessments, predictive models, etc. were used in our study [1, 14–16]. On the basis of the conducted research and analysis of statistical empirical data, the conclusions were drawn and given in the results of the work [17–19]. Also, the actual regulatory legal acts are used in this chapter [20].

3.3 Results

The formation of the sixth technological stage based on the NBIC-convergence (nano-, bio-, information, and cognitive technology convergence) [21] implies the creation of products of nanotechnology and biotechnology, as well as nuclear, molecular, cellular technologies, etc. In the world economy, in general and in Russia in particular, it has long been relevant. It is obvious that for a full-fledged transition from one technological stage to another, large-scale investments are required in the sphere of the key factor of the technological stage on which, first of all, scientific research and development work will be carried out. In this regard, the transition to the sixth technological stage should presuppose large-scale investments in digital and nanotechnology [22], as key ones within the framework of this stage.

It should be noted that in recent years, these technologies have been successfully developing in the Russian economy. However, it is too early to talk about a full-fledged transition to the sixth technological stage, which responds to a range of challenges.

Firstly, this is the absence of federal authority, whose powers would include a technological policy of the state as a whole (development of scientific and technological tasks, inventions, and their control).

Secondly, many researchers recognize the fact that, compared with the level of science development in Russia, during the transition to the fourth stage, there was a significant decline in applied science as a whole, since there previously had existed a greater number of research and design institutes that had been engaged in various scientific, experimental, and design developments.

Thirdly, there is no required number of highly qualified personnel interested in the development of science in our country.

Fourthly, for many reasons it is confirmed that the demand of Russian manufacturers for innovative technologies is not as great as in foreign countries. And finally, there are limited resources [5].

As a result, the share of technologies of the sixth technological stage in the structure of the Russian economy remains insignificant. In particular, according to the studies of domestic scientists, technologies of the fourth stage in the Russian economy account for about 50%, the fifth technological stage—only 10%, while the remaining 40% contain technologies of the earlier stages.

As shown in Fig. 3.1, the share of the sixth stage in the Russian economy is

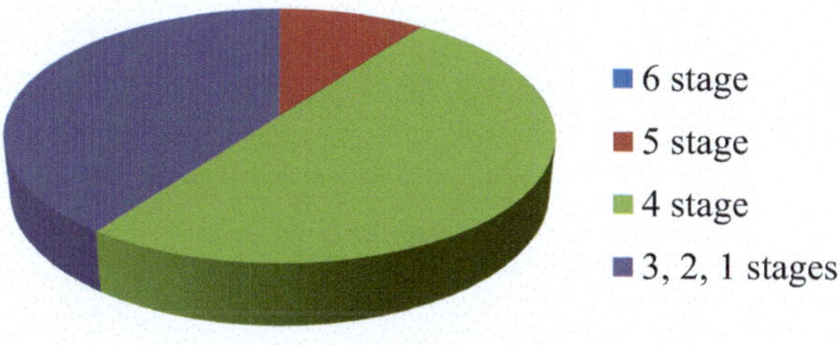

Fig. 3.1 Share of technologies of various technological stages in the structure of Russian economy, %. *Source* [6]

actually less than one percent. This fact confirms the existence of a significant technological gap between the world and domestic economies, which adversely affects the assessment of the global prospects for Russia's entry into the sixth technological stage.

However, as it was noted, this does not mean that the development of these technologies will not have prospects in the near future. Today, they are already being actively implemented and soon they may become dominant. In addition to the growth of activity on the part of business, the transition to the sixth technological stage is facilitated by the implementation of a number of national projects of our state [23].

Indeed, today Russia has sufficient potential for the growth of innovation activity of the main economic agents. If we talk about the place of the Russian Federation in the ranking of countries according to the Global Innovation Index GII, it ranks 47th out of 131 countries, wherein having prospects for further growth [17].

We may note that the Global Innovation Index is calculated as the average value of innovation resources and innovation results. The resources of innovations include institutions, human capital and science, infrastructure, the level of market, and business development, and the results of innovation imply the development of technologies and knowledge economy, the output of creative activity. These components of the index can be used to trace the dynamics of Russia's positions in the GII during 2015–2020 (Table 3.1).

Based on Russia's position in the GII 2020 compared to the GII 2019, we can draw a conclusion about the advantages and disadvantages of the innovation system in Russia, including the context of urban and rural areas [4].

Thus, the strengths of the Russian innovation system imply a high number of graduates of natural science and engineering specialties who can be employed both in industrial enterprises and in enterprises of the agro-industrial complex; the number of people employed in knowledge-intensive industries, allowing the development of industry and agriculture, etc. The weaknesses include small environmental sustainability and low energy efficiency; the lack of rapid development of clusters, despite the large number of studies in this area, etc.

In turn, the growth of innovative activity and the transition to the sixth technological stage, in our opinion, can contribute to the implementation of the model of inclusive economic growth in cities and rural settlements, as well as promote their integration into a single ecosystem.

Table 3.1 The Place of Russia in the ranking according to the Global Innovation Index (GII) during 2015–2020

Indicator	2015	2016	2017	2018	2019	2020
GII value	48	43	45	46	46	47
Innovation resources	52	44	43	43	41	42
Innovation results	49	47	51	56	59	58

Source Compiled by the authors based on [4]

At present, as it was noted, the development of modern cities in Russia and rural areas is extremely uneven. One of these problems is the underdevelopment of rural areas and their lag behind urban areas in many respects, including the level of innovation. As statistics shows, the population of cities of less than 100 thousand people is steadily declining. The number of citizens living in rural settlements is also decreasing at a faster rate.

Over the past decade, the percentage of the rural population in the country total has decreased from 26.5 to 25.26% [19]. If we talk about the regions of the Southern Federal District, it is worth noting that, for example, in the Rostov region, the share of the rural population in the total of the region has decreased from 32.1% in 2015 to 31.8% in 2021. For comparison, in Krasnodar Territory this indicator has decreased from 47.1% in 2011 to 44.4% in 2021 [19]. A similar trend is observed in the Volgograd region (Table 3.2).

Taking into account the fact that the share of the rural population of the Volgograd region in recent decades has fluctuated within a quarter of the total population of the region [24], it should be noted that this share always remains below the all-Russian level (Table 3.3).

Analyzing this dynamics, it should be noted that this is a reflection of the living standard differentiation of the population in cities and rural areas. This is manifested both in the differentiation of per capita income and in the availability of public goods, education, medicine, cultural facilities, and transportation.

Comparing urban and rural areas in terms of innovative development, as noted, the former are significantly ahead of the latter. This is due to a large number of problems in rural areas that need to be solved before the introduction of innovative technologies.

Of course, when the issue concerns the integrated development of villages, one should resort to a comparative analysis of the advantages of specific territories. Such advantages need to be used and developed, and look for points of growth. If we try to develop an unpromising industry in a particular territory, there is a high probability of wasting time and resources. Therefore, priority areas of development should be specifically identified and supported, and there is also a need to resort to improving the existing support mechanisms, including financial, for the development of municipalities' own revenue base to ensure inclusive development.

Comprehensive development of rural areas implies that it is not necessary to be limited only to one sphere of life in the village; it is important to develop all spheres. One cannot but agree with this, but it is extremely important to find an individual approach in the policy of rural development. For the optimally fast finding approach, it is proposed to divide the regions where particular rural areas are located, into 4 types in accordance with their specific problems and advantages.

Firstly, these are predominantly rural areas with agricultural specialization. Such territories also have a rather favorable climate and natural conditions for such activities. Here, social circs are also important. They will make it possible to use a certain set of universal measures in relation to such territories, allowing focusing the attention on the development of the rural economy. We can here list such measures:

Table 3.2 Dynamics of the urban and rural population of the Volgograd region from 2011 to 2021

Population, thousand people	2011	2012	2013	2014	2015	2016	2017	2018	2019	2020	2021
	2607,5	2594,8	2583,0	2569,1	2557,4	2545,9	2535,2	2521,3	2507,5	2491,0	2474,6
Urban	1982,1	1975,4	1970,5	1963,0	1957,2	1951,8	1947,2	1940,3	1933,7	1925,2	1915,8
Rural	625,4	619,4	612,5	606,1	600,2	594,1	588,0	581,0	573,8	565,8	558,8

Source Compiled by the authors based on [18]

Table 3.3 Dynamics of the share of the rural population in the total of the Volgograd region from 2010 to 2021

Year	2010	2011	2012	2013	2014	2015	2016	2017	2018	2019	2020	2021
Share,%	24,25	23,98	23,87	23,71	23,59	23,47	23,34	23,19	23,04	22,88	22,70	22,59

Source Compiled by the authors based on [18]

- diversification of the rural economy of specific territories, which subsequently makes it possible to support various types of business in the villages. Such businesses can create a great number of jobs in the village, but one should not forget about self-employment, as well as rural tourism and social development, the promotion of crafts in the particular area;
- social services should become more accessible for rural residents. Due to this, in future, the living standard of citizens and housing affordability may also increase;
- in rural areas, their own market system should be developed, allowing rural producers to have unrestricted access to the agricultural sales market, both in their own region and in other territories of the country.

Secondly, these are areas where multifunctional rural economy prevails. Typically, these rural regions are characterized by suburban agriculture. They have a fairly high population density, moderate migration outflow, and upcoming construction of housing facilities. Here, social infrastructure is developed as much as necessary, depending on the particular distance from the city. A policy in the field of development of such rural areas should take into account the already established highly productive agriculture and be developed in certain directions:

- the main feature of such territories, as already mentioned, is their close location to cities, therefore, the authorities should provide an opportunity for outdoor recreation for the townspeople and the villagers themselves. But it is essential that such a development direction should not harm the ecosystem. On the contrary, in these areas, special attention should be paid to the preservation and restoration of natural and agrarian rural landscapes. There is also an opportunity to promote sustainable agriculture, which could increase ecosystem sustainability in future;
- there is a need to stop the development of such a trend as irrational transfer of agricultural land to totally different categories. It is necessary to improve the mechanisms for regulating land relations in this area and support industrial and infrastructural suburbanization at the legislative level. It will help to diversify the rural economy, as well as create jobs for rural residents;
- it is worth paying attention to the promotion of population migration from city to village, including temporary migration, for example, recreational. If the construction of new private houses is also encouraged, it will put the suburbanization process on the path to prosperity. The growth of migration to rural areas, the qualitative and quantitative strengthening of the service sector will undoubtedly lead to the creation of new jobs for the villagers and to the development of rural entrepreneurship.

Thirdly, these are regions with unfavorable social conditions for the villages' development. In such rural areas, first of all, the main problem is depopulation. There is a set of measures for a solution to this issue:

- one cannot do without a strengthened state policy, namely a demographic one, which is carried out with an increase in the share of families with public support.

Special additional state assistance in rural areas is needed for families that can be categorized as socially vulnerable. We mean families with disabled children, incomplete, low-income, and very large ones. Single-parent families and, of course, orphans should be under state protection. For them, special complexes of measures should be developed to form and motivate a healthy lifestyle, to strengthen and maintain health, and to reduce the mortality rate in villages. A special factor for such citizens is the promotion of their employment;

- despite the depopulation in these territories, it is important to preserve people's access to social institutions, including health care and education. Of course, the development of social infrastructure should not only be supported, but also developed, and transport accessibility to villages should be improved. Problems with gasification, water supply networks, and sewerage systems should also be resolved;
- in the villages that are characterized by depopulation, agriculture is not the emphasis. However, there should be opportunities for its development; therefore, special state support for agriculture is necessary. It may imply a transition to more extensive, but less intensive, branches of agriculture in peripheral areas, where we mean crop production, grazing, etc. Access to getting credits should also be facilitated, whereby assistance will be provided in the renewal of technological facilities and the purchase of animal feed;
- to solve the problems of such territories, it seems possible to develop such a phenomenon as seasonal dacha settlement of remote villages, when at least during the warm season the villages will be filled with city dwellers for gardening and recreation on rural land.

Fourth, these are regions with poor local development of villages, partly due to unfavorable natural and climatic conditions. Here, we should pay attention to such measures:

- development and implementation of a program of unhindered access to basic life-supporting social benefits and services for villagers as healthcare and pharmaceutical organizations, cultural and educational institutions, food and non-food stores, access to the Internet, and cellular communications. It is also worth noting here the development of transport connecting with the city, both motor and water, emergency air transport;
- given the fact that focal settlements of villages rarely allow preserving their culture, an important factor is the need to stimulate the activity of rural communities in various ways and to allocate grants or other subsidies for the development of folk art and activities.

Taking into account the current state of Russian regions, it may be difficult to determine their type, since sometimes in one region there may be rural areas that require comprehensive measures for their development, belonging simultaneously to two or more types of regions. Then the question arises about the necessity to divide all rural areas into types, depending on their advantages and problems. It would also make it possible to implement integrated rural development more efficiently.

Therefore, when today we mention economic growth in Russia, the GDP growth, it should be noted that it most often concerns large cities—megalopolises, as well as oil-producing regions. In the peripheral regions, especially in rural areas negative economic growth is most often observed. In this regard, the existing model of economic growth in Russia today cannot be fully called inclusive, since the results of this growth are distributed and become available to a limited circle of economic agents.

The centric model of the geographical location of settlements in Russia is due to the processes of urbanization, which have recently tended to intensify. Along with the growth of natural population decline in rural areas, its migration to large urban agglomerations is observed. The introduction of digital technologies and elements of nanoindustry makes it possible to observe the reverse process, deurbanization, which was especially manifested during the period of the pandemic and the need for remote work of a great part of the employed.

Even if it is too early to talk about long-term deurbanization in the country at this stage, it is possible to gradually maintain this trend and limit the growth of urban agglomerations. This will also be facilitated by many disadvantages of living in large cities, which citizens often pay attention to. These are, for example, environmental degradation in cities, traffic jams on highways and long city roads, and a change in consumer preferences in favor of natural and organic food. A trend to the values of rural lifestyle began to appear, which is manifested in the less intense rhythm of life, in the possibility of living in a private house with a backyard, etc. [1].

It is worth noting that deurbanization can manifest itself in various forms: the move of urban residents to a permanent place of living in the countryside, and the construction or purchase of a "second housing" in the suburban areas, and even agritourism, which has become more and more popular since 2015. Taking into account the proposed measures, development of rural areas and introduction of deurbanization trend will be a significant step toward stabilizing the socio-economic situation in the post-pandemic period, upon condition of the introduction of innovative technologies into the aspects of living of rural areas citizens [14].

These tendencies are aimed at the integration of economic and social processes taking place separately in urban agglomerations and rural settlements. This integration is an important condition for the implementation of the concept of inclusive economic growth in our country.

Indeed, at present, large urban agglomerations, medium and small cities, as well as rural settlements in Russia can be considered as isolated ecosystems. Of course, there are forward and backward linkages between these systems, but today it is not necessary to talk about a single ecosystem, which organically includes cities of various sizes and geographic locations, as well as rural areas. As a result, economic growth is extremely uneven. Many resources are not fully used, which significantly affects the GDP growth rates of the Russian economy as a whole in the near future.

Large-scale introduction of technologies of the sixth technological stage creates conditions for the growth of integration processes between urban and rural areas within a single ecosystem.

As shown in Fig. 3.2, a single ecosystem includes large urban agglomerations, small and medium-sized cities (C), and rural settlements (V), the interaction between which is mediated by the use of technologies of the sixth technological stage (T).

Within this single ecosystem, economic growth, on the one hand, becomes more sustainable due to the possibilities of using a diversified portfolio of resources. On the other hand, it is possible to achieve the inclusiveness of economic growth, which will make its results available to a wide range of economic agents, final consumers, regardless of the territorial location of their residence.

Thus, the balanced development of urban and rural areas can be ensured through the diffusion of innovations within the framework of the transition to the sixth technological stage. Development of a single ecosystem that includes both cities of various sizes and rural settlements will make economic growth more sustainable in the long term and make it inclusive.

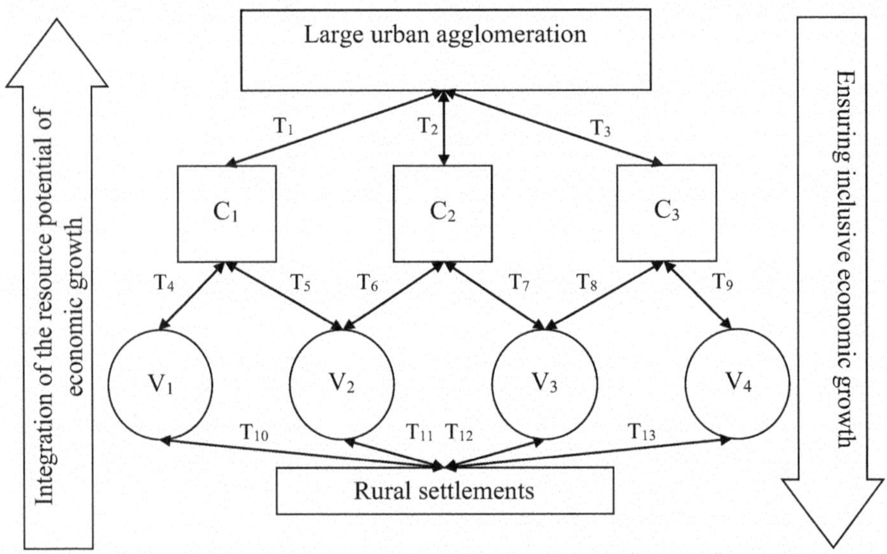

Fig. 3.2 An integrated ecosystem of urban and rural areas in the context of inclusive economic growth. *Source* Developed by the authors

3.4 Conclusion

Summing up the conducted research, some conclusions and generalizations can be drawn. Currently, economic growth in Russia is unstable and uneven. This is due to commodity dependence of the country's economy, uneven distribution of production structures, and insufficient level of technological development of the productive forces. In this regard, for many years, there has been a slight increase in the GDP, a high-level inflation, occasional crisis phenomena in the economy. In addition, the results of economic growth are distributed extremely unevenly toward large urban agglomerations, while rural areas are stagnating. This gives rise to unidirectional migration flows from villages to large cities, intensification of urbanization processes, and growing differences in the living standards of the urban and rural populations.

In the struggle for world leadership in new technologies, Russia needs to move to the new sixth technological stage, the core of which is information-, nano- and biotechnology. The introduction of these technologies will allow large-scale diffusion of innovations in all key elements of the integrated ecosystem in urban and rural areas. This will lead, on the one hand, to economic growth acceleration by mobilizing new resources, and on the other hand, it will make the processes associated with this growth more inclusive.

Consequently, at present, the state faces the most important strategic objective of establishing a roadmap for the development of innovative technologies as a factor of urban and rural area growth, taking into account the difficulties of transition to the sixth technological stage and the specific features of these territories.

References

1. Petrikov, A.: Coronavirus as an incentive for deurbanization. Will the COVID-19 pandemic catalyze population movement from cities to villages? Agroinvestor. https://www.agroinvestor.ru/analytics/article/34499-koronavirus-kak-stimul-dlya-deurbanizatsiya-stanet-li-pandemiya-covid-19-katalizatorom-peremeshcheni (2020). Last accessed 09 Mar 2021
2. Roco, M.C., Harthorn, B., Guston, D., et al.: Innovative and responsible governance of nanotechnology for societal development. J. Nanopart. Res. **13**, 3557 (2011). https://doi.org/10.1007/s11051-011-0454-4
3. Insch, A.: Positioning cities: Innovative and sustainable strategies for city development and transformation. Place Brand Public Dipl **10**, 249–252 (2014). https://doi.org/10.1057/pb.2014.30
4. Gokhberg, L., Gershman, M., Rud, V., Streltsova, E.: Global Innovation Index–2020. Institute for Statistical Research and Economics of Knowledge, National Research University Higher School of Economics. https://issek.hse.ru/news/396120793.html (2020). Last accessed 18 July 2021
5. Mishina, M.: The current technological state of Russia and the difficulties of the transition to the sixth technological pattern. In: Science and Education: Problems and Prospects: Materials of the II International Scientific and Practical Conference, pp. 91–94 (2015)
6. Prokhorova, I.: Technological potential for the development of an innovative economy in Russia: the limits and limitations of the sixth technological stage. Bull. Univ. **2**, 68–75 (2020). https://doi.org/10.26425/1816-4277-2020-2-68-75

7. Bourgeois, R.: What Future for Rural Areas? Seven plausible rural transformations. Development **58**, 177–186 (2015). https://doi.org/10.1057/s41301-016-0017-1
8. Medinskaya, O., Randau, H.R.: Centralization and urbanization. In: Medinskaya, O., Randau, H.R., Altmann, C. (eds.) Russia Business. Springer, Cham (2021). https://doi.org/10.1007/978-3-030-64613-4_20
9. Brindley, T.: The social dimension of the urban village: a comparison of models for sustainable urban development. Urban Des Int **8**, 53–65 (2003). https://doi.org/10.1057/palgrave.udi.9000093
10. Karnaev, M.: Improvement of urban areas: concept and meaning. Issues Sustain. Dev. Soc. **7**, 207–210 (2020). https://doi.org/10.34755/IROK.2020.56.12.184
11. Kurchenkov, V., Koneva, D.: The value of the type of region in the integrated development of rural areas. In: Economics and Management of Industries, Complexes Based on an Innovative Approach: Materials of the International Scientific Conference Dedicated to the 50th Anniversary of the Kalmyk State University. B. B. Gorodovikov "and the 100th Anniversary of the Autonomy of the Republic of Kalmykia, pp. 265–268 (2020)
12. Shibaeva, S., Makurina, S., Tsukarev, S.: The essence and concept of rural areas: an integrated approach. Far Eastern Agrar. Bull. **2**(46), 199–207 (2018)
13. Bibri, S.E., Krogstie, J.: Environmentally data-driven smart sustainable cities: applied innovative solutions for energy efficiency, pollution reduction, and urban metabolism. Energy Inform. **3**(29) (2020). https://doi.org/10.1186/s42162-020-00130-8
14. Kurchenkov, V., Koneva, D.: Prospects for the development of rural areas in the post-pandemic period. In: Economy and Management of Industries, Complexes Based on an Innovative Approach: Proceedings of the International Scientific Conference, Elista, pp. 235–237 (2021)
15. Mardenskaya, E.: Features of the transition to sustainable development of rural areas of the Pskov region during a pandemic and economic crisis. Electron. Sci. J. **8**(37), 247–252 (2020)
16. Employment problems in rural areas amid the coronavirus pandemic: Official site of the Public Chamber of the Russian Federation. https://www.oprf.ru/press/news/2617/newsitem/53556 (2020). Last accessed 03 Sept 2021
17. Global Innovation Index 2020: Key Findings. Who Will Fund Innovation. Cornell University, INSEAD, WIPO. https://www.wipo.int/edocs/pubdocs/ru/wipo_pub_gii_2020_keyfindings.pdf (2020). Last accessed 13 July 2021
18. Official site of the Territorial Authority of the Federal State Statistics Service for the Volgograd Region. https://volgastat.gks.ru/storage/mediabank/. Last accessed 18 Jan 2020
19. Federal State Statistics Service: https://www.gks.ru/folder/12781/. Last accessed 06 Jul 2021
20. Federal Law of 06.10.2003 No. 131-FZ: On the General Principles of Organization of Local Self-Government in the Russian Federation. http://www.consultant.ru/document/cons_doc_LAW_44571/6d3b1321c4f9966d07ca33533fc7ca347581c3a8/ Last accessed 20 July 2021
21. Inshakov, O.V., Fesyun A.V.: Industry as a form of approval, expansion and domination of economic mode in economy. Sci. J. Volgograd State Univ. Glob. Econ. Syst. **2**(25), 44–53 (2014)
22. Inshakov, O.V.: Strategy and tactics of nanoindustry development in the Russian Federation. Econ. Reg. **3**, 246–250 (2011)
23. National projects are on the mend: The official website of the Russian newspaper. https://rg.ru/2020/06/22/biudzhetnye-dengi-pomogut-bystree-preodolet-posledstviia-pandemii.html (2020). Last accessed 03 Apr 2021
24. Demographic Yearbook of Russia: https://www.gks.ru/storage/mediabank/Dem_ejegod-2019.pdf/ (2019). Accessed: 18 Mar 2020

Chapter 4
Adaptation Potential of Inclusive Growth of the Regions of the South of Russia in the Context of the COVID-19 Pandemic

Inna V. Mitrofanova, Olga A. Chernova, Henrietta Nagy, and Marina V. Pleshakova

Abstract The ongoing economic crisis caused by the COVID-19 pandemic and its asymmetric impact on some territories evoke research interest in identifying factors that determine the resilience of regional economies to external shocks. Solving this problem implies the need to study the specifics of the territory adaptation capabilities as well as measures taken by regional authorities aimed at entering the trajectory of inclusive growth. The research goal was to assess the adaptation potential of regions to the coronavirus crisis and determine the development trajectory as a result of this potential realization. The authors determined directions of the development of adaptation potential of regions ("bounce forward" or "bounce backward"). For this purpose, the analysis of regional changes in financial results of organizations, volumes of industrial production and provision of services, and the unemployment rate was carried out. As a result, it was revealed that the "bounce forward" is characteristic of the South of Russia regions with a high level of socio-economic potential and a diversified economy. For the regions with a low level of socio-economic development, adaptation opportunities were expressed in a "bounce backward". This research contributes to the study of behavioral trajectories of regional economies in the conditions of the coronavirus crisis.

I. V. Mitrofanova (✉)
Federal Research Centre the Southern Scientific Centre of the Russian Academy of Sciences (SSC RAS), Rostov-on-Don, Russia
e-mail: mitrofanova@volsu.ru

Volgograd State University, Volgograd, Russia

O. A. Chernova
Southern Federal University, Rostov-on-Don, Russia
e-mail: chernova.olga71@yandex.ru

H. Nagy
Kodolányi János University, Budapest, Hungary
e-mail: nagy.henrietta@kodolanyi.hu

M. V. Pleshakova
Moscow City University, Moscow, Russia
e-mail: pmv23@list.ru

E. I. Inshakova and A. O. Inshakova (eds.), *New Technology for Inclusive and Sustainable Growth*, Smart Innovation, Systems and Technologies 287,
https://doi.org/10.1007/978-981-16-9804-0_4

4.1 Introduction

Today, at the level of major international organizations, the need for inclusive growth is recognized, and in the European Strategy for Smart, Sustainable, and Inclusive Growth ("Europe 2020", 2010), inclusive growth is considered as an instrument of social and territorial consolidation. The concept of inclusive growth is considered by some experts as "coordinated growth", which for national economies means their involvement in improving the quality of life of the population not only within the country, but also in a global context. This concept can be applied to regions and macro-regions in order to assess their involvement in the development of the country.

The research currently underway emphasizes that the inclusiveness of regional development in the context of the coronavirus pandemic is determined not so much by the region's ability to reflect shocks as by its adaptation potential, that is, the ability to adapt to changing conditions. Moreover, these shocks can not only have negative effects, but also have a positive impact on economic development, leveling inefficient practices and developing promising business processes and institutional structures. Under these conditions, the task of the regional authorities is to support positive transformational trends by helping citizens and businesses to adapt to these transformations. Given the significant differentiation of regions by the level of use and development of socio-economic potential, by the type of resource and geographical conditions, as well as by the nature of the pandemic processes, there are no uniform recommendations for preventing the negative impact of external effects. However, the analysis of patterns of regional development, the identification of factors that would contribute to an increase in the adaptation potential of the economy, is of great theoretical and practical importance for the country's development [1–3].

Southern Russian regions, most of which have a pronounced agro-industrial specialization, are most vulnerable to shocks due to the low level of development of high-tech industries that stimulate innovative activity in the region; the high proportion of small- and medium-sized businesses most affected by the coronavirus crisis; the low level of economic diversification, which generates excessive dependence of regions on individual industries; lack of highly qualified personnel; and underdevelopment of transport and engineering infrastructure [4, 5]. Therefore, the search for ways to increase the sustainability of the Southern Russian economies through the development of the adaptation potential of the region is of great theoretical and practical importance.

Therefore, the purpose of this study is to analyze the adaptation potential of the regional economy in the conditions of the coronavirus crisis. The object of the study is the regions of the Southern Federal District. The hypothesis of the study is that the adaptation potential of the region is determined by specific regional factors and conditions, as well as the nature of state restrictive and supportive measures.

The economic crisis caused by the spread of COVID-19 has significant differences from the so-called "traditional" economic crises. For example, R. Cortes and Johnston W. identify the following distinctive features of the coronavirus crisis, which should be taken into account when developing measures to overcome it: (1) formation, (2) focus, (3) temporality, (4) government jurisdiction, (5) preparedness, (6) normality, (7) business, and (8) operational deployment [6].

Analyzing and observing the consequences of the COVID-19 crisis, the researchers note that one of its most significant characteristics determining the response of national and regional economies to the crisis from the point of view of sustainability is temporality. The specificity of the manifestation of this characteristic is that the COVID-19 crisis is characterized by a higher rate of deployment than "traditional" economic crises. The sequential deployment of the first, second, and possibly the next waves of the coronavirus crisis makes the recovery period uncertain and cyclical. At the same time, there are significant differences in the speed, duration of individual stages, and trajectory of development in different countries and regions, largely determined by how their economies reacted to the "first blow" of the pandemic. In this context, it is of particular importance to study the impact of the pandemic precisely at the first stages of the COVID-19 crisis, when the adaptation potential of the region's economy manifests itself. The change in the socio-economic indicators of the region's development at this stage can be either insignificant or multiple, and thus largely predetermine the trajectory of the development of crisis processes.

The concept of "adaptation potential of the region" was introduced into scientific circulation as part of the study of evolutionary transformations in socio-economic systems. As noted by V. Muštra, B. Šimundić et al., adaptive changes in regional socio-economic and institutional structures are necessary: (a) to preserve or restore the former path of development of the region, or (b) to transition to a new sustainable path. At the same time, it is assumed that the regional system does not lose the ability of allocating resources at its disposal efficiently [7].

In the works of W. Bonß, X. Hu., R. Hassink, R. Martin et al., the idea of "bouncing back" and "bouncing forward" as possible options for implementing the adaptation potential of the region is reflected [8–10]. A rollback occurs when the economic recovery involves a return to the original pre-crisis parameters. For example, the economy of the region, having survived the shock, retains the same structural proportions, traditional business processes, and business strategies. With a leap forward, there are significant changes in the industrial base, enabling the region to return to its usual business, as well radically new business models appear. The direction of adaptive changes is determined by the current state of the regional socio-economic system, as well as a complex set of proactive and reactive factors and conditions, which we define as the adaptation potential of the region.

Despite the fact that a lot of attention is paid to the adaptive capabilities of the territory in modern research, there is no definition of what is meant by the adaptation potential of the region. The concept of adaptation potential is most frequently used to denote the conceptual phenomenon of the adaptation process, characterizing the degree of ability of the socio-economic system to adapt to environmental conditions [11, 12]. The term "adaptation potential" is usually identified with the term "adaptive capabilities/abilities".

The set of factors that are defined as determinants of the adaptive capabilities of the territory, according to the researchers, as a rule, includes characteristics of the socio-economic potential of the local economy and institutional conditions [13, 14]. Along with this, X. Hu, L. Li, K. Dong, C. Kakderi et al. note that in the early stages of the development of the crisis, the state influence is the most important and implies the introduction of various kinds of restrictive and supportive measures [15, 16].

The adaptive cycle model developed by R. Martin, P. Sunley, B. Gardiner B., and P. Tyler suggests that the adaptation potential of the region includes market innovations, training, and technological changes [see 10]. At the same time, a number of scientists identify social factors (social cohesion, social values and rules, and social trust) as the most important determinants contributing to the realization of adaptive opportunities, which enable the pooling of resources to counteract economic shocks [17].

4.2 Methodology

Despite the differences in approaches to assessing the adaptation potential of the region, the following main features can be identified by researchers to determine the ability of the region's economy to withstand external shocks and adapt to new conditions: the presence of a financial buffer, economic diversification, and government support. This ability has already manifested itself at the initial stage of the crisis, the so-called "first strike" stage. The nature of the reactions of the regional economy to emerging challenges at this initial stage determines the direction of further changes, launching the adaptation mechanism itself.

In this study, we will focus on the first wave of the pandemic in Russia, which began in March 2020. To analyze the adaptation potential of the regions, we will consider the dynamics of changes in the following indicators as the pandemic develops: financial results of organizations; indices reflecting changes in production volumes and the provision of services in areas of activity; the rate of unemployment.

The study of the trajectory of changes in these indicators will enable us to conclude to what extent the economy of the regions was able to adapt to the new realities and to assume in which direction the adaptation is taking place: a "leap forward" (growth of financial indicators, change in the structure of the economy, and low unemployment) or a "leap backward" (decline in financial indicators, preservation of the structure of the economy, and high unemployment).

4.3 Results

The uniqueness of the southern Russian regions consists in an extremely high degree of polarization of the socio-economic space, significant differentiation by industry structure, by the level of development of innovative potential, by the characteristics of natural resource potential, by the configuration of regional settlement systems, and many other features [18]. This causes differences in the South Russian regional economies in the levels of their sensitivity to the factors of the coronavirus crisis.

The official start date of the coronavirus crisis is March 12, 2020, when the first cases were recorded in Russia. Small businesses were the most sensitive to the "first blow". Thus, according to the Federal Tax Service, 66,820 individual entrepreneurs ceased their activities in Russia in March–April 2020. This is 77% more than in March 2019 (37.718). Including 64,237 entrepreneurs themselves decided to terminate the business. 1563 enterprises ceased to exist due to the death of the owner, and 158 enterprises were declared bankrupt. In the South of Russia, the largest number of small business entities that stopped operating was recorded in the Rostov region, which ranks first in the region in terms of the share of people employed in small business (15.5% of the employed population in the region).

As the results of the analysis show, according to the results of January–December 2019, the highest financial results were typical for the Krasnodar Territory. The Rostov region demonstrates a negative balance of financial results. However, according to the results of January–December 2020, we can note a significant improvement in financial indicators in the Rostov region (the second position in the region in the absolute value of the indicator), despite the massive closure of enterprises. The growth of financial results is also typical for the Republics of Kalmykia and Adygea (Fig. 4.1). The impact of the coronavirus crisis was manifested in the fact that outdated and inefficient business models left the market, and the remaining, more effective practices were spread and developed.

Paradoxically, the best indicators of adaptability were demonstrated by enterprises in those regions that had a low financial buffer. Whereas the regions with a high

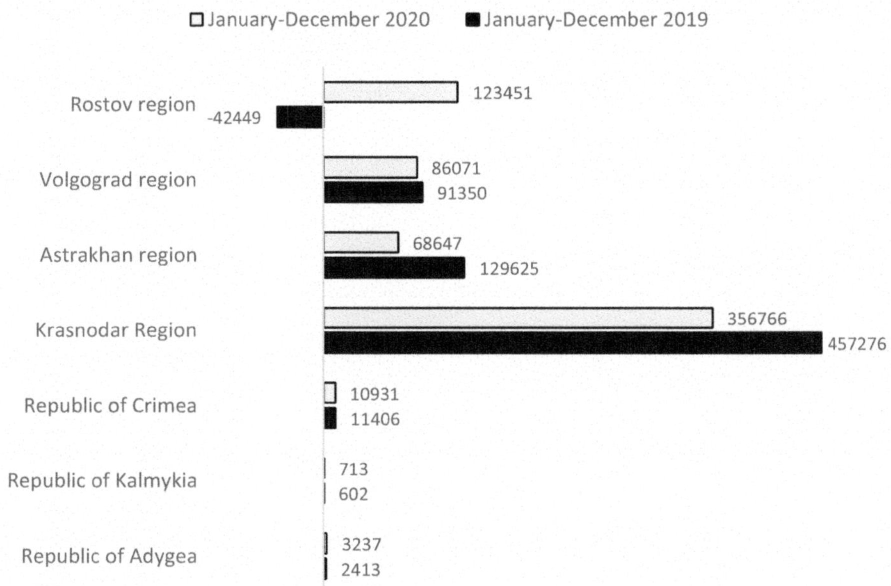

Fig. 4.1 The net financial result of the activities of organizations in the southern regions of Russia, million rubles. *Source* Developed and compiled by the authors based on [19]

value of financial results showed the highest level of their decline as a result of the coronavirus crisis.

Considering the change in structural proportions, we will focus on Rosstat (Russian Statistical Bureau) monitoring data, according to which the following areas of activity were identified: industrial production, agriculture, construction, transport, trade, and services. Indices of production volumes (rendering of services) by the studied regions for the period January–December 2020 in comparison with the corresponding period of 2019 are shown in Table 4.1.

From the data presented in Table 4.1, it can be seen that the conclusion is confirmed that the least adaptation potential was realized in the regions with the highest indicators of socio-economic development before the pandemic. Thus, in the Krasnodar Region and the Volgograd Region, there is a decrease in production volumes in all spheres of activity, the only exception is agriculture in the Volgograd region. At the same time, regions that do not have a high level of socio-economic development, nevertheless, showed growth in certain sectors of the economy. For example, in the Republic of Adygea, there is an increase in transport services, industrial production, and agriculture; in the Astrakhan region, there is an increase in construction; and in the Rostov region, despite the negative balance of the financial result of enterprises at the beginning of 2020, by the end of 2020, there was an increase in industrial production.

However, it cannot be said that it is regions with low indicators of socio-economic development that have a higher adaptability potential compared to developed regions. For example, in the Republics of Crimea and Kalmykia with low socio-economic potential, the indices of production volumes also significantly decreased in almost all

Table 4.1 Indices of production volumes (rendering of services) in the southern regions of Russia, percentage

Region	Industrial production	Agriculture	Construction	Transport	Trade	Rendering of services
Republic of Adygea	105.1	114.1	77.8	120.0	97.1	92.3
Republic of Kalmykia	92.1	91.5	84.3	54.8	101.1	92.0
Republic of Crimea	99.8	85.0	90.0	100.6	92.0	87,5
Krasnodar Region	97.0	91.3	98.2	97.7	91.7	90.1
Astrakhan Region	97.2	102.3	126.1	87.4	93.7	87.7
Volgograd Region	99.3	101.7	96.2	90.3	93.3	86.5
Rostov Region	101.7	97.1	95.2	118.8	93.0	89.2

Source Developed and compiled by the authors based on[19]

spheres of activity. These data show that the level of socio-economic development of the territory is not the key determinant for the realization of its adaptive production potential.

Our research enables us to conclude that the measures taken by the regional authorities in the conditions of the coronavirus crisis played a significant role in realizing the adaptation potential of the region. The authors believe that it was the state that influenced the adaptive capabilities of regional economies through the restrictions imposed, as well as supportive measures.

The strongest state support to business in the conditions of the pandemic was provided to the Rostov region. In this region, income tax rates were reduced, direct financing channels were significantly expanded, and preferential loans with deferred payment were provided. Whereas in the Krasnodar Region, tax support measures included only postponing the payment of taxes, and financial and credit measures consisted of issuing loans for urgent needs and salary payments, loan restructuring. Moreover, only in the Krasnodar Region among the southern Russian regions, strict quarantine measures were introduced by means of special passes for movement. In this region, restrictions on the activities of a number of enterprises were eliminated later than in other regions. This was due to the desire of the authorities to prevent the flow of tourists to the region to prevent the spread of the coronavirus crisis. As a result, the economy of the Krasnodar Region showed the lowest adaptability indicators in the South of Russia.

In general, despite the business support packages being introduced, the economic situation in the regions remained quite difficult. The deferral of lease payments extended only to state and municipal property, leaving entrepreneurs renting a private property in a very vulnerable position. Facilitating access to finance during the pandemic did not solve the long-term problems of business development. The sector of small- and medium-sized businesses remains unattractive for lenders. Moreover, a sharp decline in demand and services led to the situation when it was easier for an entrepreneur to declare bankruptcy than to take loans from a bank to pay wages to his employees.

Considering the indicators of production and economic activity in the regions in the sectoral context, it can be noted that agriculture has demonstrated the greatest resistance to the factors of the pandemic. This is due to the fact that agricultural production did not suspend its activities during the coronavirus crisis due to its importance for ensuring the food security of the national economy. In addition, outdoor work restrained the spread of the virus, as a result of which the number of cases in this field of activity was lower. Industrial production as a whole was also able to adapt to the new realities and in most regions return to the previous pre-crisis level.

The largest decrease in performance indicators is typical for retail trade, rendering services to the population, as well as the catering sector. It was during the first wave of the pandemic that restrictions were imposed on the activities of enterprises in this area, as well as on the movement of citizens, which led to a sharp decline in business activity. However, as restrictions were lifted, a period of recovery growth began, when enterprises and organizations were able to begin realizing their adaptation potential. As a result, despite the existing differences in the pace of development, most of the southern Russian regions have maintained positive dynamics of indicators, practically bringing them to the value of the pre-crisis period.

Analyzing the change in the unemployment rate in the regions, it can be noted that its dynamics is directly related to the level of economic diversification. The number of unemployed increased to a greater extent due to the suspension and termination of activities during the lockdown of enterprises in the service sector, catering, hotel business, etc. However, in economically developed economies with a differentiated industry structure, the unemployment rate turned out to be significantly lower, since laid-off workers were able to find work in other areas. Thus, the highest unemployment rate is registered in the Republics of Kalmykia and Adygea, in the Astrakhan Region, where, despite the gradual recovery growth of the regional economies, many citizens who lost their jobs could not find a job, the lowest—in the Volgograd Region and Krasnodar Region. At the same time, it is in these regions that the rate of unemployment reduction is the highest.

According to Rosstat data (Fig. 4.2), in all regions, the peak of unemployment occurred in the period August–October.

To characterize the trajectory of adaptation changes in the regions, the regions are grouped into particular types in accordance with a combination of the following features: a change in the net result of financial activity; a change in structural proportions in growth indices; a change in the unemployment rate. Possible types of scenarios are presented in Table 4.2.

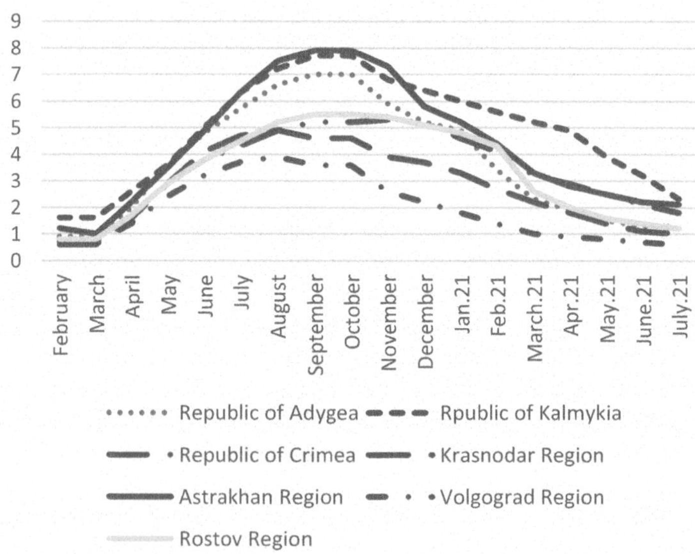

Fig. 4.2 Unemployment rate in the regions of the South of Russia, %. *Source* Developed and compiled by the authors based on [20]

Table 4.2 Types of the trajectory of adaptive changes

Changes in indicators	Level of change							
Change in the net result of financial activity	Growth				Decrease			
Changes in structural proportions of growth indices	High		Low		High		Low	
Changes in the rate of unemployment	High	Low	High	Low	High	Low	High	Low
Type of scenario	Bounce forward	Bounce forward	Bounce backward	Bounce forward	Bounce backward	Bounce forward	Bounce backward	Bounce backward

Source Developed and compiled by the authors

In cases where there is an increase in financial activity indicators in the region, then with a decrease in the level of employment, we can talk about the presence of a "leap forward". At the same time, the structural proportions of changes in growth indices may or may not be significant. Structural changes will indicate that some industries and spheres of activity receive an impetus for development in new conditions when other industries and spheres of activity (as a rule, these are traditional industries) reduce their business activity. If effective practices are consolidated in traditional activities, then significant changes in the industry structure may not occur.

A decrease in the financial result with an increase in the unemployment rate and a low change in structural proportions suggests that adaptation is following a "leap backward" trajectory. A decrease in the unemployment rate with high structural changes, even with a decrease in the financial result, may indicate the formation of a "leap forward" trajectory—the gradual consolidation of new business models in the economy.

The characteristics of the features and the type of trajectory of adaptive changes determined by the combination of these features are presented in Table 4.3.

According to the results of the analysis presented in Table 4.3, the "leap forward" occurred in most regions with higher indicators of socio-economic development and was characterized by a diversified economy: the Rostov region, the Volgograd Region, and the Krasnodar Region. In the Republics of Adygea, Kalmykia, Crimea, and the Astrakhan Region, the adaptation trajectory is developed along the path of a "leap back". The conclusions, that diversified economies are more resistant to the coronavirus crisis, are generally consistent with studies conducted by other authors.

Nevertheless, in order to conclude exactly what changes contributed to the formation of such a direction of the adaptive trajectory of development, additional research is needed, which determines the further tasks of the scientific search for authors.

Table 4.3 The type of trajectory of adaptation changes in the regions of the South of Russia

Regions	Change in the net result of financial activity		Changes in structural proportions of growth indices		Changes in the rate of unemployment		Type of scenario
	Growth	Decrease	High	Low	High	Low	
Republic of Adygea	+			+	+		Bounce backward
Republic of Kalmykia	+			+	+		Bounce backward
Republic of Crimea		+	+		+		Bounce backward
Krasnodar Region		+	+			+	Bounce forward
Astrakhan Region		+		+	+		Bounce backward
Volgograd Region		+	+			+	Bounce forward
Rostov Region	+		+		+		Bounce forward

Source Developed and compiled by the authors

4.4 Conclusion

The consequences of the coronavirus crisis have different manifestations in the southern Russian regions. This is due both to the margin of safety that the regional economy had before the outbreak of the pandemic and to the level of state support and the rigidity of the restrictive measures introduced.

The availability of a financial buffer was not a determining factor for ensuring a higher level of adaptability of regional economies. Moreover, regions with high socio-economic potential were among the most affected by the coronavirus crisis. This is largely due to the fact that regions with a high level of business activity in the conditions of the imposed restrictions had numerous connections destroyed, which put them in a vulnerable position at the very beginning of the coronavirus crisis. Additional problems for these regions were connected with the lack of sales in the conditions of the border closure.

Most of the southern Russian regions have a weak ability to withstand the factors of the pandemic. Their adaptation potential is realized mainly along the trajectory of a "leap backward", which is expressed in a return to previous business models and business strategies as restrictions are lifted. The sectoral structure of the economy and the degree of development of interregional ties were of decisive importance in realizing the adaptation potential. Regions in which, along with industrial production, there were agricultural enterprises (regions of agro-industrial specialization) demonstrated a higher level of adaptability compared to regions with an undiversified economy, such as the Republics of Kalmykia and the Crimea.

State support significantly contributed in realizing the adaptation potential of the regions. It is the differences in the anti-crisis measures taken at the regional level that explain the differences in the adaptability of their economies. In this regard, it can be said that the level of adaptability of regional economies is largely determined by the extent to which regional authorities are able to make effective management decisions [21, 22]. That is, the quality of regional governance is more important for increasing the ability of the economy to resist external shocks than structural factors. This is of particular importance for the initial stage of the development of the crisis. The further ability to adapt and create conditions for inclusive growth is determined by how effectively economic agents will be able to use the resources provided to them, and consequently, the importance of the social trust factor increases in the long term. Further research of the authors will be devoted to the study of the role of this factor in increasing the adaptability of regional systems.

Acknowledgements The publication was prepared as part of the implementation of the State Assignment of the Southern Scientific Centre of the Russian Academy of Sciences for 2022, project "Strategic vectors of development of the socio-economic complex of the south of Russia taking into account regional resilience (economic and demographic aspects)".

References

1. Kazakova, S.M.: Sustainability Assessment of Russian Macro-Regions by Means of the Index of Inclusive Development. Regional Economy. South of Russia, 8 (4), 30–38 (2020). https://doi.org/10.15688/re.volsu.2020.4.3
2. Europe 2020: A European strategy for smart, sustainable and inclusive growth (2020). http://ec.europa.eu/research/era/docs/en/ec-understanding-era-13.pdf. Accessed 01 Sept 2021
3. Ljaskovskaja, E.A., Grigorjeva, K.A.: The Rating of Inclusive Development of Macro-Regions of the Russian Federation. Bulletin of the South Ural State University. Ser. Econ. Manag. **12**(2), 45–53 (2018)
4. Mitrofanova, I.V., Chernova, O.A., Pyankova, S.G., Batmanova, V.V.: Innovative impact of import substitution processes in the economy of the South of Russia. In: Inshakova A., Inshakova E. (eds) Competitive Russia: Foresight Model of Economic and Legal Development in the Digital Age. Lecture Notes in Networks and Systems. 110, 49–58 (2020). Springer, Cham. https://doi.org/10.1007/978-3-030-45913-0_6
5. Mitrofanova, I.V., Shkarupa, E.A., Batova V.N.: Export of agricultural products: trends and directions of state support in modern conditions. Reg. Econ. **9**(2), 131–140 (2021). https://doi.org/10.15688/re.volsu.2021.2.13. (South of Russia)
6. Cortez, R.M., Johnston, W.J.: The Coronavirus crisis in B2B settings: crisis uniqueness and managerial implications based on social exchange theory. Ind. Mark. Manag. **88**, 125–135 (2020). https://doi.org/10.1016/j.indmarman.2020.05.004
7. Muštra, V., Šimundić, B., Kuliš, Z.: Effects of smart specialization on regional economic resilience in EU. First SMARTER Conference on Smart Specialisation and Territorial Development: Changing Patterns of Territorial Policy: Smart Specialisation & Innovation in Europe, Sevilla, Španljoska (2016). https://www.bib.irb.hr/842329

8. Bonß, W.: The notion of Resilience: Trajectories and social science perspectives. In: Mauer A. (ed.) New Perspectives on Resilience in Socio-Economic Spheres, pp. 9–24. Springer VS, E-book (2016)
9. Hu, X., Hassink, R.: Exploring adaptation and adaptability in uneven economic resilience: a tale of two Chinese mining regions. Camb. J. Reg. Econ. Soc. **10**(3), 527–541 (2017). https://doi.org/10.1093/cjres/rsx012
10. Martin, R., Sunley, P., Gardiner, B., Tyler, P.: How regions react to recessions: resilience and the role of economic structure. Reg. Stud. **50**(4), 561–585 (2016). https://doi.org/10.1080/00343404.2015.1136410
11. Zemtsov, S., Barinova, V., Semenova, R.: The risks of digitalization and the adaptation of regional labor markets in Russia. Foresight STI Gov. **13**(2), 84–96 (2019). https://doi.org/10.17323/2500-2597.2019.2.84.96
12. Ziablitckaia, N.V., Ishniyazova, A.R., Prokopev, A.V., Prokopeva, T.V.: Adaptation potential assessment of oil and gas enterprises. Qual. Access Success **20**(170), 15–18 (2019)
13. Korolev, V.A., Toroptsev, Y.L., Matveeva, L.G., Chernova, O.A.: Modeling of dynamic properties of national economic systems. Serbian J. Manag. **13**(1), 133–143 (2018). https://doi.org/10.5937/sjm13-12991
14. Martin, R.: Regional economic resilience, hysteresis and recessionary shocks. J. Econ. Geogr. **12**(1), 1–32 (2012). https://doi.org/10.1093/jeg/lbr019
15. Hu, X., Li, L., Dong, K.: What matters for regional economic resilience amid COVID-19? Evidence from cities in Northeast China. Cities, p. 103440, ISSN 0264-2751 (2021). https://doi.org/10.1016/j.cities.2021.103440
16. Kakderi, C., Tasopoulou, A.: Regional economic resilience: The role of national and regional policies. Eur. Plan. Stud. **25**(8), 1435–1453 (2017). https://doi.org/10.1080/09654313.2017.1322041
17. Huggins, R., Thompson, P.: Local entrepreneurial resilience and culture: The role of social values in fostering economic recovery. Camb. J. Reg. Econ. Soc. **8**(2), 313–330 (2015). https://doi.org/10.1093/cjres/rsu035
18. Nikitaeva, A.Yu., Kiseleva, N.N.: Reconfiguration of business models of industrial enterprises: vectors of increasing sustainability in the new realities. J. Volgogr. **23**(1), 110–120 (2021). https://doi.org/10.15688/ek.jvolsu.2021.1.9. (State University. Economics)
19. Information for Monitoring Social and Economic State of the Subjects of the Russian Federation. Rosstat (2020). https://rosstat.gov.ru/folder/11109/document/13259. Accessed 20 Aug 2021
20. Report “Social and Economic State of Russia”. Rosstat (2020). https://rosstat.gov.ru/compendium/document/50801. Accessed 01 Sept 2021
21. Inshakov, O.V.: Strategic ideas from regions–potential of effective consolidation (chief editor’s deliberations). Reg. Econ. **1**(15), 4–12 (2017). https://doi.org/10.15688/re.volsu.2017.1.1. (South of Russia)
22. Inshakov, O.V.: Strategic planning of the social and economic development of a region: scientific substantiation and updating of a model. Reg. Econ. **1**(19), 23–43 (2018). https://doi.org/10.15688/re.volsu.2018.1.3. (South of Russia)

Chapter 5
The Success Model to Manage the Cross-Border Infrastructure Projects

Hans-Christian Brauweiler, Aida Yerimpasheva, Dina Alshimbayeva, and Aida Myrzakhmetova

Abstract Cross-border cooperation (CBC) projects are essential for both sustainable development and creating an environment for entrepreneurship and investment. There is a strong correlation between the successfulness of implemented projects and the sustainable development of border regions. Because of the One Belt One Road (OBOR) initiative, the management of cross-border projects is becoming increasingly important, as Kazakhstan is located at the intersection of transcontinental corridors between Europe and China. A feature of cross-border interaction is that projects are implemented by different countries together, increasing the risks of cross-border projects. In this regard, it was essential to analyze successful examples of CBC projects to understand better how they affect and contribute to regional development and, based on the information collected, build a successful model of the cross-border and infrastructure projects. This paper explores the issues to build the model to manage cross-border projects successfully. In so doing, the study contributes to the search for critical success factors that underpin the model. There are two contributions to cross-border project management knowledge. The first contribution is in revealing the critical success factors. The second contribution demonstrates how the success model has been built. Our findings are a call for more research connecting the OBOR and project management issues.

H.-C. Brauweiler (✉)
WHZ Zwickau University of Applied Sciences, Zwickau, Germany
e-mail: christian.brauweiler@fh-zwickau.de

A. Yerimpasheva · A. Myrzakhmetova
al-Farabi Kazakh National University, Almaty, Kazakhstan
e-mail: aida.zakirova@kaznu.kz

A. Myrzakhmetova
e-mail: aida.myrzakhmetova@kaznu.kz

D. Alshimbayeva
Satbayev University, Almaty, Kazakhstan

E. I. Inshakova and A. O. Inshakova (eds.), *New Technology for Inclusive and Sustainable Growth*, Smart Innovation, Systems and Technologies 287,
https://doi.org/10.1007/978-981-16-9804-0_5

5.1 Introduction

Cross-border cooperation (CBC) projects are essential for sustainable development and creating an environment for entrepreneurship and investment [2]. From the beginning, the European Union (EU) promoted the idea that an increase in CBC "contributes to enhanced European integration," "supports sustainable development" along borders, and "helps reduce differences in living standards" [11]. More than a third of EU citizens live and work in border regions. Borders, thus, both directly and indirectly affect their lives [7]. European Territorial Cooperation (INTERREG programs) "plays an important role in removing barriers and promoting sustainable cross-border cooperation" [24]. European researchers state about a strong correlation between implemented projects and the sustainable development of border regions [9, 19–21, 24, 26].

It is crucial to analyze successful examples of CBC projects to understand how they influence and contribute to regional development and construct the success model of cross-border projects based on the data collected.

According to Joanna Kurowska-Pysz [15], cross-border projects strengthen relationships between partners and are aimed at "achieving mutual objectives" that are important both for cooperating organizations and for the development of borderland where these projects are being implemented. Studying all factors that influence the success or failure of cross-border projects is critical "to reaching long-lasting territorial sustainability" [5]. At the same time, Liu et al. [18] highlight the importance of building "institutions for achieving effective cross-border governance," paying much attention to the development of CBC in technological innovation. Due to economic integration, "a growing number of related projects are under planning or construction" [16].

Sanotska et al. [23] believe that market failures in border areas should be resolved at the state level; in particular, the programs of CBC developed by the governments of neighboring states should become the primary tool for moving toward economic equality in the regions of both countries.

The sources of the "innovation-entrepreneurial ecosystem's growth" are the internal and external factors of development. External factors include "a proactive innovation policy" at the regional level with the support of "science-intensive manufacturers" [12].

There is no universally accepted definition of a cross-border project, although this term is popular and widely used. Two or more countries are implementing a cross-border project. On the other hand, any national infrastructure project can also be considered a cross-border project if it significantly influences the cross-border region. Examples of national projects with significant cross-border impact are transport infrastructure projects (land, sea, and air transport) that facilitate international transport, energy projects, the sale of electricity to neighboring countries, and the construction of cross-border telecommunication networks.

The Asian Development Bank's Regional Cooperation and Integration Strategy do not clearly define the concept of "cross-border infrastructure," but its distinctive feature is the presence of cross-border external effects not created by national projects. The cross-border infrastructure creates "additional" benefits that affect participating countries. It means that the net benefits from cross-border projects are greater than those obtained only in the framework of national projects [10]. There can be no doubt that infrastructure is critical to managing international trade across international borders. Transport infrastructure "has a long-run positive impact on economic development" [1]. At the same time, cross-border projects are "accompanied by high complexity, difficulty, and uncertainty" [17].

The analysis of [10] has shown that cross-border projects do not attract sufficient attention from relevant research, and there is no systematic study of cross-border project management models. Fujimura M. and R. Adhikari identified essential factors that contribute to the successful preparation and implementation of cross-border infrastructure projects [10]. They dwell upon the next ones: clear justification and formulation of the corridor's concept; using distribution analysis (as a component of situational analysis and input data for SWOT analysis) as a guide for a win–win option; fair public–private partnership; overcoming the limitations of transport infrastructure.

However, from the point of view of project management, these factors are not fundamental, although we do not deny them and understand that the more positive external effects of cross-border projects lead to better management and more stable cooperation between countries that implement the project and, therefore, the project itself can be more successful. We have decided to build and propose the success model of managing cross-border projects.

In general, there are two concepts in project management: (i) project success and (ii) project management success. Project success is the ultimate goal of any project, regardless of industry affiliation and sources of resource support. The project success characterizes the product's properties: its quality, its implementation on time, and the PR perception of the project product. In turn, the project management success is associated with compliance with the agreed parameters of the project, the level of professionalism, the skills of the project team, and the capability to interact with all stakeholders.

5.2 Literature Review

In recent decades, CBC has reached an unprecedented level not only because of its potential for territorial integration but also taking into account its role in supranational processes such as infrastructure construction, activity planning, and project implementation [3]. This interaction is becoming increasingly important when we live in a globalized world [4]. This scenario is evident in the case of the EU, the EAEU, and the OBOR initiative. In this regard, identifying success factors in managing cross-border projects is considered critical for achieving sustainable development through CBC strategies that lead to a consistent improvement in the quality of life of the living population in these regions. From our point of view, it was essential to determine the set of these factors for managing cross-border projects successfully.

CBC, becoming one of the most critical cooperation priorities within the EAEU and the OBOR initiative, aims to strengthen cooperation between the EAEU member states and partner countries within the Silk Road Economic Belt.

CBC projects "can play a significant role in developing the local community" [13]. In addition, [13] argues that the cross-border nature of projects not only promotes mutually beneficial "transfer of knowledge and experience" but also "gives weight to a broader perspective" that a particular local need acquires through CBC.

At the same time, CBC has some disadvantages in situations of economic and institutional asymmetries. In this case, cross-border projects are subject to a higher risk. Peña Medina [22] believes that a central actor or network coordinator would allow the two administrative systems to be brought together and efficient cooperation. Although the public–private partnership (PPP) model is more effective in providing public goods, any CBC project case should be addressed separately [22]. Korneliuk et al. [14] have determined that in the conditions of strengthening integration processes between the border states, the need "to expand and deepen international relations between local self-government bodies grows."

De Soyres, Mulabdic, and Ruta estimated the effects of the OBOR. According to them, the OBOR projects "increase GDP by up to 3.35% and 2.61% for BRI and non-BRI economies" [8]. On the other hand, for Kazakhstan, cross-border cooperation within the framework of OBOR should be carried out through the EAEU. In addition, the EU's experience might come in handy.

Castanho et al. [6] have identified critical factors for success in the EU CBC projects. The findings of their study have coincided with our results on three critical factors: (i) General goals and plans of projects implemented by countries, strong territorial strategy; (ii) Political transparency and commitment; (iii) Good relations between countries (cities) to communicate. It suggests that the problems of CBC projects are typical for any region.

CBC can be divided into three distinct stages of development. At the first stage, the member states participating in the EAEU receive expert knowledge and undergo training, organize study tours to member states, and conduct feasibility studies on the potential of the border regions. It would create essential links between the member states and China. At the second stage, plans should be drawn up for socio-economic convergence between the border regions. Among other things, the planning of investment projects facilitates convergence. At the third stage, the regions jointly recognize the importance of regional development. Possible investments should be aimed at the formation of cross-border infrastructure to remove barriers at a border crossing. Regional bodies develop joint projects, and direct contacts between citizens and communities across the border are maintained.

Each country has its specific priorities to solve the common problems that it faces in CBC that is aimed at achieving common strategic goals as: to promote economic and social development; to solve common problems in the field of environment, health, safety, and security; and to ensure mobility of people, goods, and capital. The strength of integration of neighboring countries lies in cooperation across national borders. By combining the resources of individual countries, it is possible to achieve more significant and better results that benefit everyone.

The OBOR Initiative consists of two projects: the twenty-first Century Marine Silk Road and the Silk Road Economic Belt, whose goals are to deepen economic integration, eliminate barriers to investment and trade, and create a unified transport infrastructure. The Initiative supports infrastructure projects that go beyond national borders to improve cargo mobility and the ability of people to travel throughout Eurasia and beyond.

Dozens of cross-border projects involving at least two countries, either completed or under implementation, among hundreds of similar projects implemented worldwide, have outstanding achievements in creating a faster, safer, and more environmentally friendly transport network.

The lessons learned from successful cross-border projects can and should be used by other projects implemented within the EAEU and OBOR.

Cross-border infrastructure is necessary to support the economic development of states and the well-being of people. Cross-border infrastructure, located on the border of two or more states, having a beginning on the territory of one state and continuation on the territory of another, has specific characteristics. Cross-border projects based on interstate cooperation and within the framework of international corridors and international initiatives aim to create cross-border infrastructure to ensure the integration of countries in various industries. These projects solve problems that go beyond national development.

The first stage helps identify potential projects that meet the requirements of donors. For selecting projects that meet the requirements of funding organizations, a clear justification of cross-border and infrastructure projects is very important. The cross-border and infrastructure project compliance is determined by the requirements and procedures established by the funding organizations or source of funding. When analyzing the project's compliance with the requirements of donor organizations, it is essential to identify both problems and opportunities.

The second stage is aimed at preparing the project's documentation. In order to meet the requirements of the financing organization, the project's design must be developed under the established requirements for the selection, rules for filing applications, and all necessary procedures. At this stage, financial institutions can assist in preparing documentation and determining the financing source of the project.

Another critical step in preparing project documentation that has to meet the requirements of the financing organization is to understand the project cycle of the financial institution where the project proposal will be sent since any organization investing in a project has its project cycle. Particular attention should be paid to risk management because any discrepancies between the legislative framework and project management practice can create specific difficulties and thereby increase the investment risks of projects. In order to ensure that the cross-border project infrastructure project meets the requirements of the financing organizations, it is crucial to determine the protocol for sharing risks between the stakeholders at the early stage. It is essential to pay attention to the feasibility study, which should include an assessment of various scenarios and the cross-border risks. One of the ways to manage risks should be insurance.

Another effective tool aimed at ensuring compliance of projects with the requirements of the financing organization is marketing research. Marketing research helps to establish a complete list of all stakeholders and identify and evaluate their interests concerning the project before its initiation.

The demand for cross-border infrastructure in Asia far exceeds available funding [10]. Development banks provide funds to support many cross-border projects to promote the social and economic development of the participating countries. OBOR has become a "catalyst…for facilitating transnational transport infrastructure projects" [25].

At the policy level, there are two main economic rationales for regional cooperation between two or more countries: (i) The need to take into account additional opportunities associated with the cross-border project and its positive and negative externalities; (ii) Defining the potential for achieving economies of scale to pursue national goals.

By doing this, all participating countries benefit from regional cooperation. However, the removal of physical and non-physical barriers to realize these benefits requires investment, as well as harmonization and simplification of relevant policies and procedures.

As for the first part of the primary economic rationale for regional cooperation, cross-border projects may attract additional preferential and non-preferential funds. Positive externalities (for example, benefits such as saving time and money, protecting the environment, and facilitating trade) and negative externalities (such as costs such as environmental pollution, human trafficking, and the spread of infectious diseases) occur when the consequences of one or several countries go beyond national borders. If the countries do not make joint agreements, there will be too few positive external factors and too many negative ones. As for the second rationale, regional programs and cross-border projects can provide economies of scale in providing public or private (market) goods and services beyond what any country can achieve alone. Thus, regional cooperation can help achieve national goals.

Cross-border infrastructure projects have an essential role in optimizing resource allocation, promoting sustainable development, and enhancing regional security.

5.3 Methods

To build the success model for managing cross-border projects (project management success model), we have determined the composition of the criteria taking into account the specifics of CBC. To simplify the model, we decomposed criteria into elements. When choosing the criteria, we were guided by the fact that the criteria for successful project management reflect the context and content components. The context (external environment of the project) determines criteria related to CBC. The criteria related to the internal environment of the project are attributed to the content

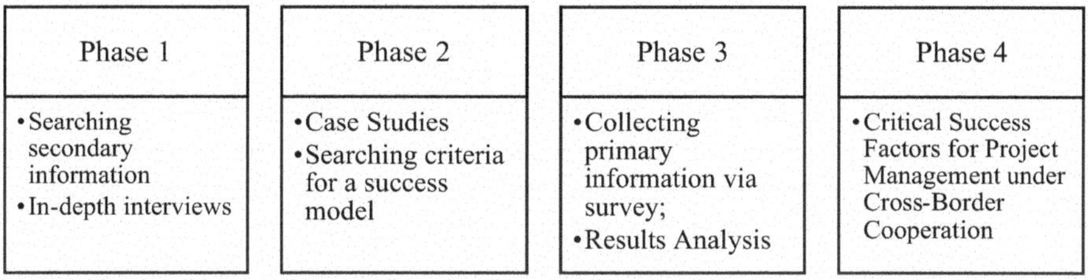

Fig. 5.1 Phases of the research. *Source* Compiled by the authors

and relate to any project. Therefore, the task of building the success model has been reduced to searching: (a) Context criteria related to CBC, (b) Content criteria related to project management issues, and (c) A number of criteria included in the success model.

To simplify the managerial processes of the project, we have narrowed down the number of criteria to a vital number of indicators using the Pareto chart. Based on field studies, both qualitative and quantitative ones, we have compiled a list of critical criteria for the success model and then identified the most critical ones.

The study aimed to identify and analyze the critical factors that determine the success of project management in cross-border projects and find out where the primary efforts should be applied to achieve sustainable development. As successful cross-border projects represent the critical step toward the prosperity of border areas, demand for these projects helps to create the basis to identify the main structural changes and policies aimed at solving the problems of regional sustainability.

However, developing a complete universal list of criteria for all projects seems impossible or at least difficult. In this regard, to achieve the study's objectives, we have decided to choose such a design of the study that would include searching both secondary and primary information, research methods such as desk and field, preliminary and conclusive, including descriptive cross-sectional ones. The study has consisted of the phases shown in Fig. 5.1.

During the first phase, qualitative secondary and primary studies were used. Qualitative research in the form of in-depth interviews was conducted with project managers, civil servants, and experts. Interviews were also conducted with experts and specialists from the Institute of Project Management of Satbayev University.

During the second phase, we studied information directly related to cross-border projects within the framework of the OBOR initiative. The third and most active phase included a survey conducted online and the analysis of the results, which made it possible to determine the critical factors for the success of managing projects and to build a model for the success of managing cross-border projects.

5.4 Results

Thus, a literature review, combined with interviews with technical specialists, experts, key participants, and stakeholders of the CBC process and case study analysis of cross-border projects within the framework of the OBOR, revealed preliminary criteria for successful project management. In the case study analysis, information was analyzed on 20 projects implemented in Kazakhstan under the OBOR, which allowed identifying the 14 critical factors that determine the success of managing cross-border projects shown in Table 5.1.

The list of critical success factors for managing cross-border projects has become the deliverables of phase 2. Moving to the next phase, we compiled a questionnaire for online-offline surveys. The circle of respondents for this field research included specialists, project management experts, and employees of the performing organizations as team members and other stakeholders. To conduct a survey, we chose non-probability sampling; that is, we used convenience and snowball sampling methods. This type of sample was determined by the specifics of the subject of study. The sample size was N = 77 respondents related to project management and CBC.

The questionnaire for the survey was made both in electronic form (Google Form) and in paper form. The central part of the questionnaire was devoted to the critical success factors of managing cross-border projects (Table 5.1) and the general 12 critical success factors of managing any projects shown in Table 5.2, also determined by using case study analysis.

Because of the quarantine that took place in Spring 2020, the sample size turned out to be less than planned, but this did not prevent us from drawing certain conclusions that were necessary to build the success model for managing cross-border projects. The survey results are presented in Figs. 5.2 and 5.3.

A survey conducted among specialists, project management experts, as well as employees of the performing organizations, including members of project teams, as well as stakeholders, showed that the following factors are the most crucial ones for the success of managing cross-border projects:

Table 5.1 Critical success factors for managing cross-border projects identified using case study analysis

(1) Relations between countries (cities)	(8) Ensuring quality of life standards
(2) Attracting young and talented people	(9) General goals and plans of projects implemented by countries;
(3) Strengthening political commitment	(10) Citizen participation in decision making
(4) Strong territorial strategy	(11) Prevention of duplication of infrastructure
(5) Enhancing a sense of belonging	(12) Availability of diverse infrastructure
(6) Access to investment	(13) A stronger economy
(7) Political transparency and commitment	(14) Marketing and advertising

Source Compiled by the authors

Table 5.2 Critical success factors of project management, identified using case study analysis

(1) Clear goal setting	(7) Competent project management
(2) Proper decomposition of project objectives	(8) Support for project curators
(3) Promoting transparency	(9) Competent members of the project team and its permanent membership
(4) Commitment to project-related decisions	(10) Adequate information support
(5) Promoting communication within the project	(11) Search and correction of deviations according to the schedule and budget
(6) Promoting communications by stakeholders	(12) Feedback and customer focus

Source Compiled by the authors

Fig. 5.2 Survey results on critical success factors for managing cross-border projects. *Source* Compiled by the authors

1. (1) Relations between countries (cities);
2. (6) Access to investment;
3. (4) Strong territorial strategy;
4. (9) General goals and plans of projects implemented by countries;
5. (7) Political transparency and commitment;
6. (12) Availability of diverse infrastructure;
7. (13) A stronger economy.

The listed factors from the most important to the less important were selected under the Pareto 80/20 law (Fig. 5.3). Thus, using field studies of online/offline surveys, seven key variables were selected, which formed the basis for the success model of cross-border projects under the OBOR initiative.

The following are the results of a survey on critical success factors for project management (Fig. 5.4).

The survey results showed that project management experts put in the first place such criteria for the success of project management as a clear goal setting, facilitating communication within the project, and finding and correcting deviations on the schedule and budget (which is essentially an analysis of the mastered volume).

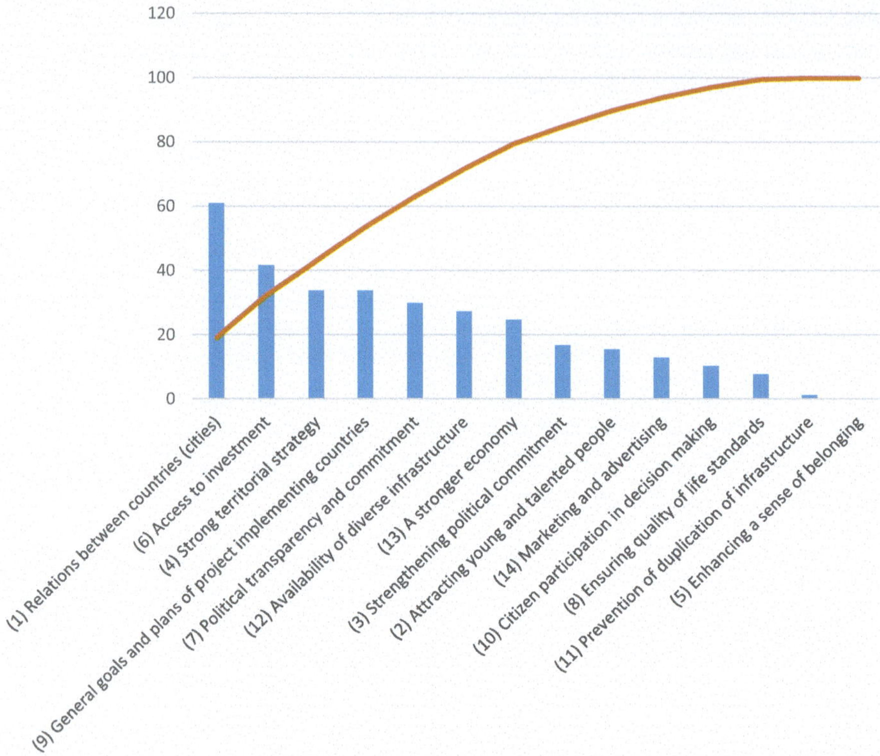

Fig. 5.3 Pareto diagram for selecting the most important success factors for managing cross-border projects. *Source* Compiled by the authors

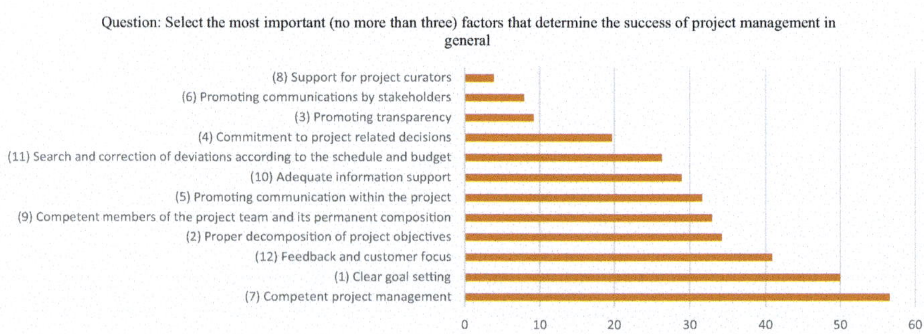

Fig. 5.4 Survey results on critical success factors for managing cross-border projects. *Source* Compiled by the authors

Further, as in the case of the criteria for the success of managing cross-border projects, we conducted a Pareto study to select criteria for a model of success in project management (Fig. 5.5).

Critical success factors for project management mainly depend on the types, attributes, and characteristics of an effective work team. Modern concepts and methods of personnel management allow us to evaluate the project team's effectiveness and develop criteria for the success of project management. Some authors consider it necessary to form a project team taking into account the interpersonal relationships of employees and offer a criterion characterizing the contribution of

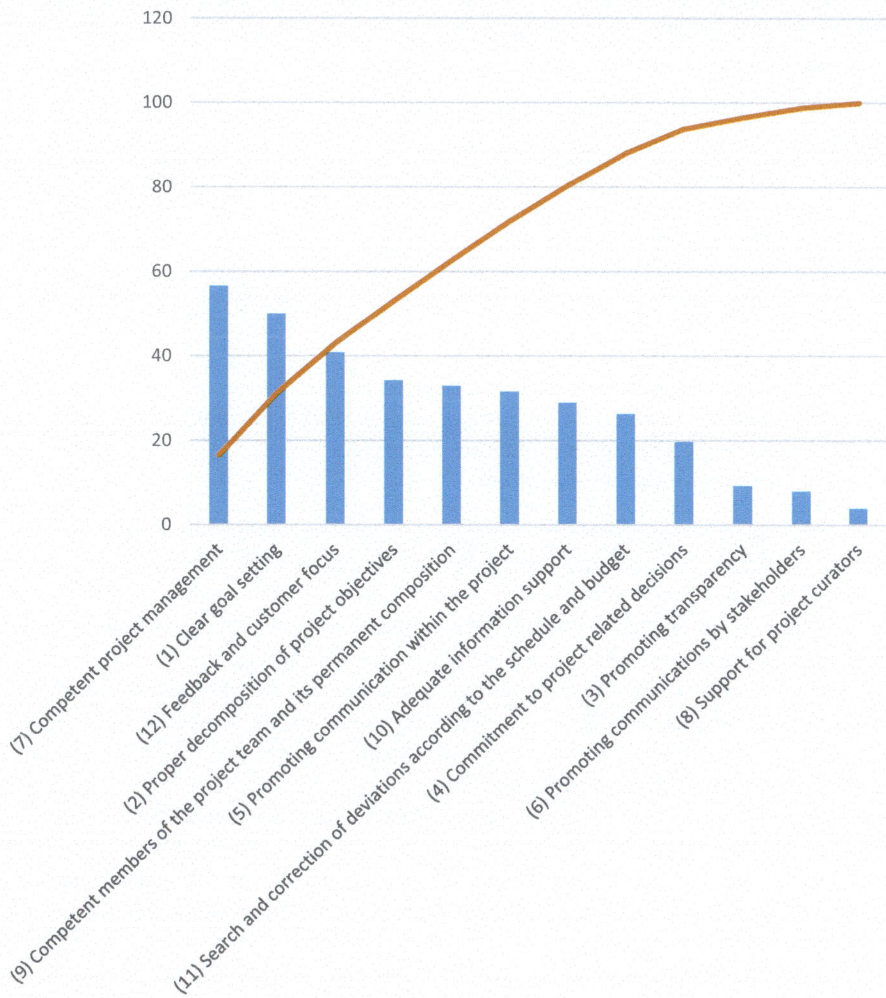

Fig. 5.5 Pareto diagram for selecting the most important factors for the success of project management. *Source* Compiled by the authors

Table 5.3 Content and context critical success factors of the cross-border projects

Content of the cross-border projects	Context of the cross-border projects
(1) Clear goal setting (2) Proper decomposition of project objectives	(9) General goals and plans of projects implemented by countries (4) Strong territorial strategy
(7) Competent project management	(12) Availability of diverse infrastructure (13) A stronger economy
(5) Promoting communication within the project (10) Adequate information support (12) Feedback and customer focus	(6) Access to investment (1) Relations between countries (cities)
(9) Competent members of the project team and its permanent membership	(7) Political transparency and commitment

Source Compiled by the authors

employees to group interaction [27]. However, we do not take into account the interpersonal relations of team members and take into account only managerial factors, which then will form the basis of the model for successful management of cross-border projects:

1. (7) Competent project management;
2. (1) Clear goal setting;
3. (12) Feedback and customer focus;
4. (2) Proper decomposition of project objectives;
5. (9) Competent members of the project team and its permanent membership;
6. (5) Promoting communication within the project;
7. (10) Adequate information support.

To form the success model, we analyzed, combined, and collated the content (internal) and context (external) critical factors shown in Table 5.3.

Thus, the case study results, interviews, and online/offline surveys allowed us to identify the most important critical factors for successful project management in the context of CBC and form the success model, as shown in Fig. 5.6.

In the model's center, we have presented the content factors that generally correlate with routine project management. On the outer petals of the model (a, b, c, d), we have presented contextual factors associated with the external environment necessary for successful project management in CBC.

It is important to note that internal and external factors are related. For example, if C factors as communications, information, and feedback are essential for any project. In the context of CBC, communications between countries and cities and access to investments are crucial. This part of the model is responsible for the communications of the cross-border projects.

If we talk about the analytical model, then it looks like this:

$$A = f\left\{x_i;\ y_j\right\}, \tag{5.1}$$

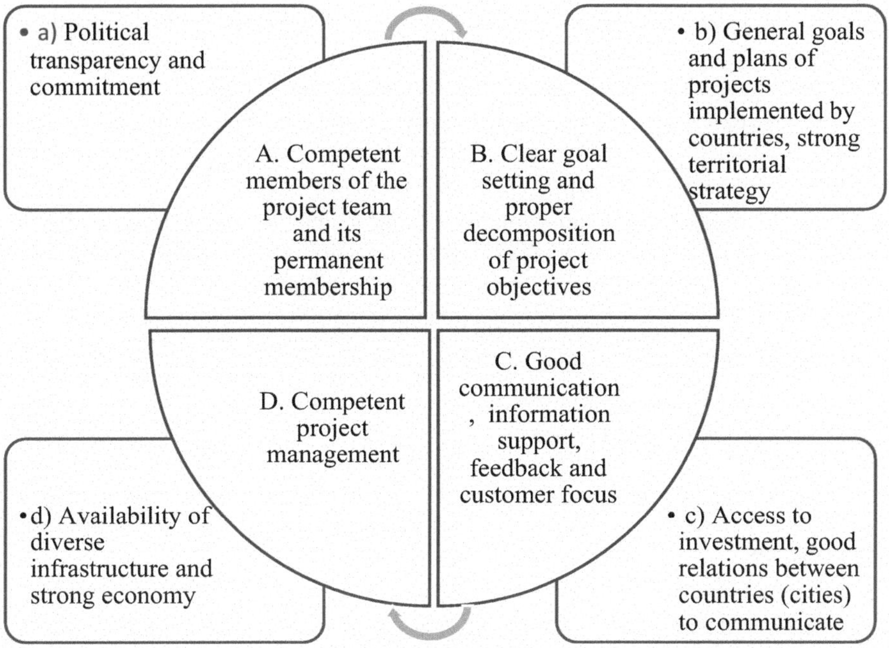

Fig. 5.6 Project management success model in a CBC. *Source* Compiled by the authors

where A is the success of project management in the conditions of CBC, which depends on the function f;

x_i—critical success factors for project management (content);

y_j—critical success factors for managing cross-border projects (context).

It is vital to ensure the quality of the project at the initial stage. It is about providing the quality characteristic of cross-border projects at the project preparation phase, on which the subsequent stages of the project cycle depend. Uncertainty of the purpose of the project usually leads to inadequate analysis.

5.5 Conclusion

In recent decades, CBC has reached an unprecedented level not only because of its potential for territorial integration but also taking into account its role in supranational processes such as infrastructure construction, activity planning, and project implementation. This interaction is becoming increasingly important when we live in a globalized world. This scenario is evident in the case of the EU, the EAEU, and the OBOR initiative. In this regard, identifying success factors in managing cross-border projects is considered critical for achieving sustainable development through

CBC strategies that lead to consistent improvement in the quality of life of the living population in these regions. From our point of view, it was essential to determine the set of these factors for managing cross-border projects successfully.

The strength of the integration of neighboring countries lies in cooperation across national borders. By combining the resources of individual countries, it is possible to achieve more significant and better results that benefit everyone.

As for the first part of the primary economic rationale for regional cooperation, cross-border projects may attract additional preferential and non-preferential funds. Positive externalities (for example, benefits such as saving time and money, protecting the environment, and facilitating trade) and negative externalities (such as costs such as environmental pollution, human trafficking, and the spread of infectious diseases) occur when the consequences of one or several countries go beyond national borders. If the countries do not make joint agreements, there will be too few positive external factors and too many negative ones.

References

1. Alam, K., Baig, S., Li, X., Ghanem, O., Hanif, S.: Causality between transportation infrastructure and economic development in Pakistan: an ARDL analysis. Res. Transp. Econ. 100974 (2020). https://doi.org/10.1016/j.retrec.2020.100974
2. Alshimbayeva, D., Myrzakhmetova, A., Erimpasheva, A.: Cross-border relations of neighboring states in the context of the development of border territories: theoretical aspect (on the example of Kazakhstan and China). Econ. Strateg. Pract. **15**(1), 153–163 (2020). https://doi.org/10.51176/jesp/issue_1_t11
3. Brauweiler, H.: Innovationen im peripheren Raum. Gabler/Dt. Univ.-Verlag, Wiesbaden (2002)
4. Brauweiler, H.: Internal audit in a multinational perspective. In: 2nd International Research Symposium: "Global Challenges of Management Control and Reporting", pp. 25–32. Wroclaw University of Economics, Wroclaw (2017). https://www.dbc.wroc.pl/Content/37298/Brauweiler_Internal_Audit_a_Multinational_Perspective_Problems_2017.pdf. Accessed 26 Oct 2021
5. Castanho, R.: The Relevance of Political Engagement and Transparency in Cross-Border Cooperation Environments: Analyzing Border Cities in Europe. Lex Localis–J. Local Self-Gov. **18**(3), 487–502 (2020). https://doi.org/10.4335/18.3.487-502
6. Castanho, R., Loures, L., Fernández, J., Pozo, L.: Identifying critical factors for success in Cross Border Cooperation development projects. Habitat Int. **72**, 92–99 (2018). https://doi.org/10.1016/j.habitatint.2016.10.004
7. Cross-border cooperation in the European Union: an opportunity for Europe's border regions (2021). https://ec.europa.eu/regional_policy/en/newsroom/news/2015/09/cross-border-cooperation-in-the-european-union-an-opportunity-for-europe-s-border-regions. Accessed 27 Oct 2021
8. De Soyres, F., Mulabdic, A., Ruta, M.: Common transport infrastructure: a quantitative model and estimates from the Belt and Road Initiative. J. Dev. Econ. **143**, 102415 (2020). https://doi.org/10.1016/j.jdeveco.2019.102415
9. Frątczak-Müller, J., Mielczarek-Żejmo, A.: Networks of cross-border cooperation in Europe–the interests and values. The case of Spree–Neisse–Bober Euroregion. Eur. Plan. Stud. **28**(1), 8–34 (2019). https://doi.org/10.1080/09654313.2019.1623972
10. Fujimura, M., Adhikari, R.: Critical evaluation of cross-border infrastructure projects in Asia. SSRN Electron. J. (2010). https://doi.org/10.2139/ssrn.1653699

11. Grad-Rusu, E.: Exploring cross-border cooperation in Eastern Europe: What kind of initiatives have developed in the Romanian-Hungarian border area? Észak-Magyarországi Stratégiai Füzetek **18**(2), 35–43 (2021). https://doi.org/10.32976/stratfuz.2021.33
12. Hofmeister, G., Mukhtarova, K., Abdykalikova, M., Yerimpasheva, A., Abikenov, A.: Ecosystem of technological business: methods of analysis and development factors. Central Asian J. Soc. Sci. Humanities **5**(1) (2019). https://doi.org/10.26577/cajsh-2019-1-s1
13. Jurilj, I.: Cross-border cooperation projects as a community development generator–"Adriatic Canyoning" project. E-Zbornik **10**(19), 63–73 (2020). https://doi.org/10.47960/2232-9080.2020.19.10.63
14. Korneliuk, O., Zavadska, Y., Chapko, R.: Forms of implementation of cross-border cooperation in the European Union. Innov. Econ. (3–4), 12–18 (2021). https://doi.org/10.37332/2309-1533.2021.3-4.2
15. Kurowska-Pysz, J.: The process of joint learning as a determinant of cross-border project management. East. J. Eur. Stud. **11**(Special Issue), 47–76 (2020)
16. Li, J., Gao, G., Ma, L., Zhao, T., Qu, H., Chen, F.: Analysis of profit models for cross-border power interconnection projects. Global Energy Interconnect. **2**(5), 457–464 (2019). https://doi.org/10.1016/j.gloei.2019.11.021
17. Li, J., Zhang, J., Suo, W.: Risk assessment in cross-border transport infrastructure projects: a fuzzy hybrid method considering dual interdependent effects. Inf. Sci. **488**, 140–157 (2019). https://doi.org/10.1016/j.ins.2019.03.028
18. Liu, J., Lo, K., Mah, D., & Guo, M.: Cross-border governance and sustainable energy transition: the case of the Guangdong-Hong Kong-Macao Greater Bay Area. Current Sustain. Renew. Energy Rep. **8**(2), 101–106 (2021). https://doi.org/10.1007/s40518-021-00178-4
19. Nagy, I.: Cross-border cooperation on the external borders of the EU and the impact of the received EU CBC funds on AP Vojvodina/Serbia. Belgeo (2) (2020). https://doi.org/10.4000/belgeo.3873
20. Nave, E., Franco, M.: Cross-border cooperation to strengthen innovation and knowledge transfer: an Iberian case. Innov. Eur. J. Soc. Sci. Res. 1–19 (2021). https://doi.org/10.1080/13511610.2021.1964354
21. Palmowski, T., Fedorov, G.: The potential for development of Russian-Polish cross-border region. Geogr. Environ. Sustain. **13**(1), 21–28 (2020). https://doi.org/10.24057/2071-9388-2019-70
22. Peña Medina, S.: Cross-border spaces and environmental planning: the Border 2012 Program at the U.S.-Mexico border. Sociedad Y Ambiente, 6 (2015). https://doi.org/10.31840/sya.v0i6.1573
23. Sanotska, C., Saienko, O., ilyk, R.: Ukrainian-Poland cross-border cooperation: current situation. Sci. J. Yuriy Fedkovich Chernivtsi Natl. Univ. Econ. 820 (2020). https://doi.org/10.31861/ecovis/2019-820-5
24. Vaňková, L., Kocourková, G., Krejza, Z.: Cross-border cooperation in the field of investment activity with regional development implications. In: IOP Conference Series: Earth and Environmental Science, vol. 656, pp. 012009 (2021). https://doi.org/10.1088/1755-1315/656/1/012009
25. Wang, J., Yau, S.: Case studies on transport infrastructure projects in Belt and Road Initiative: an actor network theory perspective. J. Transp. Geogr. **71**, 213–223 (2018). https://doi.org/10.1016/j.jtrangeo.2018.01.007
26. Wróblewski, Ł. (2021). Where there is a will, there is a way: some remarks on institutional distance in the Polish-German borderland. Pogranicze. Polish Borderl. Stud. **9**(1), 39–62. https://doi.org/10.25167/brs3408
27. Zakirova, A., Soloaga, A.: Managing employees in organizations via leadership skills. J. Econ. Res. Bus. Adm. **123**(1), 96–105 (2018). https://doi.org/10.26577/be-2018-1-2066

Chapter 6
Digital Transformation of Managing Business Entities Development in Agricultural Production

Sergey A. Korobov, Ilya V. Pshenichnikov, and Veronica S. Epinina

Abstract The level of agricultural entrepreneurship development is one of the defining conditions for the sustainable economic growth, increasing GDP, providing employment, and reducing risks to food security. In the conditions of uncertainty and unpredictability, the digital transformation of planning and forecasting systems for managing the business entities' development in agricultural production becomes the objective necessity. The authors have analyzed the branches of the agricultural sector, which, in the pandemic context, received a strong momentum for the development (industrial biotechnology and Foodnet) and defined the main threats to the activities of agricultural enterprises: low digital maturity of business organizations, the lack of effective business models of digital transformation, the high level of resistance to changes on the part of personnel, and the low level of competence in implementing digitalization. The contributors have searched effective tools for implementing the development strategy of an agricultural enterprise, taking into account the change in the technological order, the active digital transformation of the competitive environment. The study substantiates the feasibility of forming a digital system for managing the business entities' development in agricultural production, based on the integrated use of machine learning, Big Data technology, blockchain technologies, virtualization of contractual relations, and transactions using smart contracts.

S. A. Korobov (✉) · I. V. Pshenichnikov · V. S. Epinina
Volgograd State University, Volgograd, Russia
e-mail: econmanag@volsu.ru

I. V. Pshenichnikov
e-mail: pshenichnikov-iv@volsu.ru

V. S. Epinina
e-mail: epininavs@volsu.ru

E. I. Inshakova and A. O. Inshakova (eds.), *New Technology for Inclusive and Sustainable Growth*, Smart Innovation, Systems and Technologies 287,
https://doi.org/10.1007/978-981-16-9804-0_6

6.1 Introduction

Nowadays, the business institute is going through a transformation stage in a completely new environment, which is primarily characterized by the emergence of opportunities for them to use digital tools in order to carry out their economic activities and to interact with other market participants.

The active development of information and communication technologies within the framework of the fourth scientific and technological revolution and the need to search the innovative ways for developing working relationships between economic entities create the preconditions for the business entities' virtualization and the digital transformation of their activities. For business organizations to adapt successfully, it is necessary that digital transformation encompasses the process of identifying consumer preferences; organizational and managerial sphere of activity of economic agents; transfer of material flows within the enterprise; and means of interaction and contracting between an economic agent and market participants. Among the basic problems of digital transformation of managing a business entity development, one can single out:

- the lack of the strategy for developing the stack of digital technologies necessary for business entities development;
- high level of resistance to digital transformation by the business entities personnel.

The fourth industrial revolution gave rise to opportunities to expand the market potential of business entities by forming new market segments which are subjected to a high degree of digital transformation. Agriculture is one of the most susceptible industries to digital transformations in the market economy. In this industry, both complex digital technologies for enterprise management are reflected, which allow the virtualization of the management system for end-to-end business processes, and a new market niche is being reproduced with the active involvement of the scientific and technological progress achievements and the results of implementing the national technological initiative.

The trend analysis of developing the digital transformation indicators of agricultural enterprises and biotechnological enterprises made it possible to identify the promising areas of business entities' digital development in these industries.

6.2 Methodology

Nowadays, many researchers [4–6, 8, 9, 11–13, etc.] recognize that the digital transformation of managing the business entities development in the field of agricultural production is necessary to increase the efficiency and effectiveness of their functioning through the fundamentally new change in the quality of managing the technological processes and decision-making processes at all levels of the hierarchy as well. This fact, in turn, predetermines the active introduction of modern production methods and the further intensification of using the information data on the state and forecasting of controlled elements and subsystems possible transformations and also economic conditions in agriculture.

The experience of successful agricultural producers has proved that the use of modern digital technologies allows creating optimal soil-agrotechnical and organizational-territorial conditions that ensure the significant increase in productivity and labor efficiency; the decrease in material costs for fuels and lubricants, electricity, funds plant protection, wages, and other types of expenses, soil fertility preservation; and environmental protection during the entire life cycle of agricultural products. Therefore, in the territories' digital agenda, one of the priorities is the accelerated introduction of digital technologies in managing the agricultural enterprises' activities.

The industry evolution is only at the very beginning of the digital chain now. At the same time, according to the majority of researchers [1, 2, 10, 13, 14, 16–18], the fourth industrial revolution and scientific and technological progress make it possible to form the methodology for a digital system for managing business entities in the field of agricultural production. This methodology should be based on using the following information technologies: automation of business processes in the information flow transfer; work with big data; machine learning; blockchain technology. These digital transformation areas of agriculture predetermine implementing the projects of a complete innovative complex scientific and technical cycle of end-to-end digital systems based on modern competitive technologies, methods, and algorithms.

In this work, the authors have used general scientific research methods: analysis and synthesis, deduction and induction, methods of scientific abstraction and expert assessment, historical and logical analysis, as well as the methods of statistical, systemic, structural–functional, and comparative analysis. The trend analysis of developing the indicators of digital transformation of agricultural enterprises and biotechnological enterprises has made it possible to identify the promising areas of business entities' digital development in these industries.

6.3 Results

The driver for agricultural enterprises' digital transformation is the development of the following industries: foodnet, aeronet, and industrial biotechnology.

The market "Foodnet" is a dynamically developing market segment that includes many start-up projects of the corresponding direction, which propose various innovative ways of organizing economic activities. New efficient forms of management will substitute the traditional enterprises on the market, which will fail to restructure. Taking into account the forecast of developing the Foodnet market, the global consumption of products in this market will amount to 30 trillion dollars in the world market, and for the Russian geographic market, the projected capacity will be about 2.5 trillion dollars (Fig. 6.1).

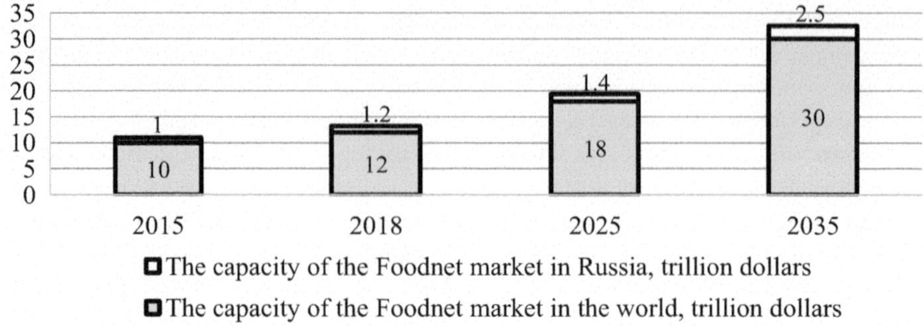

Fig. 6.1 The capacity of the Foodnet market in the world and in Russia. *Source* Compiled by the authors based on [7]

According to the forecast data of the analytical agency «J'son & Partners Consulting», the segmentation of the Foodnet market will be distributed as follows: «smart agriculture»—32%; available organic—28%; new sources of raw materials—15%; personalized nutrition—13%; accelerated selection—12% (Fig. 6.2). At the same time, biotechnology, innovative food products, and the Internet of Things (IoT) will become the fastest-growing food tech areas, and the leader among them will be innovative food products, including artificial meat and fish.

«Smart Agriculture» is one of the fastest-growing innovative segments of crop production. Five years ago, the technology market in it made up 2.3 billion euros, with the average market growth being about 12% per year in recent years—this is three times faster than the growth of the traditional equipment market.

The main directions of developing the projects for this segment of the Foodnet market are automation and robotization of production and technological processes in agriculture and the food industry; geolocation; and Big Data.

According to analytical data, the annual volume of investments in developing this market niche amounted to $21 billion in 2019; compared to the figures of 2016, it

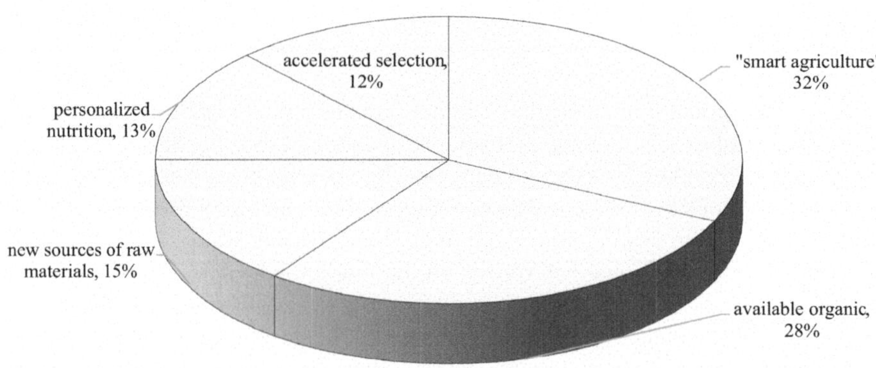

Fig. 6.2 The segmentation of the Foodnet market. *Source* Compiled by the authors based on [7]

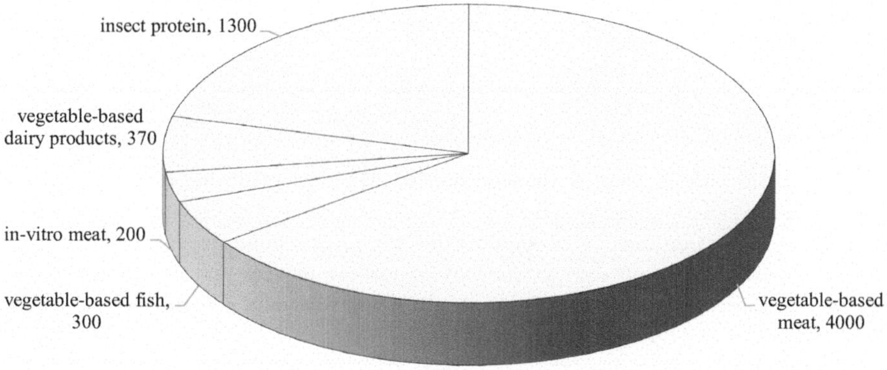

Fig. 6.3 The capacity of the market segment "new sources of raw materials", in billions of dollars. *Source* Compiled by the authors based on [7]

was $14 billion more. This fact also confirms the dynamic development of the "smart agriculture" market segment.

According to forecasts, by 2035, the market of solutions for "smart agriculture" will have been able to reach 15.8 billion dollars with the average annual growth of profitability being at the level of 10.4%. Besides, the market segment "new sources of raw materials" has significant market potential (Fig. 6.3).

The advanced demand for food products, especially for protein foods, forces people to look for new sources of raw materials. New sources of raw materials are designed to solve the problem of obtaining the artificial protein and the protein necessary for solving the food security problems of national economic systems.

Today these developments are reaching a new level. First of all, this is artificial meat, processing of algae, and pseudo-cereal structures. This also includes the recycling of production waste and "smart packaging" as they allow to reduce losses, which means that they can act as the equivalent of new sources of raw materials on the principle "saved resources = reproduced results".

It is difficult to single out the common leaders in this segment as the technologies differ too much, and, as a rule, each market participant specializes in its own direction. In general, we can say that by 2035 the volume of this segment will have been able to reach $218 billion with the average annual growth rate being 8.2%.

The capacity of the Russian market for developing and applying unmanned flying vehicle technology, including for agriculture, has the following growth dynamics (Fig. 6.4).

The growth dynamics of the market "Aeronet" capacity is provided by the following market segments: logistics operations and cargo transportation; support of technological operations in agriculture; remote sensing and farmland monitoring (Fig. 6.5). The use of unmanned flying vehicle technology significantly reduces transaction costs for technological operations and increases the cargo delivery efficiency, sown areas monitoring, control of germinating grain crops, and implementing chemical treatment of sown areas.

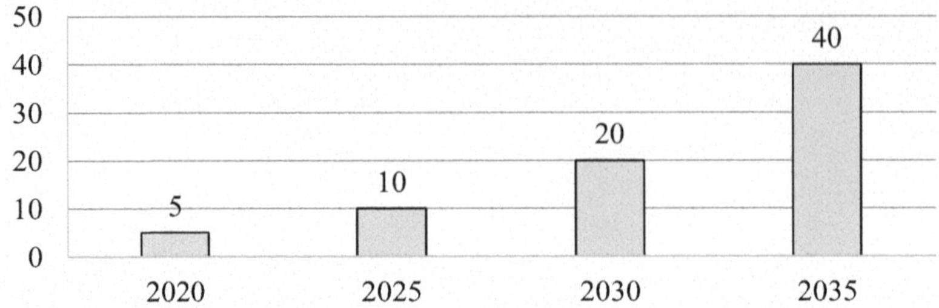

Fig. 6.4 The capacity of the market "Aeronet" in Russia, in billions of dollars. *Source* Compiled by the authors based on [3]

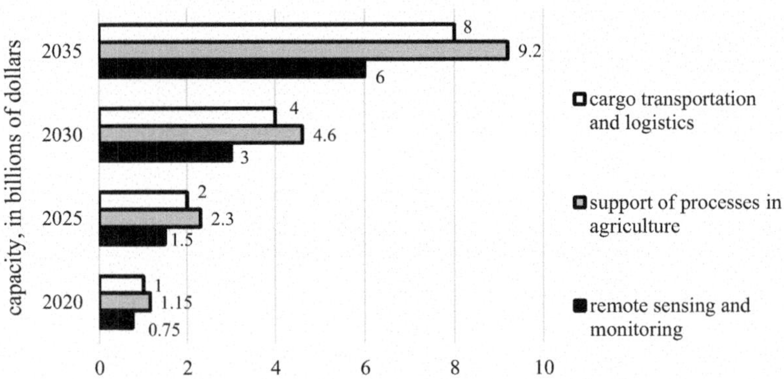

Fig. 6.5 The segments capacity of the market "Aeronet" in Russia, in billions of dollars. *Source* Compiled by the authors based on [3]

The biotechnology used by agricultural enterprises can significantly increase efficiency indicators, as well as it can reduce environmental damage from production. According to the estimates of the interdepartmental working group on the control over the biotechnology introduction under the Government of the Russian Federation, the total economic effect of using biological products in crop and livestock production in Russia may amount to more than 100 billion rubles per year at a cost of 10.5 billion rubles.

The capacity of the Russian market for developing and applying industrial biotechnologies in agriculture has the following growth dynamics (Fig. 6.6). The state program for developing agriculture for 2013–2020 assumes financing of biotechnology in 2015–2020 in the amount of 2 billion 226 million rubles from the federal budget and 780 million rubles from the budgets of the constituent entities of the Russian Federation.

The agrobiotechnology market is based on three segments:

– biotechnologies used in crop production (biological plant protection products, including plant growth stimulants and microbiological fertilizers);

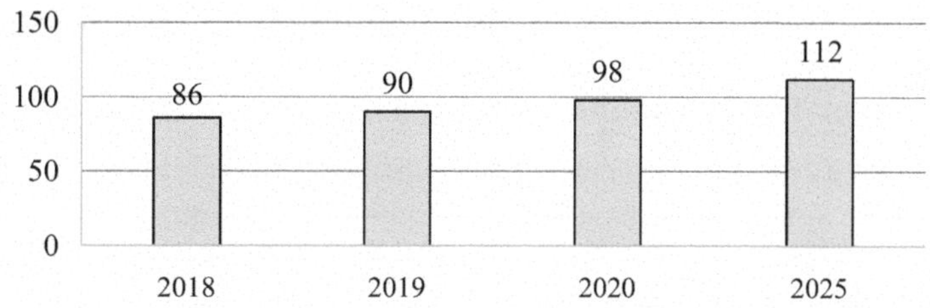

Fig. 6.6 The capacity of the market "industrial biotechnology in agriculture" in Russia, in billions of roubles. *Source* Compiled by the authors based on [15]

- creating new types and varieties of plants by the method of genetic engineering;
- biotechnologies used in animal husbandry (vaccines, therapeutic and feed antibiotics, diagnosticum, probiotics, and biological components of feed additives).

From the point of view of the potential for developing new technologies and projects, the following segments of this market are of greatest interest: biotechnology for crop production; production of enzymes and biopolymers; immunobiologicals; feed antibiotics.

The identified promising market sectors require reforming the management system of an economic agent and, above all, the inclusion of a preventive marketing analysis system in management.

For the successful integration of business entities in agriculture into new market niches, it is required to create a digital tool that allows people to form a prototype of the future market and to determine its development strategy by obtaining the reliable dynamic analysis of the indicators of the financial and economic activities of an economic agent with the possibility of predictive analytics.

Business processes automation during the information flow transfer within the digital management system of agricultural enterprises is designed to solve the following tasks:

- reception of primary documents required for implementing financial and management accounting of the business entities activities;
- sorting primary documents and checking the correctness of their filling;
- distribution of documents between necessary specialists and analytical modules of the digital management system of an economic agent;
- data distribution from primary documents between business transactions into the accounting system;
- forming financial records and sending them to government agencies;
- data distribution from the generated financial statements into special forms for analyzing financial and economic activities;
- creating the data array about the internal environment state of a business entity for the purpose of its further processing by Big Data technology for subsequent predictive analytics necessary to predict the financial and economic activities of business entities.

It is advisable to process the obtained data array, which characterizes the activities of business entities, with the help of Big Data technology tools in order to solve the following tasks of digital transforming the organization's management system:

- identifying the criterion parameters for comparative analysis of incoming data on the economic, financial, and marketing activities of a business entity;
- identifying the parameters of the business entity, which characterize its competitive advantages and weaknesses;
- forming the database of a business entity activity parameters, which allows using the theory of weak signals and elements of machine learning technology to determine the forecast for developing economic and financial activities;
- searching the optimal activity parameters of a business entity user for the subsequent formation of a hypothetical sectoral model of a reference economic agent necessary for benchmarking and determining the statistical data required by the government bodies in order to make effective decisions to support the economic sectors development.

Digital transformation of collecting and analyzing data on the business entities activities will allow cutting down the risks of obtaining inaccurate information and making ineffective strategic decisions, which will strengthen the business entities market positions significantly, increase the profitability of their activities, and also create the preconditions for the more transparent establishment of performance indicators of personnel activity in an organization. The establishment of these indicators supposes reducing the resistance of personnel to the process of digital transformation of a business entity management system.

The cause-and-effect relationships of establishing the indicators values identified with the help of big data tools become the basis for machine learning, which is necessary to solve the following problems of digital transformation of the business entity management system:

- creating a new market image based on the theory of weak signals, which consists in the need to detect and to rank the weak signals from current and potential consumers of services or products, to build their development forecast, and to highlight significant needs and qualitative assessment, which contributes to forming the potential forecast for the development of market niches corresponding the business entity potential;
- forming a preventive marketing system, consisting of the following stages: establishing constant monitoring of all trends in the external environment; forecasting the consumer preferences development using weak signals; describing the upcoming strategic and tactical implications for an enterprise; forecast results assessment and possible risks analysis; flexible change in the assortment of goods and services, taking into account the constructed forecast;
- developing the forecast of changes in economic and financial indicators of business entities, taking into account the peculiarities of the macro and microenvironment development;

- making forecasts of the financial and tax burden of business entities, taking into account the formed prototype of the market and changes in the macro environment, as well as taking into account the changes in the internal environment of an economic entity.

The use of digital transformation of the business entity management system also implies the inclusion of blockchain technologies necessary to create smart contracting between market participants and to build a transparent interaction system, taking into account reliable information about the financial and economic situation of business entities and market prospects for the business entities activities. The use of smart contracts will allow both to expand the geographical boundaries of the organizations' activities and to create a new digital format for the economic integration of entrepreneurial structures.

The comprehensive use of digital transformation of the business entities' management system in agriculture will increase the level of their digital maturity by including digital competencies and technologies in business processes, which will increase their competitiveness significantly and create the preconditions for adapting to the conditions of Industry 4.0.

As a result of forming the digital system for managing agricultural enterprises, it will be possible to achieve an increase in the economic efficiency of these economic agents' activities by reducing the cost of agricultural production and increasing the enterprises' profitability.

The digital system use for managing agricultural enterprises will simultaneously reduce both variable and fixed costs of economic agents. Reducing variable costs for agricultural enterprises using a digital control system occurs by means of: reducing the cost of fertilizing; selecting the placement of agricultural crops between the fields with different germination conditions; adjusting business processes when using agricultural machinery in the process of sowing and harvesting campaigns.

The basic tool for reducing the variable costs of a digital system for managing business entities in the field of agriculture is processing big data on production, technical and technological, financial, and economic components of economic activity.

The use of big data tools in the digital management system of a livestock business can reduce variable costs in such a «problem area» as the feeding process. Thanks to the data analysis, a schedule is drawn up for the acquisition, storage, and supply of feed to the livestock complex, which increases the efficiency of managing an economic agent significantly. The reduction in fixed costs in the livestock complex is due to the use of tools for processing big data and machine learning for controlling the temperature regime and supporting water supply and energy supply to the livestock complex. Blockchain technologies, as well as in the case of an agricultural enterprise, play a significant role in reducing the cost of the process of transactional interaction within the enterprise and with the external environment. The cumulative reduction in the total cost of the livestock complex using the tools of the digital control system can be 35–40%.

The share of variable costs of agricultural enterprises varies from 60 to 80% in the structure of the total cost, and the use of tools of the digital agricultural enterprise management system can reduce these costs by 20–30%.

A decrease in the fixed costs of agricultural enterprises takes place by means of increasing the transparency of interaction between the structural divisions of an economic agent and its counterparties and the formation of new effective contracting conditions by reducing transaction costs. Blockchain technologies will allow making effective smart contracts, both in the interaction of an agricultural enterprise with third-party organizations and within an economic agent, which will significantly affect the objectivity and transparency of business processes valuation and the efficiency of financial and material flows redistribution at the enterprise.

Reducing fixed costs by 10–15% due to the use of a digital system in managing an agricultural enterprise will allow an economic agent to reduce the total cost and increase the competitiveness of its activities.

With the simultaneous use of machine learning elements that contribute to the reliability of the crop yields forecast and the attractiveness of their implementation on the market, an agricultural enterprise receives a balanced picture of income and expenses and the ability to budget its activities effectively.

The digital system for managing agricultural enterprises by reducing costs and predicting the receipt of financial results of their activities allows increasing the economic viability of these economic agents.

6.4 Conclusion

The results obtained in this study allow us to make the following conclusions:

- the fourth industrial revolution has formed new perspective market niches, the development of which is largely predetermined by innovative technologies and digital services: foodnet, aeronet, and industrial biotechnology;
- developing a digital system for managing business entities in the field of agriculture should be based on an integrated approach to the digitalization of business processes for collecting and analyzing data and predictive analytics, which will allow adapting the activities of an agricultural enterprise effectively to new market niches and business conditions;
- the comprehensive digital transformation of the system for managing business entities in the field of agriculture will be carried out more effectively on the RPA technology stack (automation of business processes in data transfer), Big Data, and Machine Learning;
- the use of blockchain technologies for making smart contracts with business entities in the field of agriculture will increase the efficiency of interaction between economic actors and expand the possibilities of agglomeration in rural cooperation, which will lead to a decrease in transformation and transaction costs, and, as a result, will reduce the cost of manufactured products significantly.

References

1. Abdullah, R.S., Faizal, M.A.: Block chain: cryptographic method in fourth industrial revolution. Int. J. Comput. Netw. Inf. Sec. **10**, 9–17 (2018). https://doi.org/10.5815/ijcnis.2018.11.02
2. Alias, N., Al-Rahmi, W.M., Yahaya, N., Al-Maatouk, Q.: Big data, modeling, simulation, computational platform and holistic approaches for the fourth industrial revolution. Int. J. Eng. Technol. **7**, 3722–3725 (2018). https://doi.org/10.14419/ijet.v7i4.21759
3. Analytical data from the National Technology Initiative. http://nti-aeronet.ru/RABOCHAJA-GRUPPA/. Accessed 14 Aug 2021
4. Etri, E., Karblank, E., Geerten, D., Lesher, M., Pilate, D., Wykoff, E., Keihin, B.: Vectors of digital transformation. Bull. Int. Organ. **3**, 7–50 (2020). https://doi.org/10.17323/1996-7845-2020-03-01
5. Fedotova, G.V., Gontar, A.A., Kulikova, N.N., Kurbanov, A.K.: Threats to food security of the Russia's population in the conditions of transition to digital economy. Adv. Intell. Syst. Comput. **622**, 542–548 (2018). https://doi.org/10.1007/978-3-319-75383-6_68
6. Fedotova, G.V., Gorlov, I.F., Glushchenko, A.V., Slozhenkina, M.I., Natyrov, A.K.: Trends of scientific and technical development of agriculture in Russia. Lect. Netw. Syst. **87**, 193–200 (2020). https://doi.org/10.1007/978-3-030-29586-8_23
7. ICT Analytics and Digital Media. Analytics. https://json.tv/. Accessed 10 Aug 2021
8. Inshakov, O.V.: Strategy and tactics of nanoindustry development in the Russian Federation. Econ. Reg. **3**, 246–250 (2011). https://doi.org/10.17059/2011-3-29
9. Khasanov, A.: The Impact of predictive analytics on the activities of companies. J. Strateg. Decis. Risk Manag. **3**, 108–113 (2018). https://doi.org/10.17747/2078-8886-2018-3-108-113
10. Kimani, D., Adams, K., Attah-Boakye, R., Ullah, S., Frecknall-Hughes, J., Kim, J.: Blockchain, business and the fourth industrial revolution: whence, whither, wherefore and how? Technol. Forecast. Soc. Chang. **161**, 120254 (2020). https://doi.org/10.1016/j.techfore.2020.120254
11. Kondratyuk, T.: The fourth industrial revolution: what competencies do employees need? J. Strateg. Decis. Risk Manag. **3**, 66–79 (2018). https://doi.org/10.17747/2078-8886-2018-3-66-79
12. Korobov, S.A., Moseiko, V.O., Novoseltseva, E.G., Epinina, V.S., Marusinina, E.Yu .:Application of the tools of cognitive analysis in formation of regional system of development of small and medium entrepreneurship Contrib. Econ. 105–112. https://doi.org/10.1007/978-3-319-55257-6_15
13. Lee, K., Malerba, F., Primi, A: The fourth industrial revolution, changing global value chains and industrial upgrading in emerging economies. J. Econ. Policy Reform. 1–12 (2020). https://doi.org/10.1080/17487870.2020.1735386
14. Li, G., Hou, Y., Wu, A.: Fourth industrial revolution: technological drivers, impacts and coping methods. Chin. Geogra. Sci. **27**, 626–637 (2017). https://doi.org/10.1007/s11769-017-0890-x
15. Research by Abercade Consulting. Industry materials. http://abercade.ru/research/. Accessed: 11 Aug 2021
16. Shkalenko, A.V., Tymchuk, Yu.A.: Business entities in the context of technological modernization of the economy: priority areas of change taking into account digital transformation. J. Volgogr. State Univ. Econ. **22**, 74–87 (2020). https://doi.org/10.15688/ek.jvolsu.2020.4.7
17. Tarasov, I.: Technologies of Industry 4.0. The Impact on increasing the productivity of industrial companies. J. Strateg. Decis. Risk Manag. **2**, 62–69 (2018). https://doi.org/10.17747/2078-8886-2018-2-62-69
18. Trachuk, A., Linder, N.: The fourth industrial revolution: how does the Internet of Things affect the interaction of industrial companies with partners. J. Strateg. Decis. Risk Manag. **3**, 16–29 (2018). https://doi.org/10.17747/2078-8886-2018-3-16-29

Part II
Determinants of Inclusive Growth: A Statistical and Econometric Analysis

Chapter 7
Decarbonization Trends in the Largest Post-soviet Countries and the Specifics of Their Inclusion in the Global Climate Agenda

Lyudmila Yu. Bogachkova, Lidiya S. Guryanova, and Nadezhda Yu. Usacheva

Abstract The research purpose is to identify the features and problems of the inclusion of Kazakhstan, Russia, and Ukraine in the global climate agenda through the analysis of aggregated indicators of decarbonization of these countries' economies. The final goal is to support the adoption of managerial decisions aimed at ensuring the sustainable development of these largest post-Soviet countries. The study outlines a modern vision of the technological conditions for the green transformation of world energy (energy transition) and describes the common economic tools of the global climate agenda. The ambiguity of the concept of "inclusiveness" of this policy has been revealed. It is argued that the consistency in achieving the entire set of sustainable development goals including not only economy's decarbonization, but also inequality reduction and improvement of the population welfare is of utmost importance for emerging markets which, according to the UN, include the post-Soviet countries under consideration. A comparative analysis of the dynamics of aggregated indicators of the carbon intensity of the economies of Kazakhstan, Russia, and Ukraine for 1990–2020 has been carried out. Prospects for improving their climate and energy policy have been discussed to support decision-making aimed at the coordinated achievement of energy policy goals.

L. Yu. Bogachkova (✉) · N. Yu. Usacheva
Volgograd State University, Volgograd, Russia
e-mail: bogachkova@volsu.ru

N. Yu. Usacheva
e-mail: n.y.usacheva@volsu.ru

L. S. Guryanova
Kharkiv National University of Economics, Kharkiv, Ukraine

E. I. Inshakova and A. O. Inshakova (eds.), *New Technology for Inclusive and Sustainable Growth*, Smart Innovation, Systems and Technologies 287,
https://doi.org/10.1007/978-981-16-9804-0_7

7.1 Introduction

A new annual report of the UN Intergovernmental Panel on Climate Change [18] published in August 2021 provides impressive data indicating the development of the global warming process and negative changes in the Earth's climate system, including those that seem to be irreversible. According to the authors of the report, 234 experts from 66 countries, the apparent driver of these processes is the anthropogenic impact on the environment, emissions of greenhouse gases (GHG) into the atmosphere, including the emission of carbon dioxide (CO_2), formed mainly in the field of traditional energy when burning hydrocarbons.

A purposeful transition (energy transition) to green energy, which exploits renewable energy sources (RES), such as hydroenergy, solar energy, and wind energy, is the main way to prevent global warming recognized by the governments of most countries and cross-country associations. The Global Energy Transition Strategy has been developed and implemented under the auspices of the UN with the aim of creating a low-carbon, energy-efficient, and inclusive green economy, characterized by the involvement of previously unclaimed natural and labor resources in production processes [30].

In the modern world, a broad consensus has been reached on measures for preventing climate change, despite the fact that there are climate skeptics among scientists and political and public leaders who question the anthropogenic nature of global warming [2, 7, 16, 24, 26] and the feasibility of the conception of energy transition [5].

In 1992, heads of a number of states adopted the Framework Convention on Climate Change [4]. Later in 1997, the leaders of 159 developed countries signed the Kyoto Protocol thus committing themselves to the reduction of greenhouse gas emissions during 1997–2012 (the first stage) and 2013–2020 (the second stage). As a result, tens of billions of dollars were invested in renewable energy. Finally, in September 2015, the heads of 193 states agreed on the 2030 Agenda for Sustainable Development (hereinafter referred to as the Climate Agenda), which included 17 Sustainable Development Goals (SDGs) [31]. The goals are as follows: cutting off subsidies for the use of fossil fuels and creating conditions in which the perpetrators of environmental pollution would pay for the negative consequences of their activities (SDG 13); application of environmentally friendly technologies, reduction of greenhouse gas emissions (SDG 9); increasing the share of renewable energy in the global energy balance (SDG 7); poverty elimination (SDG 1); inequality reduction (SDG 10) [29], etc.

In December 2015, the Paris Agreement was signed to secure the obligations of countries to implement the Climate Agenda. The main goal of the Paris Agreement is to keep the rise in mean global temperature to well below 2 °C above pre-industrial levels and preferably limit the increase to 1.5 °C by reducing global greenhouse gas emissions down to zero level by 2050, i.e., to achieve such a level of harmful emissions, which will be balanced by the absorbing capacity of forests and oceans. Currently, 191 countries have joined the Paris Agreement [29, 31].

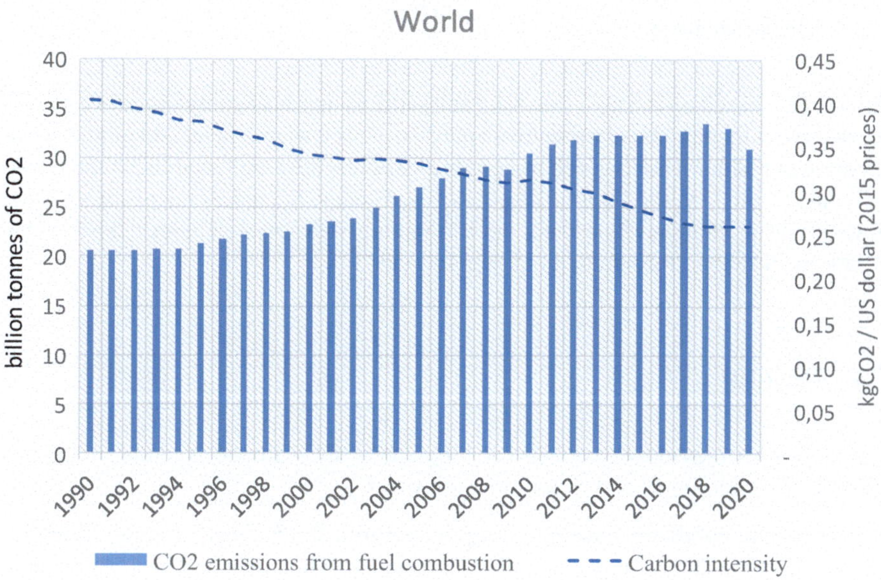

Fig. 7.1 Carbon dioxide emissions (on the left) and carbon intensity (on the right) of the world economy. *Source* Compiled by the authors based on [8, 11]

The effectiveness of international climate policy can be judged by the dynamics of the total volume of CO_2 emissions on the planet and by the aggregate indicator of the carbon intensity of the world economy, which is calculated as the ratio of global CO_2 emissions to the world gross product (Fig. 7.1). Figure 7.1 shows that from 1990 to 2018 carbon intensity had been declining every year by an average of 7 g. of CO_2 per one dollar of value added (at constant prices of 2015). However, the gross volume of carbon dioxide emissions had been increasing on average by 542.09 million tons per year [12]. In the last pre-pandemic year of 2018, global CO_2 emissions made 33.5 billion tons.

Despite the fact that from 2018 to 2020, the volume of global CO_2 emissions decreased from 33.5 to 31.1 billion tons (see Fig. 7.1), this is rather explained by the negative impact of the COVID-19 pandemic on global industrial production than by climate policy efficiency.

Therefore, CO_2 emissions are expected to show growth trends in the course of the post-pandemic recovery of the global economy. This means that the peak in CO_2 emissions has not yet been passed, and more work remains to be done to achieve zero CO_2 emissions—the main goal of the Paris Agreement.

The research purpose is to identify the features and problems of the inclusion of Kazakhstan, Russia, and Ukraine in the global climate agenda through the analysis of aggregated indicators of decarbonization of these countries' economies. The final goal is to support the adoption of managerial decisions aimed at ensuring the sustainable development of these largest post-Soviet countries.

7.2 Methodology

The study is based on the use of general scientific methods of analysis and synthesis, descriptive statistics, and comparative analysis, as well as graphical and tabular visualization of results. The research materials are represented with analytical and statistical data of the UN, the International Energy Agency, the independent information and consulting company Enerdata, the BloombergNEF research center, the Russian Center for Strategic Research, and other sources.

7.3 Results

The system of measures developed to implement the Climate Agenda includes a number of technologies and economic policy tools.

Technological conditions of the energy transition. While discussing the technologies required for the planet's transition to a carbon–neutral state, it is reasonable to compare two mid-2021 expert reports on the development of world energy until 2050. One of them was issued by the International Energy Agency (IEA) [20], and the other by the BloombergNEF research center [21].

The IEA report [20] contains the traditional vision of energy decarbonization based on the development of renewable energy sources (RES), solar and wind energy in particular. Let us note that in this case, the massive introduction of nanotechnology can play an important role in increasing the efficiency of RES generation and energy storage [14, 19]. At the same time, according to the IEA experts, the possibilities of achieving a zero balance of emissions by 2050 through the use of RES generation are very limited.

Contrary to the existing green tradition, the report prepared by the BloombergNEF analysts for the first time announces the ambiguity of low-carbon development and outlines three scenarios—"green", "gray", and "red", which are not necessarily alternative [21]. All scenarios envisage an increase in the share of electricity in final energy consumption from the current 19% to 49% but the generation of electricity is carried out in different ways. The green scenario, as always, is based on the powerful development of renewable energy sources; the gray scenario assumes the preservation of gas and coal generation and a large-scale implementation of carbon capture and storage (CCS) technology; and according to the red scenario, decarbonization is provided primarily by means of nuclear energy production. Thus, it is increasingly recognized by the global expert community that not only renewable energy sources, but also hydrogen, nuclear energy, and carbon capture can play an important role in achieving zero emissions in the world.

From the viewpoint of the climate agenda, special attention should be paid to nuclear energy, since direct CO_2 emissions from nuclear power plants are practically zero. Today, there are 443 operating nuclear reactors in 34 countries around the world. On a global scale, the operation of all nuclear power plants in the world prevents the emission of greenhouse gases in the amount of 2 billion tons of CO_2 equivalent per year, which is commensurate with the absorbing capacity of the entire forest area of the planet [22].

Scientists at the Joint Research Center under the European Commission performed a comparative analysis of various methods of generating electricity in terms of material consumption, emissions of pollutants into the atmosphere, impact on human health, and life expectancy [27]. As a result, it was found that in normal operation, nuclear energy is highly competitive with the green methods of energy generation. However, within the framework of the European climate strategy, the issue of recognizing nuclear energy as a sustainable source of clean generation has not yet been resolved.

The Russian standpoint on the peaceful atom is unambiguous: it will be impossible to solve the tasks of the climate agenda in the medium term without the use of nuclear energy [22].

Economic tools of the climate agenda. The climate agenda includes a wide range of economic tools [23]. International financial institutions, investment funds, and insurance companies are in transition toward de-investment in projects related to the exploitation of hydrocarbon energy resources and large volumes of CO_2 emissions. Stock exchanges and shareholders require companies to disclose their greenhouse gas emissions. In the field of carbon reporting, both voluntary initiatives (for example, the Carbon Disclosure Project [3]) and mandatory requirements (Directive 2014/95/EU44) have been developed. ESG (environmental, social, and governance impacts) reporting is becoming one of the factors that predetermine the success of large- and medium-sized enterprises. Energy efficiency standards and requirements are becoming more widespread around the world [13].

The International Civil Aviation Organization and the International Maritime Organization have developed and are currently implementing guidelines aimed at compliance with mandatory measures to improve energy efficiency and reduce GHG emissions from international air travel and shipping.

Today, green finance, green certificates in particular, have become widespread in the world. Initially, they were used to record and confirm the production and consumption of electrical energy based on alternative renewable energy sources. Later, a market for these financial instruments was formed. This market was used to attract private capital to the RES sector allowing companies to fulfill their environmental or social obligations, as well as to enhance a "climate responsible" image of the business. However, green certificates usually only apply to alternative RES (wind, solar, and small-scale hydropower), which limits support for low-carbon energy sources such as nuclear and large-scale hydropower [23].

Carbon pricing is one of the most drastic measures to ensure the reduction of greenhouse gas emissions. This economic tool implies financial obligations for companies-producers of CO_2 emissions and other harmful substances. Both mandatory and voluntary carbon pricing is applied in practice. Three types of carbon pricing can be distinguished: carbon tax; emissions trading system; mixed schemes.

In the Summer of 2021, it became known about the plans of the European Union to tighten the European climate agenda which includes the introduction of a cross-border tax on goods and services with a large carbon footprint imported into the EU [6]. Thus, it is planned to equalize the competitive conditions of European and foreign companies. The fact is that since 2005, European producers have paid for greenhouse gas emissions, while importers from other countries do not usually incur such expenses. Based on expert estimates, Russia and Ukraine will bear the greatest losses associated with the introduction of this carbon tax. For Russia in particular, losses can amount to €1.1 billion per year [28]. This makes urgent the problem of inclusiveness of the climate agenda in relation to countries with emerging market economies and the problem of adaptation to the energy transition of countries that export goods to the EU.

Understanding the inclusiveness of the global climate agenda. The concept of inclusiveness is multifaceted and is widely used in the energy transition literature. In a broad sense, inclusiveness is the involvement of countries in the global climate agenda. In the UN program aimed at the preservation of the environment, the term "inclusiveness" is used to denote the widespread production use of previously unclaimed natural and labor resources based on the development of renewable energy sources [30, p. 16].

The concept of inclusiveness has one more meaning which is important for developing countries and countries with economies in transition—conformity of climate change prevention (SDG 13) with other sustainable development goals, such as poverty elimination (SDG 1), improving the well-being of population (SDG 3), inequality reduction (SDG 10), and others [17]. The fact is that the climate policy tools developed for the leading countries (carbon pricing and promotion of alternative RES generation) lead to an increase in prices for energy and other goods. For this reason, relatively low-income populations in the developing countries and countries with economies in transition may have to spend a larger share of their income on electricity and gasoline. This leads to the deterioration of already low living standards and to exacerbated inequality in these countries and consequently contradicts SDGs 1–3 and SDG 10. In our opinion, ensuring the inclusiveness of climate policy in relation to these countries, understood as achieving consistency among all the goals of sustainable development, requires a synthetic approach to the choice of means of the energy transition. This implies the need to go beyond the traditional climate policy, which is limited by the regulation of price levels, the choice of specific methods of carbon pricing, and one or another form of compensation for losses of the most vulnerable segments of the population [17]. A set of additional measures is also required, including the development of safe nuclear energy, the introduction of carbon capture and storage (CCS) technologies, etc.

On the inclusion of the largest post-Soviet countries in the climate agenda: obligations and economies' carbon intensity dynamics. Kazakhstan, Russia, and Ukraine are the largest CO_2 emitters in the post-Soviet space. These countries are characterized by the largest population and production scale [1]. Table 7.1 shows the climate goals and obligations of the countries under consideration as parties to

Table 7.1 Basic, expected, actual, and target levels of GHG emissions in Russia, Kazakhstan, and Ukraine (exemplified by CO_2)

Year		1990	2018 (pre-COVID)		2030
Level Country		Basic	Expected	Actual	Target
Russia	mln tons	2,163.50	1,747.03	1,587.00	1,568.54
	%	100	80.75	73.35	70–75
Kazakhstan	mln tons	237.3	204.08	214.0	189.84
	%	100	86	90.18	75–85
Ukraine	mln tons	688.6	495.79	181.8	No more than 413.16
	%	100	70	26.4	no more than 60

Source Compiled by the authors based on [31]

the Paris Agreement to reduce greenhouse gas emissions. Thus, in a broad sense, the inclusion of Kazakhstan, Russia, and Ukraine in the climate agenda is currently proceeding.

If we assume that the movement of countries toward the set goals will take place through a uniform reduction of emissions (both by years and by types of GHGs), then the estimation of volumes of CO_2 emissions in 1990 will allow us to easily calculate the expected levels of emissions that should have been achieved by the last pre-COVID year of 2018. Comparison of actual CO_2 emissions with their expected values demonstrates the effectiveness of achieving the set goals (Table 7.1).

As shown in Table 7.1, Russia is reducing CO_2 emissions ahead of schedule: in 2018, the actual level of emissions (1,587 million tons) was lower than the expected level (1,747.03 million tons). Kazakhstan demonstrates the opposite trend: the expected level (204.8 million tons) is lower than the actual one (214 million tons), which can be considered as a sign of deficiency of the corresponding state policy.

Finally, Ukraine is fulfilling its commitments to reduce CO_2 emissions much ahead of the curve: actual emissions (181.8 million tons) are well below the expected level (495.79 million tons). Thus, against the background of growing global emissions (Fig. 7.1), the volumes of CO_2 emissions in the largest post-Soviet countries are decreasing (Table 7.1).

There is also a decrease in the relative shares of Kazakhstan, Russia, and Ukraine in the global volume of CO_2 emissions (Fig. 7.2).

Figure 7.2 shows that in 1990 the total contribution of the three countries to global emissions was about 15%, and by 2018, it decreased to 6% and mainly due to Russia's green policy. The shares of Russia and Ukraine in global emissions have more than halved. Kazakhstan's contribution decreased slightly. Thus, Kazakhstan, Russia, and Ukraine are not responsible for the increase in global CO_2 emissions (Fig. 7.1).

The carbon intensity of the economies of Kazakhstan, Russia, and Ukraine is also decreasing at a faster pace as compared to the global dynamics (Fig. 7.3). According to the data of the International Energy Agency, over the period of 1990–2018, CO_2 emissions per unit of value added decreased by 35% on a global scale, by 40%—in Russia, by 59%—in Ukraine, and by 58%—in Kazakhstan.

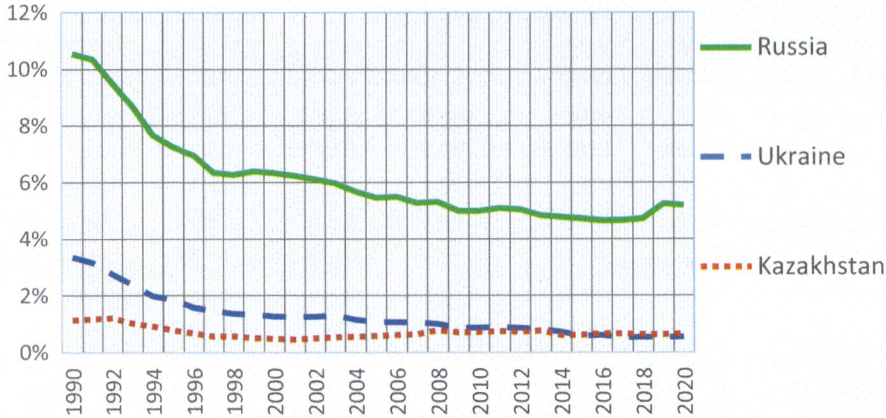

Fig. 7.2 Contributions of Russia, Kazakhstan, and Ukraine to global greenhouse gas emissions over the period of 1990–2020. *Source* Compiled by the authors based on [8, 11]

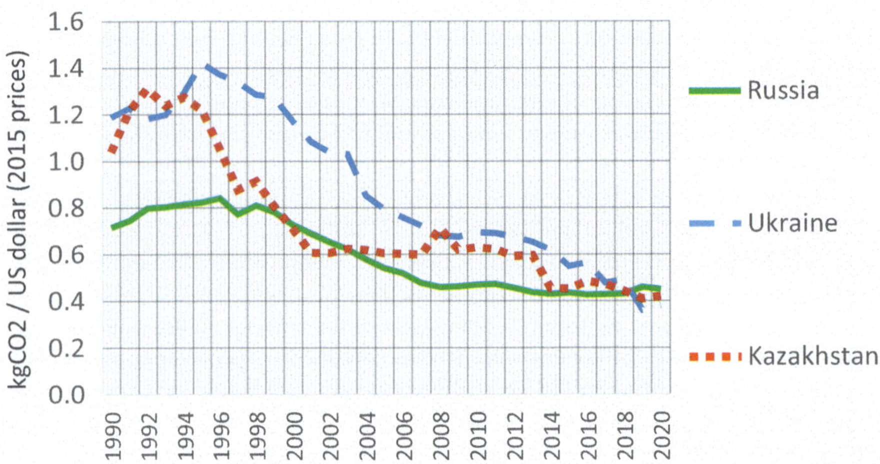

Fig. 7.3 Carbon intensity or CO_2 emissions per GRP unit at purchasing power parity in 2015 US dollars. *Source* Compiled by the authors based on [8, 11]

Reduction of CO_2 emissions and downward trends in the dynamics of relative contributions of Kazakhstan, Russia, and Ukraine to global emissions, as well as the decrease in the carbon intensity of their economies, give the impression of a high degree of these countries inclusion in the climate agenda.

Factors reducing the economy's carbon intensity and the problem of consistency of various sustainable development goals. The achievement of the described results in Kazakhstan and Ukraine was facilitated by the high rates of RES development, stimulated by green tariffs. In 2018, Kazakhstan ranked first among countries under consideration in terms of contribution of RES generation facilities, including hydroelectric power plants, to electricity generation. At the same time, a slight decrease in Kazakhstan's contribution to global CO_2 emissions (Fig. 7.2) can be explained by the relatively high rates of economic growth in this country and the structure of energy balance, in which coal has the largest share [1].

Russia's economy is characterized by a relatively low level of carbon intensity as compared to Kazakhstan and Ukraine (Fig. 7.3) due to the large shares of hydro and nuclear generation in electricity production. High rates of nuclear energy development have a positive effect on the reduction of CO_2 and other greenhouse gas emissions.

As shown in Table 7.1, over the period of 1990–2018, Ukraine had ranked first among the countries under consideration in terms of reduction of CO_2 emissions (in % to the level of 1990). In 2020, Ukraine made the smallest contribution to global emissions (Fig. 7.2) and reached the lowest level of the economy's carbon intensity (Fig. 7.3). It is also known that over the past 5 years, the country had achieved the greatest success in increasing the contribution of solar and wind power plants to the total volume of electricity generation. This was facilitated by large-scale foreign investment [1], which testifies to a high degree of inclusiveness understood as involvement in the process of wind and solar energy generation. However, the main factors behind these results also include a deep economic recession, a sharp decline in the real disposable income of the population, deindustrialization, industrial decline, and the war in Donbas [1]. Thus, Ukraine's inclusion in the climate agenda, understood as consistency in achieving sustainable development goals, is highly problematic.

In all post-Soviet countries, the systemic economic crisis that followed the collapse of the USSR and its negative socio-economic manifestations played a significant role in reducing CO_2 emissions. Real GDP growth rates in 1995–2018 were lowest in Ukraine (128.51%), medium in Russia (197.57%), and the highest in Kazakhstan (345.14%) [1]. In the course of the economic development of these countries, other conditions being equal, the trends in CO_2 emissions may be reversed. In this regard, despite the reduction in the contributions of Kazakhstan, Russia, and Ukraine to global CO_2 emissions, coordinators of the United Nations Environment Program (UNEP) assess the progress of Russia and Ukraine in the field of decarbonization of the economy as "critically insufficient" and the achievements of Kazakhstan as "insufficient" to achieve the goals of the Paris Agreement [23, p. 21].

Enhancement of national energy policies for coordinated achievement of sustainable development goals. The problem of modernizing energy policy in Kazakhstan, Russia, and Ukraine has become urgent and aggravated against the background of the anticipated introduction of cross-border carbon regulations by the European Union [25]. All three countries are major suppliers of goods with a large carbon footprint to the EU, and their losses from the introduction of the cross-border carbon regulations are estimated by analysts at billions of dollars. To reduce and neutralize the expected damage, it is planned to use national systems for monitoring CO_2 emissions and internal trade in CO_2 emissions.

Kazakhstan already has an exchange for trading CO_2 emissions, and now this republic has announced plans to introduce its own cross-border carbon regulation similar to the EU to reduce the tax burden on Kazakh exporters to the EU. Experts admit the possibility of creating a unified carbon regulation system in the EAEU after the unification of the gas and energy markets in 2025 [25].

Today, Russia does not have an internal carbon pricing system. To exclude payments of the carbon tax to the EU countries, Russia is developing the mechanisms of internal regulation of emissions with payments to the state budget. It is also planned to achieve recognition of this regulation at the international level so as to replace payments to the EU budget with internal payments for CO_2 emissions in Russia [10, 15].

Ukraine also plans to launch an internal emission trading market as part of the country's obligations under the Agreement of Association with the European Union. Ultimately, the EU's position on the introduction of the cross-border carbon regulations in relation to Ukraine will depend on the degree and quality of implementation of this agreement [9].

To solve the problem of the inclusion of post-Soviet countries in the coordinated achievement of various sustainable development goals, it is advisable to use nuclear power, which is characterized by a low-carbon nature and high technological and economic efficiency. The most promising areas for the development of this industry are represented with two-component nuclear power with a closed fuel cycle, hydrogen power, and mass commercial production of small-scale nuclear power plants for power supply to remote regions [22].

7.4 Conclusion

Thus, the inclusion of the post-Soviet countries in the climate agenda and sustainable development policy can be ensured through the enhancement of national energy strategies in the following areas: an integrated approach to the choice of technologies for the transition to low-carbon energy, including the elements of the gray and red scenarios in addition to the green trend; introduction of national systems for monitoring greenhouse gas emissions and CO_2 emissions trading; development of transparent competitive market mechanisms for stimulating renewable energy generation; and creation of the conditions for attracting investments in the capital market by means of green finance.

References

1. Bogachkova, L.Yu., Guryanova, L.S., Usacheva, N.Yu., Usacheva I.V.: Conditions and trends of green energy development in the largest economies of the post-Soviet space. Lect. Notes Netw. Syst. **155**, 152–163 (2021).https://doi.org/10.1007/978-3-030-59126-7_18

2. Carter, R.M.: Climate: The counter Consensus. Stacey International, London (2010)
3. CDP Worldwide Charity (2021). https://www.cdp.net/en/ Accessed: 28 Sept 2021
4. Climate change (2021). Global Issues. Official website of the UN. https://www.un.org/ru/global-issues/climate-change Accessed: 28 Sept 2021
5. Compston, H., Bailey, I.: Turning Down the Heat: The Politics of Climate Policy in Affluent Democracies. Palgrave Macmillan, London. (2008). https://doi.org/10.1057/9780230594678
6. European Green Deal (2021). Commission proposes transformation of EU economy and society to meet climate ambitions. https://ec.europa.eu/commission/presscorner/detail/en/ip_21_3541 Accessed: 30 Sept 2021
7. Giaever, I.: Global warming revisited. Speech at 65th Lindau Nobel Laureate Meeting (2015). https://www.mediatheque.lindau-nobel.org/videos/34729/ivar-giaever-global-warming-revisited/meeting-2015 Accessed 23 Sept 2021
8. Global Energy Statistical Yearbook 2021 (2021). https://yearbook.enerdata.net/ Accessed: 4 Oct 2021
9. Grigorenko, Yu.: Smoke of the motherland: Ukraine is ready to launch monitoring of CO_2 emissions (2020). https://gmk.center/posts/dym-otechestva-ukraina-gotova-k-zapusku-monitoringa-vybrosov-so2/. Accessed: 15 Sept 2021
10. Grinkevich, D., Milkin, V.: The Russian government is preparing its own version of carbon collection. Vedomosti (2021). https://www.vedomosti.ru/economics/articles/2021/09/22/887822-vlasti-uglerodnogo-naloga. Accessed 28 Sept 2021
11. IEA (2021). CO_2 emissions from fuel combustion highlights. https://www.iea.org/data-and-statistics/data-product/co2-emissions-from-fuel-combustion-highlights. Accessed: 30 Sept 2021
12. IEA (2020). Energy efficiency indicators database. http://data.iea.org/payment/products/120-energy-efficiency-indicators.aspx Accessed 15 Sept 2021
13. Inshakov, O.V., Bogachkova, L.Yu., Popkova, E.G.: Energy efficiency as a driver of global competitiveness, the priority of the state economic policy and the international collaboration of the Russian Federation. Lect. Notes Netw. Syst. **44**, 119–134 (2019).https://doi.org/10.1007/978-3-319-90966-0_9
14. Inshakov, O.V., Inshakova, E.I., Orlova, A.A.: Global and Russian nanotechnology product market development: comparison of trends and impact of sanctions. Int. J. Trade Global Mark. (IJTGM) **10**(2/3) (2017). https://doi.org/10.1504/IJTGM.2017.086077
15. Kanishchev, M.: CO_2 emission control system as a national idea. Energy Policy **9**(163), 56–74 (2021)
16. Lindzen, R.S.: Climate physics, feedbacks, and reductionism (and when does reductionism go too far?). Eur. Phys. J. Plus. **127**, 52 (2012). https://doi.org/10.1140/epjp/i2012-12052-8
17. Makarov, I.: BRICS and the formation of a new climate agenda. National Committee for BRICS Research (Russia) (2020). https://www.nkibrics.ru/posts/show/5ea7f6ca62726974caf70000. Accessed 18 Sept 2021
18. Masson-Delmotte, V., et al. (eds.): Climate Change 2021: The Physical Science Basis. Summary for Policymakers. Cambridge University Press, Cambridge (2021).in press)
19. Nanotechnology will help improve the efficiency of the energy sector (2014). http://www.energosovet.ru/news.php?ver=full&zag=1394525152. Accessed 30 Sept 2021
20. Net Zero by 2050: A Roadmap for the Global Energy Sector. IEA Publications. https://www.iea.org/reports/net-zero-by-2050 (2021). Accessed 4 Oct 2021
21. New Energy Outlook 2021 (2021). Executive summary. https://assets.bbhub.io/professional/sites/24/NEO-Executive-Summary-2021.pdf Accessed: 4 Oct 2021.
22. Novak, A.: Development of nuclear energy—a necessary condition of the global climate agenda. Energy Policy **9**(163), 6–11 (2021)
23. Russia's climate agenda (2021). Responding to international challenges. Center for Strategic Research. https://www.csr.ru/upload/iblock/8f3/8f34a0e7c6cc6b8af39986ae8e71f3ad.pdf Accessed: 4 Oct 2021
24. Sacchetti, D.: Contrarian views on climate change. IAEA Bull. **49**(2), 20–21 (2008)
25. Smertina, P., Konstantinov, A.: Decarbonization: Kazakhstan can introduce a carbon tax by analogy with the EU. Kommersant (2021). https://www.kommersant.ru/doc/4966783. Accessed 30 Sept 2021

26. Soon, W., Baliunas S.: Lessons & limits of climate history: Was the 20th century climate unusual. The George C. Marshall Institute, Washington, D.C. (2003)
27. Technical assessment of nuclear energy with respect to the ‘do no significant harm’ criteria of Regulation (EU) 2020/852, EC Joint Research Centre, Petten, JRC124193 (2021). https://ec.europa.eu/info/sites/default/files/business_economy_euro/banking_and_finance/documents/210329-jrc-report-nuclear-energy-assessment_en.pdf. Accessed 4 Oct 2021
28. Tkachev, I., Kotchenko, K.: Russia will pay € 1.1 billion a year of carbon tax to the EU (2021). RBK. https://www.rbc.ru/economics/26/07/2021/. Accessed 30 Sept 2021
29. UN General Assembly: Transforming our world: the 2030 Agenda for Sustainable Development (2015). https://www.un.org/ga/search/view_doc.asp?symbol=A/RES/70/1&Lang=E. Accessed 18 Sept 2021
30. UN Environment Programme: Towards a green economy (2011). http://web.unep.org/greeneconomy/sites/unep.org.greeneconomy/files/field/image/green_economyreport_final_dec2011.pdf. Accessed: 15 Aug 2021
31. UN Framework Convention on Climate Change. Intended Nationally Determined Contributions (INDCs) (2015). https://www4.unfccc.int/sites/NDCStaging/Pages/All.aspx. Accessed 15 Aug 2021

Chapter 8
Ensuring the Openness of Environmentally Relevant Information as an Environmental Component of Inclusive Growth of Russia's Regions

Elena A. Ivantsova, Elena A. Zaliznyak, Anna A. Matveeva, and Anna V. Kholodenko

Abstract The system of environmental indicators of inclusive development of any region should be aimed not only at economic stimulation but also at the development and promotion of human well-being, including improving the environmental aspects of the quality of life of the population. The right to access information about real indicators of the quality of the environmental components is one of the priorities in the social responsibility development of business. As part of the research, the evaluation of the official websites of companies was carried out, which revealed that the open-access information does not meet the requirements for the preparation of "optimal" non-financial reports. The enterprises of the region were offered the opportunity to show their interest in disclosing information by filling out the developed checklists reflecting the indicators of non-financial reports in the field of sustainable development and the Global Reporting Initiative (GRI). The contributors gave recommendations on the main areas of interaction between business representatives and stakeholders that will enhance the effectiveness of state environmental supervision and public control in the field of environmental protection, as well as act as one of the ways to achieve the environmental component indicators of inclusive growth of Russia's regions.

E. A. Ivantsova (✉) · E. A. Zaliznyak · A. A. Matveeva · A. V. Kholodenko
Volgograd State University, Volgograd, Russia
e-mail: ivantsova@volsu.ru

E. A. Zaliznyak
e-mail: zaliznyak@volsu.ru

A. A. Matveeva
e-mail: matveeva@volsu.ru

A. V. Kholodenko
e-mail: kholodenko@volsu.ru

E. I. Inshakova and A. O. Inshakova (eds.), *New Technology for Inclusive and Sustainable Growth*, Smart Innovation, Systems and Technologies 287,
https://doi.org/10.1007/978-981-16-9804-0_8

8.1 Introduction

The emphasis on the inclusive growth concept in the search for adequate models of socio-economic development is justified from the perspective of the key challenges facing modern society. One of the five leading global risks reflected in the report "Global Risks—2015" of the World Economic Forum is environmental. Even at the initial formation stage of the green inclusive growth concept, it was emphasized that in this context, focusing only on standard indicators is impractical. Thus, GDP per capita does not reflect the real well-being of citizens and does not take into account income inequality and the damage caused to the environment, including the depletion of natural resources.

Existing approaches to measuring inclusive growth are usually associated with quality of life and living conditions elements. They are often equated with issues such as healthcare, the environment, education, access to basic services, security, infrastructure, and financial affordability. In order to measure and understand inclusive growth, it is necessary to take into account its impact on inequality [15, 22].

Inclusion issues are related to the accumulation of net income per capita, providing incentives for environmental protection and guaranteeing access to environmentally relevant information. At first glance, these issues may seem unrelated, but the placement of data on the impact of economic entities on environmental components contributes to the transition to business openness, and the openness of any information allows identifying risks and improving the risk management system. In turn, this activates environmental protection activities, the effective implementation of environmental innovations, which contributes to the creation of corporate and environmentally responsible brands and increases the competitiveness of business, both in the domestic market and in external economic affairs. These directions affect the total income of the population. Improvements may be followed by an increase in GDP per capita if these directions are implemented in practice.

Measures taken in each area (economic, social, and environmental), in terms of stimulating inclusive growth, should be quantified according to a certain set of indicators, given that the basis of inclusive growth is, among other things, ensuring equal opportunities for all groups of the population in various spheres, including equal access to information.

Information openness of business in the field of environmental safety of the region is one of the components of the public administration quality assessment system formed by experts of the Analytical Center under the Government of the Russian Federation and the World Bank as part of the program for monitoring the quality of public authorities and mechanisms for improving public administration in the regions [10, 13].

The main problem for the development of voluntary mechanisms of environmental responsibility is the closeness of information on the impact of companies on the environment.

Table 8.1 Limitations and problems of non-financial reporting formation of Russian companies

<table>
<tr><td colspan="4">The "Trilemma" of restrictions</td></tr>
<tr><td>Difficulty in compiling the CSR scalability</td><td colspan="2">An integral part of non-financial reporting</td><td>Company expectations</td></tr>
<tr><td colspan="4">Types of report categories</td></tr>
<tr><td>Optimal</td><td colspan="2">Non-core</td><td>Technical</td></tr>
<tr><td colspan="4">Problems</td></tr>
<tr><td colspan="2">Legal</td><td colspan="2">Economic</td></tr>
<tr><td colspan="2">The lack of commitment by companies to the "objective" assessment of their economic activities consequences</td><td colspan="2" rowspan="2">Methodological inaccuracy of indicators for environmental reporting of NEI entities, legally established and corresponding to the Sustainable Development Goals (SDGs)</td></tr>
<tr><td colspan="2" rowspan="3">The lack of clear regulation of the procedure for disclosure of corporate non-financial reporting for the SDGs</td></tr>
<tr><td colspan="2">The absence of an acceptable methodology for the cost assessment of environmental damage and the impact of risk occurrences, both for the company and for the region</td></tr>
<tr><td colspan="2">The lack of incentive to non-financial reporting publicity</td></tr>
</table>

Source [2, 3, 5, 9, 11, 19, 20]

According to the research of Russian scholars, such as [4, 11, 12, 14, 19], the system of non-financial reporting implementation faces many limitations and problems (Table 8.1).

However, despite some difficulties, summarizing observations of the non-financial reporting process development among Russian companies, it should be noted [14] that the non-financial reporting of large companies acquires maturity features and creates a good basis for further development of data collection and analysis systems that will be crucial for the quality of reporting in the future.

The scientific research outlined objectives for solving the problem of environmental information openness by regional business communities and developing practical recommendations to ensure the information openness to interested parties (stakeholders). This is of particular importance in the changed conditions of the world economy transformation and the exponential development of the "green" economy. Enterprises that use natural resources should be competitive not only in the domestic market but also in the foreign market, taking into account the goals of sustainable and inclusive development.

8.2 Methodology

The environmental responsibility mechanisms are based on international environmental standards, which allow stakeholders to receive publicly available information about the state and degree of impact on environmental components. These include environmental certification and non-financial reporting containing environmentally relevant information, certified by a "third" independent party. According to a number of foreign researchers, such as [2, 3, 5, 6, 18], the implementation of such approaches to business social responsibility based on the access of stakeholders and, first of all, the population to environmental information in rapidly developing economies has shown their effectiveness and efficiency.

Environmental responsibility, first of all, is related to the existing standards of environmental safety and resource conservation adopted in a particular state, and its high-quality implementation is associated with the introduction of stricter environmental requirements by the state. M. Porter was one of the first to raise the issue of coordinating the achievement of competitive advantages by businesses based on achieving high environmental results and noted that the management effectiveness significantly depends on the institutional environment quality of entrepreneurial activity [16].

Global Reporting Initiative Standards (GRI Standards) provide a comprehensive, flexible, and adaptable structure that companies under any form of ownership, regardless of their production volume, can use to compile reports on their economic, environmental, and social impact. For example, in Switzerland, a report in the field of sustainable development is provided by both a large contractor All real-Gruppe and a representative of a small business—a real estate manager [8].

Companies use the GRI Standards when preparing reports, as well as other guidelines and standards, including the United Nations Global Compact, the Social Charter of Russian Business, the basic performance indicators of the RSPP (Russian Union of Industrialists and Entrepreneurs), and indicators of the RSPP indices for sustainable development ("Responsibility and Openness" and "Vector of Sustainable Development") [1, 4, 10, 12, 13].

The system-forming enterprises of the Volgograd region were selected as the object of the study, most of which belong to the service sector (77.2%) and only 22.8% to the production sector, i.e., the real sector of the economy. In addition, the main part of the GRP of the Volgograd region (76% of the total volume) is created by branches of the real sector of the economy, more than 40% of this GRP volume falls on industrial production and 17% on agriculture; urban districts account for 90% of the industrial production of the region [7, 21].

One of the key areas of successful development of the region industrial complex at present is, among other things, its compliance with modern environmental safety requirements. The transition to the so-called "green economy", the implementation of the rational nature management principles, minimizing the negative impact on the environment, and stimulating environmental protection activities are one of the priorities of the Volgograd region.

In the course of the study, an analysis of the official websites of enterprises was carried out to identify environmentally relevant information in public access. The analysis reflected that out of 27 objects rated in the high-risk category (21 enterprises and 6 structural units), 16 objects (9 enterprises and 7 structural units) do not have their own official websites. Such enterprises include, in most cases, either branches of large companies, for example, industrial sites of LLC "Nizhnevolzhskaya Oil

Company" and JSC "Uryupinskaya" Poultry Farm, or small companies, for example, LLC "Spetsavtotrans".

Of all the enterprises belonging to the I category of NEI entities, only JSC "Volzhsky Pipe Plant" and LLC "LUKOIL-Volgogradneftepererabotka" partially place information on the impact of the enterprise on environmental components and environmental measures carried out in the public domain.

The National Register of Non-Financial Reports contains 176 companies, and the list does not include any enterprises operating in the Volgograd Region. Transparency of environmental and energy reporting (50.1% of disclosed indicators) for the companies under study is better than the average in Russia (48.6%). This reflects the willingness of companies to discuss their achievements and problems with society and investors [1, 7, 13].

When establishing an accurate picture of the information openness of regional companies, it is necessary to take into account not only the desire of the enterprises to provide information but also the availability of requested information from relevant ministries and departments. Special attention is paid to the consumers of this information—stakeholders (the population, public environmental organizations, and the scientific community), that is, interested parties. Regional enterprises were offered the opportunity to show their interest in disclosing environmentally relevant information also by filling out authorial checklists structured by blocks.

Official requests were sent to companies for providing information on emissions, discharges, disposal of industrial and consumer waste, as well as on energy consumption and energy conservation. The "Manage" and "Responsibility and Openness" blocks included issues on environmental policy, environmental protection costs, and the degree of readiness for business openness. The "Manage" block included two blocks: the main one (the availability of an environmental management system and an industrial environmental control program) and an additional one (current and capital costs for environmental protection; the amount of payments for negative environmental impact). The additional block focused on the results of state supervision of the company, violations of environmental legislation, and existing complaints from citizens regarding the impact of facilities on the environment. The "Responsibility and Openness" block is aimed at identifying the manager's awareness of non-financial reports' existence in the field of sustainable development and the Global Reporting Initiative (GRI). A special place in the checklist was given to assessing the potential readiness of companies to provide environmental information at the request of interested parties, placing quantitative data on the amount of pollution and the real costs of environmental protection in the public domain.

According to the results of processing the checklists, it should be noted that about 30% of companies responded to official requests. More than 70% ignored the request to participate in the study, and 21.5% of respondents refused to provide information by citing trade secrecy or the availability of the requested information on the official websites of authorities in the field of environmental protection and territorial statistical bodies.

Only six enterprises provided fully completed checklists: the oil company LLC "LLK-International", JSC "Volzhsky Pipe Plant", the KF LLC research and production enterprise, Municipal Unitary Enterprise "Plumbing and Sewer Facilities" (the city of Volzhsky), JSC "Volzhsky Orgsintez", and OJSC "EKTOS-Volga". It should be noted that JSC "Volzhsky Pipe Plant" and LLC "LLK-International" are aware of non-financial reporting and are ready to support the initiative of business openness by generating reports in the field of sustainable development. However, most companies referred to the lack of accounting for such information, for example, LLC "Ecosphere" and LLC "Volga-Business" (the city of Volzhsky).

Based on the results of the study, it can be concluded that business structures have not yet expressed their willingness to disclose detailed environmental reporting, referring to corporate policy and the provision of environmental reporting to supervisory authorities.

Under the act of Federal Law "On Providing Access to Information on the Activities of Government Bodies and Bodies of Local Self-Government", No. 8-FZ, February 9, 2009, citizens have the right to receive the requested information in accordance with the procedure established by law. Within the framework of the study, official requests were formed to the relevant local government structures to obtain information about regional companies in accordance with the developed checklist. The Finance Committee of the Volgograd Region and the Economic Policy and Development Committee of the Volgograd Region forwarded the requested information "volumes of products produced by enterprises in monetary units" to other committees and departments, with reference to the lack of authority. The Committee of Housing and Utility Services and Fuel and Energy Complex of the Volgograd region provided information on the block "Energy Saving" according to the general data of the energy passport of thermal power plants of the city and the region. Volgogradstat responded that the requested information relates to primary statistical data and is not subject to disclosure and (or) dissemination. The Committee of Natural Resources, Forestry, and Ecology of the Volgograd Region in its response referred to the annual reports on the state of the environment posted on the website of the agency and also indicated that it does not carry out supervisory and other measures at the NEI entities, since this is not stipulated by current legislation. The Interregional Department of Rosprirodnadzor for the Astrakhan and Volgograd regions provided information on the indicators of state supervision for all impact blocks in a generalized form, except for the waste management system of production and consumption.

Thus, it can be noted that at present, the lack of interest in providing environmental information to stakeholders by both enterprises and authorities forms limitations of inclusive growth.

8.3 Results

The industry of the Volgograd region is a driver of sustainable economic growth and ensures the stability of the budget system, the financial base for social and infrastructural transformations, including within the framework of the implementation of priority and national projects.

Based on the analysis of official statistics and websites of regional companies, it was concluded that there is a lack of business social responsibility, in terms of the environmental component. Therefore, the problem of environmental information disclosure cannot be solved only by third stakeholders, and it is necessary to involve municipal authorities as a "carrier" of environmental information from affiliates in the social responsibility system of regional business.

According to the state register of objects, that result in negative environmental impact (NEI), 1,589 federal agencies for environmental supervision were registered in the Volgograd Region as of 01.01.2020, of which 109 (6.9%) belong to category I—objects having a significant negative impact on the environment and related to the areas of BAT application, and 3257 objects of NEI entities of the regional supervision agencies (Table 8.2).

Table 8.2 shows that most of the federal (67.5%) and regional (96.3%) NEI entities of the supervision level belong to the III and IV categories—insignificant and minimal environmental impact, respectively.

The authors proposed an innovative model for the information system formation in the structure of environmental safety of the region (openness and transparency of the impact of economic entities on society and the environment).

In our opinion, an effective model of trilateral interaction should function as follows:

- regional authorities form an open information platform (based on a set of systematized data on the state of the environment, as well as factors affecting its quality indicators and environmental safety of the region as a whole, updated by the authorities in accordance with their competence). Interaction of the Federal Service for Supervision of Natural Resources, the Committee of Natural Resources, Forestry, and Ecology of the Volgograd Region, Rospotrebnadzor, and the Economic Policy and Development Committee of the Volgograd Region is recommended.
- the electronic database is compiled on the basis of environmental monitoring data, environmental reporting of enterprises, integrated/environmental reports of

Table 8.2 The number of NEI entities on the territory of the Volgograd region as of 01.01.2020

The level of state environmental supervision	I	II	III	IV
Federal	109	407	957	116
Regional	–	121	2.171	965

Source [17, 21]

enterprises, reports in the field of sustainable development, the results of control and supervisory activities of Rosprirodnadzor, and the Committee of Natural Resources, Forestry, and Ecology of the Volgograd Region;
- informational and methodological support of business structures on the preparation of reports in the field of sustainable development is provided by the authorities (in the pilot project, representatives of public environmental organizations and scientists of regional universities can assist in the preparation of the report);
- regional authorities receive an information request from interested parties to provide data on the impact of enterprises on the environment; bring to the attention of the stakeholder possible sources containing information of interest in the public domain (e.g., a regional open information platform, the UONVOS portal); provide data based on environmental reporting of business entities; and with the help of existing institutional levers of influence and, based on the information model created in the region, request information directly from the enterprise.

An important element is the boundaries of environmental reporting, i.e., the company must take into account the impact on environmental components not only within its activities but also go beyond the boundaries of its territorial location (suppliers and consumers of services and goods).

The achieved results provide a solution for a number of key issues:

- Identification of successful practices for improving public reporting and activities in the field of environmental responsibility and sustainable development of the region;
- Methodological assistance to interested companies in the preparation of non-financial reports and further development of the reporting process;
- Identification of potential opportunities for solving problems in the field of environmental protection by the efforts of three sectors (state, private, and public), optimizing the resources of each of them, through information interaction.

8.4 Conclusion

The experience gained will allow identifying key elements of interaction between business representatives and stakeholders (public authorities, public organizations, and the population) to increase the effectiveness of state environmental supervision and public environmental control in the field of environmental protection.

Business communities should increase the degree of environmentally relevant information openness and introduce the best available technologies at their enterprises as a tool to increase competitiveness in the global and domestic market.

Companies should be guided by the principles of completeness and reliability of the information, which contain a complete list of legally established information, including up-to-date data not only for providing environmental reporting to regulatory authorities but also the interested parties.

Undoubtedly, when introducing mandatory public non-financial reporting for companies, a necessary condition is the development and actual functioning of an innovative model for the information system formation in the structure of environmental safety of the region (openness and transparency of the impact of economic entities on the population and the environment).

Information flows should be distributed in such a way that interested parties (population, public environmental organizations, and scientific community) have access to the requested information by obtaining it only from official sources (reliability guarantee), thereby fulfilling one of the conditions for achieving sustainable and inclusive development of the region.

References

1. On approval of the Concept for the development of public non-financial reporting and an action plan for its implementation: Order of the Government of the Russian Federation dated 05.05.2017 No. 876-r. Consultant Plus. http://www.consultant.ru/cons/cgi/online.cgi?req=doc&base=LAW&n=216631&fld=134&dst=1000000001,0&rnd=0.9296204677257955#029497191440890225. Accessed 29 April 2019
2. Adhariani, D., de Villiers, Ch.: Integrated reporting: perspectives of corporate report preparers and other stakeholders. Sustain. Accou., Manage. Policy J. **10**(1), 126–156 (2019). https://doi.org/10.1108/SAMPJ-02-2018-0043
3. Baret, P., Helfrich, V.: The "trilemma" of non-financial reporting and its pitfalls. J. Manag. Gov **23**, 485–511 (2019). https://doi.org/10.1007/s10997-018-9430-z
4. Belomestnov, V.G., Ognev, D.V., Saktoev, V.E.: Environmental responsibility of business. ESSUTM. Bulletin **68**(1), 98–102 (2018)
5. Camilleri, M.A.: Case Study 2: Environmental, social and governance reporting in Europe. In: Camilleri, M.A. (ed.) Corporate Sustainability, Social Responsibility and Environmental Management, pp. 119–138 Springer, Cham (2017). https://doi.org/10.1007/978-3-319-46849-5_7
6. Crotty, J.: Corporate social responsibility in the Russian federation: a contextualized. Bus. Soc. **55**(6), 825–853 (2016). https://doi.org/10.1177/0007650314561965
7. Enterprises of the Volgograd region. Interfax. https://interfax-era.ru/predpriyatiya/?region=12. Accessed 24 March 2020
8. GRI Services (2019). Making Sustainability Reporting Accessible to All! https://www.globalreporting.org/information/news-and-press-center/Pages/kick-off-service-gri.aspx. Accessed: 21 May 2019
9. GRI Standards Global Reporting Initiative (2019). Global Reporting. Available https://www.globalreporting.org/standards/gri-standards-download-center/. Accessed 27 May 2019
10. Information partnership with GRI (2019). Russian Union of Industrialists and Entrepreneurs. https://rspp.ru/simplepage/informatsionnoe-partnerstvo-s-gri/. Accessed: 24 March 2020
11. Matova, N.I., Mishulina, S.I.: Social-ecological responsibility of business: essence, factors of formation, manifestation characteristics. J. Volgograd State Univ. Econ. **22**(3), 123–137 (2020). https://doi.org/10.15688/ek.jvolsu.2020.3.11
12. Medvedeva, O.E., Mikerin, G.I., Medvedev, P.V.: Integrated non-financial reporting and the problems of inclusion in it environmental component in Russia. Property Relat. Russian Federation **194**(11), 6–14 (2017)
13. National Register and Library of Corporate Non-Financial Reports. Russian Union of Industrialists and Entrepreneurs. http://rspp.ru/simplepage/157. Accessed 26 April 2019

14. Nikiforova, O.A., Mitrofanova, D.O.: Concepts of social responsibility of business: initial concepts and classifications. Bull. Saint-Petersburg Univ. Sociol. **2**, 214–228 (2017)
15. Pahomova, N.V., Rihter, K.K., Malyshkov, G.B.: Inclusive sustainable development: priorities, indices, international experience, potential of coordination with reindustrialization model. Problems Modern Econ. **51**(3), 15–24 (2014)
16. Porter, M., van der Linder, C.: Green and the Competitive: ending the Stalemate. In: Welford, R., Starkey, R. (eds.) Business and the Environment, pp. 61–78. Earthscan Publications Ltd., London (1996)
17. Public register of objects of negative impact on the environment. Software and technical support for the accounting of objects of negative impact on the environment https://onv.fsrpn.ru/. Accessed 03 January 2020
18. Sankara, J., Patten, D.M., Lindberg, D.L.: Mandated social disclosure: Evidence that investors perceive poor quality reporting as increasing social and political cost exposures. Sustain. Account. Manag. Policy J. **10**(1), 208–228 (2019). https://doi.org/10.1108/SAMPJ-05-2017-0046
19. Trent'ev, A.A., Korobova, N.L.: Non-financial reporting as a means of greening activities of companies. Russia: Trends Prosp. Dev.: Yearbook **10**(1), 350–355 (2015)
20. Zalizniak, E., Matveeva, A., Polovinkina, Yu., Kirillov, S.: Environmental component of nonfinancial reporting: development trends and world practice. In: Anthropogenic transformation of geospace: nature, economy, society, IV International Research to Practice Conference, Volgograd, October 2019 in AER: Advances in Engineering Research Vol. 191. Atlantis Press, Paris (2020). https://doi.org/10.2991/aer.k.200202.064
21. Zaliznjak, E.A., Matveeva, A.A., Polovinkina, Ju.S., Kirillov, S.N.: Environmental responsibility of business in ensuring technosphere safety of the region. VolSU Publishing Office, Volgograd (2020)
22. Zavodskih, A.A.: Building an index of inclusive development of regions. Pskov J. Regional Stud. **42**(2), 19–41 (2020)

Chapter 9
The Development of Inclusive Economy Based on Digital Technologies: Econometric Assessment of Formation

Farida G. Alzhanova, Mikhail M. Guzev, Elena V. Loginova, and Aleksander A. Polkovnikov

Abstract The contributors propose an econometric model to analyze the correlation between the indexes that evaluate digital technologies adoption in economic activities and social interactions and the indexes that quantify the attributes of an inclusive economy. Firstly, open-access values were collected for the two groups of indicators, namely, the exogenous indexes measuring the effect of digital technologies on socio-economic systems, and endogenous indexes characterizing the economic development of society. The values were used to estimate the degree of the economy inclusiveness. For this purpose, with the aid of Python programming and the linear and Bayesian linear (ARD) regression methods, an econometric model was constructed to establish the correlation between the exogenous and endogenous indexes, which are crucial for inclusive economic development. The econometric model was used to rank in descending order the endogenous indicators that show a degree of inclusive economic development. The indexes were ranked in the following order: The Legatum Prosperity Index; The Index of Economic Freedom; The Doing Business Index; GDP per capita; The Corruption Perceptions Index. The ranking allowed us to identify the most significant effects of digitalization processes on the development of an inclusive economy.

F. G. Alzhanova (✉)
Institute of Economics SC MES RK, Almaty, Kazakhstan

M. M. Guzev · E. V. Loginova · A. A. Polkovnikov
Volgograd State University, Volzhskiy branch, Volzhsky, Russia
e-mail: guzev@vgi.volsu.ru

E. V. Loginova
e-mail: elena.loginova@vgi.volsu.ru

A. A. Polkovnikov
e-mail: polkovnikov@vgi.volsu.ru

E. I. Inshakova and A. O. Inshakova (eds.), *New Technology for Inclusive and Sustainable Growth*, Smart Innovation, Systems and Technologies 287,
https://doi.org/10.1007/978-981-16-9804-0_9

9.1 Introduction

The urgent matters for modern society include 'work on climate change, debt, and inequality, working towards green, resilient and inclusive development' [1]. The proposed agendum makes the study of an inclusive economy one of the intensively growing fields of research. Rapidly, the ideas of an inclusive economy have found their way into programs and strategies developed for countries and regions.

The key topics discussed at the International Conference on Technology Policy and Innovation (1998) were devoted to 'knowledge for inclusive development, the role of science and technology policy in promoting inclusive development, bridging the gap through the exchange of knowledge between countries with different levels of development [2]. In 2010, following the Lisbon Strategy in Europe, the Europe 2020 'Strategy for Smart, Sustainable and Inclusive Growth' was proclaimed [3].

However, there are relatively few publications on this topic. Whereas the search query 'digital + inclusion' in the Web of Science database yielded as many as 1080 works published since 2004 the search query 'inclusive + economy' retrieved only 61 publications dedicated to the topic under discussion.

Studies devoted to inclusive economic development address a range of issues. They seek to define attributes of an inclusive economy [4], adapt capitalist institutions to strategies that focus on empowerment [5]. Some discuss how to form a dynamic and socially inclusive economy, develop an inclusive social security system, and define measures that need to be taken to overcome inequality in the labor market [6] and health care system.

It is vital to say that to achieve the goals of an inclusive economy, policies and measures have to be developed that would ensure economic inclusion. As such, studies point out equal access to infrastructure for all economic agents regardless of their geographical location. Others point out the necessity of anti-discrimination legislation adoption, implementation of inclusive labor and employment policies, adoption of inclusive labor standards, the launch of subsidized schemes and programs for employers to encourage inclusive employment, development of inclusive urban planning technologies, and construction of inclusive transport networks. Inclusive development targets can be estimated against the GDP per capita index, the unemployment rate of the economically active population, gender equality among the unemployed, accessibility of transport infrastructure, accessibility of digital infrastructure, labor productivity, and greenhouse intensity of GDP.

It is obvious that social inclusion can be reached through some interrelated policies and programs. They ought to include the creation of a barrier-free environment (technical compensatory facilities, composite services, assessing the accessibility of facilities and services) for vulnerable and disadvantaged social groups [7], introduction of anti-discrimination norms, inclusion of disadvantaged groups into social processes, providing the availability of specific social benefits and inclusive services in health care, implementation of inclusive education technologies and enhancement of financial inclusion for all the categories of citizens and all the sectors of the financial market [8].

With the rapid growth and spread of information technology, researchers' interest in the issue of inclusion in the digital world has increased. A separate field of research on the inclusive economy is associated with research on the digital gap, including the gender digital gap [9], basic or advanced digital skills [10], digital dividends [11]. The World Bank report notes that digital technologies will neither boost productivity nor reduce inequality unless society creates conditions to improve the business climate and governance, and increase investment into education and health. However, we have to say that facilitated access to digital technologies provides new opportunities for economic development, including an inclusive one, which currently is considered a guarantor of stability in society. Experts at the World Economic Forum in Davos (2018) emphasized that an inclusively developing economy is more resilient to external threats and economic crises.

9.2 Materials and Method

Digitalization of society contributes to a great extent to the transformation of the market economy into an inclusive one. This transformation suggests that society ensures a transition from an exclusive to an inclusive system of distribution of benefits and creates conditions to improve the quality of life [12]. An inclusive economy focuses on an individual's abilities. However, based on the theory of production factors developed by O. Inshakov [13, 14], the production function of an inclusive digital economy prioritizes the technical and technological factors but not the human [12], which causes controversy in the formation of new economic relations manifesting itself in the contradiction between 'the dominant of growth' and 'the dominant of development'. To tackle this contradiction, it is necessary to carry out a factor analysis of the inclusive 'digital' economy and define the impact of digitalization on economic growth and social development. This target can be fulfilled using econometric analysis methods.

The econometric analysis that measures the impact of digital technologies on economic practices and social interactions and the development of an inclusive economy consisted of several stages. First, we defined endogenous indicators as the quantitative values of the variables that characterize an inclusive economic development in the digital environment (Table 9.1).

Then, we collected data for the exogenous indicators that characterize the readiness of countries for digitalization (Table 9.2), and endogenous indicators of the proposed econometric model that estimate the impact of digitalization on the development of an inclusive economy.

Further, we obtained correlation matrices for the indices and sub-indices of the exogenous indicators of the proposed model, the analysis of which enabled us to select those characterized by the least multicollinearity. The data remained was normalized, and regression analysis was performed using LINEAR models and Bayesian linear (ARD) regression. Then, we ordered the exogenous indicators according to the impact they have on the endogenous indicators. Finally, by evaluating the quality

Table 9.1 The endogenous indicators of an inclusive economy

No.	Index	Characteristics of an inclusive economy indicators
1	GDP per capita	The positive dynamics of the indicator characterizes economic growth that provides opportunities for economic development
2	The doing business index	Characterizes the existence of transparent and effective business regulations that ensure equal access to business opportunities
3	The corruption perceptions index	The high value of the index characterizes the openness and low corruption in the state and economic sectors
4	The economic freedom index	The high value of the index characterizes a low level of government intervention in the production, distribution, and consumption of goods and services, which means an insignificant amount of government preferences for the business sector
5	The Legatum prosperity index	Measures the parameters of public welfare (economy, entrepreneurship, management, education, health care, security, personal freedoms, social capital, ecology), which taken together characterize opportunities for inclusive growth

Table 9.2 The exogenous indicators of an inclusive economy

No.	Index	Characteristics
1	Network readiness index	Measures the level of development of information and communication technologies (ICT) and the network economy in the countries of the world
2	Global digital competitiveness ranking index	Scores the capabilities (knowledge, technology) of countries and their readiness for digital transformation
3	Global connectivity index	Scores digital transformation of countries according to the criteria: investment in ICT development, ICT maturity, and indicators of the digital economy
4	The inclusive internet index	Measures the quality and breadth of available infrastructure required for access and levels of Internet usage and the capacity to access the Internet including skills, cultural acceptance, and supporting policy
5	E-government development index	Measures the availability of public services and the openness of the government when making decisions
6	B2C E-commerce index	Measures accessibility of electronic transactions

of regressions, we ordered the endogenous indicators according to their impact on an inclusive economic development.

With the aid of the Python programming language and using its libraries Pandas, Numpy, SkiKit Learn, we constructed an econometric model to build regression lines and analyze the indexes values. Firstly, we treated the values with linear, linear Bayesian (ARD), Lasso and Ridge regression models. As the latter two models showed insignificant coefficients, we excluded them from the analysis used below.

Linear regression is based on a linear function that reflects the dependence of factors (regressors, variable values) on the predicted (resulting) variable [15].

Bayesian linear regression (ARD) is based on the conclusion made in 1992 by David Mackay [16, 17], according to which an posteriori estimate of the variance is derived for each coefficient, after which the coefficients with variance close to zero are reset to zero. In this case, each weight is compared with the so-called regularization coefficient, which is an indicator of the adaptability of the variable to the initial data.

9.3 Results

We collected the data on the values of endogenous and exogenous variables of the econometric model from open-access Internet sources to evaluate the impact of digital technologies on the development of an inclusive economy in the world's countries [18–26].

Based on the obtained data, we constructed correlation matrices, and for the exogenous indexes we excluded from the model insignificant (closely correlated) values of the correlation coefficients whose magnitude are bigger than or equal to 0.7 or less than or equal to −0.7 (Tables 9.3, 9.4, 9.5, 9.6, 9.7 and 9.8).

Due to the analysis of correlation matrices, we selected the exogenous indicators with the lowest multicollinearity (Global Connectivity Index, Network Readiness Index, E-Government Development Index, E-Participation Index, Online Service Index, Human Capital Index, Telecommunication Infrastructure Index, UPU Postal Reliability Score, B2C E-Commerce Index).

Table 9.3 Correlational matrix of the network readiness index

	Network readiness index	Technology	Humans	Management	Impact
Network readiness index	1	0.981	0.954	0.976	0.960
Technology	0.981	1	0.899	0.954	0.929
Humans	0.954	0.899	1	0.925	0.894
Management	0.976	0.954	0.926	1	0.896
Impact	0.960	0.929	0.894	0.896	1

Table 9.4 Correlational matrix of the global digital competition index

	Global digital competition index	Knowledge	Technology	Readiness for the future
Global digital competition Index	1	0.948	0.967	0.961
Knowledge	0.948	1	0.888	0.848
Technology	0.967	0.888	1	0.904
Readiness for the future	0.961	0.848	0.904	1

Table 9.5 Correlational matrix of the global connectivity index

	Global connectivity index	Supply	Demand	Algorithms of correlation	Potential
Global connectivity index	1	0.959	0.928	0.967	0.927
Supply	0.959	1	0.864	0.888	0.864
Demand	0.928	0.864	1	0.893	0.753
Algorithms of correlation	0.967	0.888	0.893	1	0.906
Potential	0.927	0.864	0.753	0.906	1

Table 9.6 Correlational matrix of the inclusive internet index

	The inclusive internet index	Availability	Accessability	Urgency	Readiness ara>
The inclusive internet index	1	0.566	0.568	0.280	0.421
Availability	0.566	1	−0.25	0.102	0.29
Accessability	0.568	−0.25	1	−0.096	−0.008
Urgency	0.280	0.102	−0.096	1	−0.061
Readiness	0.421	0.29	−0.008	−0.061	1

Table 9.7 Correlational matrix of the E-government development index

	E-government development index	Online service index	Telecommunication infrastructure index	Human capital index
E-government development index	1	0.937	0.938	0.865
Online service index	0.937	1	0.812	0.739
Telecommunication infrastructure index	0.938	0.812	1	0.716
Human capital index	0.865	0.739	0.716	1

Table 9.8 Correlational matrix of the B2C E-commerce index

	B2C E-commerce index	A proportion of population using Internet	A proportion of population at the age 0f 15 years and above having a bank account	A number of safe servers per 1 mln. population	Sub-index of postal reliability of the integrated postal development index of the Universal Postal Union
B2C E-Commerce Index	1	0.853	0.834	0.813	0.643
A proportion of population using Internet	0.853	1	0.766	0.709	0.307
A proportion of population at the age of 15 years and above having a bank account	0.834	0.766	1	0.685	0.279
A number of safe servers per 1 mln. population	0.813	0.709	0.685	1	0.237
Sub-index of postal reliability of the integrated postal development index of the Universal Postal Union	0.643	0.307	0.279	0.237	1

At the next stage of the study, we normalized the values of the selected exogenous and endogenous indicators based on the use of the preprocessing normalize function of the NumPy library of the Python programming language, to work with the indicators whose values used to have different bases before normalization.

To estimate the impact of the selected exogenous indicators values on the values of the endogenous indicators, we carried out a regression analysis using two models ARD and LINEAR. The results are shown in Tables 9.9, 9.10, 9.11, 9.12 and 9.13.

The performed regression analysis showed that the endogenous variable of GDP per capita in the ARD model is most influenced by the Network Readiness Index and the Global Connectivity Index, in the LINEAR model, it is dependent on the Human Capital Index, Online Service Index, and Telecommunication Infrastructure Index;

Table 9.9 An estimated Variable—GDP per capita

Exogenic indicators	Variable		Coefficient values	
	ARD	LINEAR	ARD	LINEAR
Global connectivity index	0.115	0.077	0.967	0.935
Network readiness index			3.631	3.142
E-government development index			−3.089	−2 327.159
E-participation index			0	0.164
Online service index			−0.785	776.649
Human capital index			0	788.762
Telecommunication infrastructure index			0	759.579
UPU postal reliability score			−0.531	−1.116
B2C E-commerce index			0	2.072

Table 9.10 An estimated variable—the doing business index

Exogenic Indicators	Variable		Coefficient values	
	ARD	LINEAR	ARD	LINEAR
Global connectivity index	0.0716	0.054	0	−0.35
Network readiness index			0	0.006
E-government development index			0	−21.476
E-participation index			0	−0.065
Online service index			0.195	−6.878
Human capital index			0.002	−7.131
Telecommunication infrastructure index			0	−7.139
UPU postal reliability score			0	−0.015
B2C E-commerce index			0.381	0.438

Table 9.11 An estimated variable—the corruption perceptions index

Exogenic Indicators	Variable		Coefficient values	
	ARD	LINEAR	ARD	LINEAR
Global connectivity index	−0.065	−0.087	0.002	0.333
Network readiness index			1.463	1.331
E-government development index			0	565.565
E-participation index			0.001	0.412
Online service index			0.001	−189.361
Human capital index			0	−191.966
Telecommunication infrastructure index			0	−185.219
UPU postal reliability score			−0.098	−0.268
B2C E-commerce index			0	0.028

Table 9.12 An estimated variable—the economic freedom index

Exogenic indicators	Variable		Coefficient values	
	ARD	LINEAR	ARD	LINEAR
Global connectivity index	0.079	0.091	0.207	0.324
Network readiness index			0.158	0.388
E-government development index			−0.001	−29.999
E-participation index			0.177	0.574
Online service Index			0	9.822
Human capital index			−0.001	10.104
Telecommunication infrastructure index			−0.011	9.334
UPU postal reliability score			0	−0.152
B2C E-Commerce index			0	0.099

Table 9.13 An estimated variable—the Legatum prosperity index

Exogenic indicators	Variable		Coefficient values	
	ARD	LINEAR	ARD	LINEAR
Global connectivity index	0.043	0.03	0	0.06
Network readiness index			0.563	0.479
E-government development index			−0.001	269.364
E-participation index			−0.001	0.069
Online service index			−0.025	−90.287
Human capital index			0	−91.434
Telecommunication infrastructure Index			0	−88.083
UPU postal reliability score			−0.062	−0.118
B2C E-commerce index			0.272	0.407

- the endogenous variable the Doing business Index in ARD and LINEAR models shows a strong correlation with the B2C E-Commerce Index;
- the endogenous variable the Corruption Perceptions Index in the ARD model correlates with the Network Readiness Index, in the LINEAR model, it shows correlation with the Network Readiness Index and the E-Government Development Index;
- the endogenous variable Index of Economic Freedom in the ARD model is most influenced by the Global Connectivity Index, in the LINEAR model, it is influenced by the Human Capital Index, Online Service Index, and Telecommunication Infrastructure Index;
- the endogenous variable, the Prosperity Index, in the ARD model is most influenced by the Network Readiness Index and B2C E-Commerce Index, in the LINEAR model, it correlates with the Network Readiness Index, E-Government Development Index, and B2C E-Commerce Index.

The constructed models were also evaluated in terms of the quality of R^2 regressions by endogenous indicators (the closer the R^2 value to 1, the better the model can predict the endogenous variable by a set of exogenous indicators):

- the values for the endogenous variable GDP per capita—ARD: 0.78155730265946, LINEAR: 0.8106852647371257;
- the values for the endogenous variable, the Doing Business Index—ARD: 0.6711404472230671, LINEAR: 0.6927878251674546;
- the values for the endogenous variable the Corruption Perceptions Index—ARD: 0.7706445586240167, LINEAR: 0.7944157164533866;
- the values for the endogenous variable, the Index of Economic Freedom—ARD: 0.5143171260865078, LINEAR: 0.6014058753656392;
- the values for the endogenous variable, the Prosperity Index—ARD: 0.9071223294784283, LINEAR: 0.9187181749180923.

9.4 Conclusion

Following the results of the econometric analysis, the exogenous indicators reflecting the use of digital technologies in economic practice and social interactions were ranked and classified according to the degree of their impact on the development of an inclusive economy:

- high impact indicators: Network Readiness Index and B2C E-Commerce Index;
- indicators of average influence: E-Government Development Index, Global Connectivity Index, Human Capital Index, Online Service Index, and Telecommunication Infrastructure Index;
- indicators, the influence of which is not revealed: E-Participation Index and UPU Postal Reliability Score.

Thus, the value of the endogenous indicator ‘Prosperity Index’ is best predicted by the exogenous baseline indicators. Both regression models have approximately equal predictive quality, but the normal linear regression is characterized by a slightly higher score. The least predicted values for the endogenous indicators are the Index of Economic Freedom and the Index of Doing Business since the quality of both models does not exceed 0.7. The values of the endogenous indexes ‘GDP per capita’ and ‘Corruption perceptions index’ are on average predicted in terms of quality, where an average quality is 0.789325711.

Even though the usual linear regression has better R^2 indicators, evaluating the obtained coefficients, we can conclude that when constructing a model for evaluating the impact of digital technologies on the development of an inclusive economy, it is more efficient to use ARD regression, since for a given dataset the coefficients of this model are in an interval $(-1; 1)$, while ordinary linear regression has coefficients far beyond its boundaries.

References

1. Malpass, D.: Building a Green, Resilient and Inclusive Recovery (2021), 29.03.2021. https://www.gks.ru/free_doc/doc_2018/region/pril-reg-sep_2018.xlsx. Accessed 15 May 2021
2. Conceiçao, P., Gibson, D., Heitor, M., Sirilli, G.: Knowledge for inclusive development: the challenge of globally integrated learning and implications for science and technology policy. Technol. Forecast. Soc. Chang. **66**(1), 1–29 (2001). https://doi.org/10.1016/S0040-1625(00)00075-5
3. EUROPE 2020 (2020). A European strategy for smart, sustainable and inclusive growth. https://eur-lex.europa.eu/legal-content/EN/TXT/PDF/?uri=CELEX:52010DC2020&from=EN. Accessed 15 Aug 2021
4. Shipton, D., Sarica, S., Craig, N.: Knowing the goal: an inclusive economy that can address the public health challenges of our time. J. Epidemiol. Commun. Health Published Online First (2021). https://doi.org/10.1136/jech-2020-216070
5. Tarko, V.: Simple rules for a more inclusive economy. European Journal of Law and Economics (2020). Accessed 15 Aug 2021. https://doi.org/10.1007/s10657-020-09674-2.
6. Etherington, D., Jones, M.B.: Contradictions of the workfare state? denmark, welfare-through-work, and the promise of job rotation. Environ. Planning C-Gover. Policy **22**(1), 129–148 (2004). doi:10.1068/c
7. Inbusiness.kz: Inklyuzivnye uslugi stanovyatsya bezbarernymi. [Inclusive services become barrier-free]. https://inbusiness.kz/ru/news/inklyuzivnye-uslugi-stanovyatsya-bezbarernymi (2021). Accessed: 21 Jun 2021
8. 8.World Bank (2020). Financial Inclusion. Europe and Central Asia Economic.https://doi.org/10.1596/978-1-4648-1409-9
9. Pawluczuk, A., Gamundani, A.: Bridging the gender digital divide: an analysis of existing guidance for gender digital inclusion programmes' evaluations. Digital Policy Regul. Gover. **23**(3), 287–299 (2021)
10. Mihai, M., Catana, A., Titan, E., Manea, D.: 14th International Scientific Conference on eLearning and Software for Education—eLearning Challenges and New Horizons Proceedings Paper. The Digital premium progress of an Inclusive Economy (2018). https://doi.org/10.12753/2066-026X-21-242
11. World Bank (2016). World Development Report 2016: Digital Dividends. https://doi.org/10.1596/978-1-4648-0671-1
12. Loginova, E.: digital coordinates of modern russian society: research methodology. Sci. J. Volgograd State Univ. Econ. **21**(3), 15–28 (2019). https://doi.org/10.15688/ek.jvolsu.2019.3.2
13. Inshakov, O.: The theory of factors of production in the context of the development economics. VolSU publ, Volgograd (2002)
14. Inshakov, O.: "DEVELOPMENT nucleus" in the light of the new factors of production theory. Econ. Contemporary Russia **1**, 11–25 (2003)
15. Vyugin, V.: Mathematical Foundations of the Theory of Machine Learning and Forecasting. Moscow (2013)
16. MacKay, D.: A practical Bayesian framework for backpropagation networks. Neural Comput. **4**(3), 448–472 (1992)
17. MacKay, D.: Bayesian Methods for Adaptive Models (1992). http://www.inference.phy.cam.ac.uk/mackay/thesis.pdf. Accessed: 06 Mar 2021
18. Abdrahmanova, G.I., Vishnevskij, K.O., Gohberg, L.M., et al.: Digital Economy Indicators of the Russian Federation: 2021: Data Book. Moscow, HSE. (2021). https://doi.org/10.17323/978-5-7598-2385-8
19. Corruption Perceptions Index. https://www.transparency.org/en/cpi/2020/index/nzl. (2020). Accessed 04 Aug 2021
20. Doing Business (2020). https://www.doingbusiness.org/en/doingbusiness. Accessed 04 Sept 2021

21. E-Government Development Index (2020). https://publicadministration.un.org/egovkb/Portals/egovkb/Documents/un/2020-Survey/2020%20UN%20E-Government%20Survey%20(Full%20Report).pdf. Accessed 04 Aug 2021
22. Global Connectivity Index (2020). https://www.huawei.com/minisite/gci/en/. Accessed: 04 Aug 2021.
23. Index of Economic Freedom (2021). https://www.heritage.org/index/ Accessed 04 Aug 2021
24. The Legatum Prosperity Index (2020). https://www.prosperity.com/. Accessed 04 Aug 2021
25. The World Bank (2020) Gross Domestic Product. https://databank.banquemondiale.org/data/download/GDP_PPP.pdf. Accessed: 04 Aug 2021.
26. UNCTAD (2020). B2C E-Commerce Index. https://unctad.org/system/files/official-document/tn_unctad_ict4d17_en.pdf. Accessed: 04 Aug 2021

Chapter 10
Digital Transformation of the National Economic System: Identification of Key Determinants

Marina E. Buyanova, Alla E. Kalinina, and Irina S. Averina

Abstract The chapter presents the results of a comprehensive study devoted to identifying the most significant determinants of digital technological development at the macro level, that allow to determine effective scenarios for national and regional economic systems. The use of correlation and regression analysis in this study made it possible to identify the most significant institutional and economic factors that affect the effectiveness of the economic space digitalization at the national level. This, in turn, made it possible to identify a set of problems arising in countries in the context of the development of this phenomenon on a global scale and to define a set of directions for their consistent resolution. The timeliness and comprehensiveness of national economy responses to such challenges and threats posed to them by the natural process of the economic system evolution can provide a state with a high level of security and bring it to a higher competitive position. The significance of this study lies in identifying and validating institutional and economic factors that have a significant impact on the pace of digital transformation of national economic systems expressing features of technological singularity in modern economic life and leading to transformation following an exponential trajectory.

10.1 Introduction

The changes occurring in the economic and institutional mechanisms of the economic systems of different countries are a response to the realities of Industry 4.0, which causes sectoral structural changes that manifest themselves at all economic levels (macro, meso, and micro). This circumstance dictates the emergence of new challenges for and threats to the national economies of all states of the world community. This, in turn, determines the need for their timely overcoming and prevention, and

M. E. Buyanova (✉) · A. E. Kalinina · I. S. Averina
Volgograd State University, Volgograd, Russia
e-mail: BuyanovaME@volsu.ru

I. S. Averina
e-mail: AverinaIS@volsu.ru

E. I. Inshakova and A. O. Inshakova (eds.), *New Technology for Inclusive and Sustainable Growth*, Smart Innovation, Systems and Technologies 287,
https://doi.org/10.1007/978-981-16-9804-0_10

creates the potential to increase the level of competitiveness, stability and security of individual macroeconomic systems. These realities provide countries with unprecedented opportunities to choose the path of further development through the prism of divergence, which allows economic entities to develop differently under initial, largely typical conditions. Identification of the most significant factors of digital technological development within the framework of the macroeconomic scale will make it possible to identify the most effective scenarios for national and regional economies, which is a significant condition within the framework of national macroeconomic policy for the further competitive development of the country in general.

It is the timely identification of fundamental determinants defining the place of national economic systems in the realities of the Fourth Industrial Revolution that will allow countries to form necessary components of public policy or transform the existing ones in order to create the necessary opportunities for their competitive, sustainable, and safe development in modern conditions.

10.2 Materials and Method

Among the researchers who can be considered founders of the theory of information society and digital economy, it is necessary to name [3, 8, 13, 24, 25, 27, 29–31, 33] and others.

The issues of theory and practice of economic system digital transformation are covered in works of [1, 2, 4–9, 12, 14–16, 19–22, 26, 28, 32, 34–43] and others.

The study analyzes the economic and institutional indicators that determine, to varying degrees, success/failure of digitalization of the economic space of the country and the extent of its involvement in Industry 4.0 realities. These indicators were included in the correlation and regression analysis to identify the most significant factors that determine the country development level within the framework of digitalization of spaces, and either stimulate or hinder its further development in different countries (for the representativeness of the sample, figures for 33 countries with a time lag of 10 years (2009–2018) were used).

10.3 Results

Thus, 33 geographical entities were included in the sample: Belgium, Bulgaria, the Czech Republic, Denmark, Germany, Greece, Estonia, Spain, France, Croatia, Italy, Latvia, Lithuania, Hungary, Malta, the Netherlands, Austria, Poland, Romania, Slovenia, Slovakia, Finland, Iceland, Norway, Switzerland, the United Kingdom, the Russian Federation, the USA, South Korea, Japan, Canada, China. These countries are presented in the most comprehensive sample according to specified indicators, including the chosen time interval (2009–2018). It is also worth noting that the sample includes countries with different levels of economy digitalization and different mechanisms of adapting to the realities of Industry 4.0.

As can be seen, these countries also differ in the level of current socio-economic development, have different resource potential (including labor and technological resources, etc.), which also justifies the significance of the calculations presented below, which made it possible to identify a number of factors that are significant for the successful implementation of strategic plans for the creation and improvement of an effective digital economy within the framework of modern transformation of economic systems by different countries.

Currently, there are various methods for assessing the national economy digitalization level, that were developed both in Russia and in other countries; among them are the following:

- Method of Moscow School of Management SKOLKOVO and Center for Financial Innovations and Cashless Economy (Digital Russia Index);
- Method of the International Telecommunication Union (ICT Development Index (IDI)): an integral indicator introduced in order to reflect the latest achievements in the field of ICT and to take into account new trends in the development of the economies of different countries);
- Method of Huawei Company reflecting the progress of the largest countries in the world in the field of transition to digital technologies (Global Connectivity Index (GCI));
- Method of the European Union countries (Digital Economy and Society Index (DESI), a composite index that summarizes corresponding indicators on the effectiveness of digital technologies in Europe and tracks the evolution of EU member states in the field of digital competitiveness) [23].

These methods take into account a number of complex components of the economic system digitalization in general, leaving the analysis of the most significant determinants affecting the speed of digital transformations occurring in national economic systems outside of the scope of research. This circumstance determines the need for researching this aspect. In this regard, a study of the determinants influencing the digital transformation of national economies was conducted and their significance was defined.

The aforementioned countries were analyzed by 7 indicators:

- GDP per capita is one of the most significant indicators characterizing the socio-economic development of a country (taken as a dependent variable)—Y;
- ICT personnel in the national economy (this indicator and the subsequent ones are presented as independent variables)—X1;
- The amount of commercial R&D expenses—X2;
- The number of patent applications—X3;
- The number of personnel engaged in R&D at enterprises and organizations in all economic sectors—X4;
- Development of employees' ICT skills and abilities at enterprises and organizations—X5;

- The amount of government budget expenditures on research and development—X6;
- Employed persons with education in ICT—X7.

This list is explained by the completeness and accessibility of the data used and can be expanded later in the process of further studies by interested researchers.

The indicators mentioned above were studied dynamically from 2009 to 2018, that is, the time series length in the model equals 10. The total number of observations is 260. The model under study contains three indicators of the financial aspect, and the logarithms of these indicators were taken due to the possibility of data contamination due to inflation-linked circumstances. The model can be represented by the following regression Eq. (10.1).

$$lnY = \beta_0 + \beta_1 X_1 + \beta_2 lnX_2 + \beta_3 X_3 + \beta_4 X_4 + \beta_5 X_5 + \beta_6 lnX_6 + \beta_7 X_7 \quad (10.1)$$

Based on statistical data, 2 models were developed:

- generalized least squares method (GLS)

$$lnY = 8.746 + 0.02X_1 + 0.114 lnX_2 + 0.0005X_3 + 0.703X_4 + 0.010X_5 + 0.201 lnX_6 + 0.325X_7 \quad (10.2)$$

- fixed effects method (FE)

$$lnY = 2.337 + 0.001X_1 + 0.112 lnX_2 + 0.00006X_3 + 0.079X_4 + 0.001X_5 + 0.022 lnX_6 + 0.023X_7 \quad (10.3)$$

The results of the evaluation of the models are given in Table 10.1.

The test statistics revealed that the aforementioned factors influence on the GDP of different countries as on one of the most significant macroeconomic indicators of the state economic development. Based on Akaike and Schwarz criteria, the GLS model is preferable to the FE model.

Considering the analysis above and greater rationality of the GLS model applicability, it follows that the most significant economic factors in the model are the following:

- the amount of commercial R&D expenses in the ICT sector (1% significance level);
- R&D personnel in total by activity sectors (1% significance level);
- development of employees' ICT skills and abilities at enterprises and organizations (5% significance level);
- government budget expenditures on research and development (1% significance level);
- employed persons with education in ICT (1% significance level).

Table 10.1 Model evaluation results

Indicator	GLS		FE	
	Coeff. (β)	St. errors	Coeff. (β)	St. errors
ICT personnel	0.020	0.033	0.001	0.003
Commercial R&D expenses	0.114***	0.019	0.012**	0.0005
Number of patent applications	0.0005	5.360	0.00006*	0.00006
Number of personnel engaged in R&D at enterprises and organizations in all economic sectors	0.703***	0.078	0.079**	0.001
Development of employees' ICT skills and abilities at enterprises and organizations	0.010**	0.004	0.001	0.0001
Government budget expenditures on research and development	0.201***	0.064	0.022*	0.0033
Employed persons with education in ICT	0.325***	0.083	0.023	0.001
Constant	8.746***	0.141	2.337***	0.003
Observations used	260		260	
Akaike criterion	179.618		190.933	
Schwarz criterion	204.543		−919.083	
Adjusted R-squared	0.708		0.879	
Significance levels: *$p < 0.1$, **$p < 0.05$, ***$p < 0.01$				

Source Compiled by the authors based on [10, 11, 18]

The significance of these factors is confirmed by statistical data on countries according to the International Digital Economy and Society Index [17] (Table 10.2).

Denmark occupies the leading position in this ranking with an index value of 0.76, followed by South Korea, which has successfully adapted to the Fourth Industrial Revolution conditions and digitalization processes. A total of 45 countries are

Table 10.2 Ranking of leading countries by the Digital Economy and Society Index [17]

Country	Index value
Denmark	0.76
The Republic of Korea	0.75
Finland	0.74
The Netherlands	0.74
The United Kingdom	0.73
Iceland	0.73
Norway	0.73
Switzerland	0.70
Russia	0.48

Source Compiled by the authors based on [17]

included in the ranking. For comparison, Russia ranks 39th with an index value of 0.48.

It should be noted that currently official statistics present a Digital Economy and Society Index country ranking 2020, however, in order to ensure data comparability, an earlier ranking was used due to a complete sample of statistical data reflected in the correlation and regression analysis. No significant changes that would violate the logic of the study were revealed in the 2020 and 2018 rankings.

A comparison of data of the ranking above (Table 10.2) and their interconnection with statistical data on the determinants that are included and highlighted as the most significant ones (Table 10.3) showed that index values generally correlate with the success of digital transformations in the countries, which indicates the correctness of the constructed models.

The correlation and regression analysis shows (this is confirmed by the statistical data included in the GLS and FE models) that the quantitative value of "the number of personnel engaged in R&B at enterprises and organizations in all economic sectors" indicator, the share of government budget expenditures on R&D, employed persons with education in ICT, enterprises that conducted training for further ICT

Table 10.3 Indicators of digital transformations of the economies in different countries, 2018

Country	Amount of commercial R&D expenses in the ICT sector as a % of total R&D expenses	Government budget expenditures on research and development, % of the total amount	Number of personnel engaged in R&D at enterprises and organizations in all economic sectors	Development of employees' ICT skills and abilities at enterprises and organizations, % of the total number of enterprises that conducted training	Employed persons with education in ICT, thousand people
Denmark	1.08	1.75	2.07	28	24.4
The Republic of Korea	13.09	3.74	1.43	N/D	N/D
Finland	0.82	1.56	1.88	34	58.0
The Netherlands	3.64	1.69	1.77	31	110.4
The United Kingdom	0.44	1.33	1.43	28	352.9
Iceland	0.43	2.09	1.64	25	2.1
Norway	1.17	2.02	1.73	42	28.3
Switzerland	7.34	2.82	1.74	28	71.3
Russia	N/D	1.53	1.09	N/D	N/D

Source Compiled by the authors based on [10]

studying/ICT competence development of their staff, as well as commercial R&D expenses in the ICT sector, are among the significant determinants of the digital transformation of the national economic systems.

10.4 Conclusion

Thus, the current realities of the macroeconomic systems development and the success of digital transformations occurrent in the conditions of the Fourth Industrial Revolution depend on the following economic and institutional factors:

- ICT personnel in the national economy;
- the number of patent applications;
- the amount of commercial R&D expenses;
- the amount of government budget allocations for research and development;
- development of employees' ICT skills and abilities at enterprises and organizations;
- the number of personnel engaged in R&D at enterprises and organizations in all economic sectors;
- employed persons with education in ICT.

However, it should be noted that important institutional factors should include the regulatory framework effectiveness, the effectiveness of the work of educational and government institutions, the degree of enterprises engagement in digital economy. A complete statistical sample of these indicators is not presented; therefore, they were not included in the econometric model presented above. However, they may be included in the model at a later stage and investigated given that there is a sufficiently comprehensive sample of necessary statistical data in official sources.

However, the absence of these data does not affect the significance of the identified factors in the model, which can only be potentially expanded and supplemented in the future.

The identification of important determinants of the economic system digital transformation will make it possible not only to identify important leverage points to accelerate the pace of their adaptation to Industry 4.0 realities, but also to improve national programs for the digital development of the economy in accordance with the most prospective scenarios, while considering the divergent development conditions in different countries.

References

1. Betancourt, M.: The Critique of Digital Capitalism: An Analysis of the Political Economy of Digital Culture and Technology. Punctum Books, Brooklyn, NY (2016)
2. Brousseau, E.: Internet and Digital Economics: Principles Methods and Applications. Cambridge University Press, Cambridge, New York (2007)
3. Castells, M.: The Informational City: Information Technology, Economic Restructuring, and the Urban–Regional Process. Basil Blackwell, Oxford, UK (1991)
4. Christensen, J. (2003). The Industrial Dynamics of the New Digital Economy. Edward Elgar Publishing Ltd.
5. Cortada, J.W. (2000). 21st Century Business: Managing and Working in the New Digital Economy. Prentice-Hall International
6. Coyle, D.: The Weightless World: Strategies for Managing the Digital Economy. MIT Press (1999)
7. Dai, Q., Shin, E., Smith, C.: Open and Inclusive Collaboration in Science: A Framework. OECD Science, Technology and Industry Working Papers. 2018/07. Paris: OECD Publishing (2018). https://doi.org/10.1787/2dbff737-en. Accessed 15 Aug 2021
8. Dizard, W.: The Coming Information Age: An Overview of Technology, Economics, and Politics. Longman, New York (1989)
9. Elder-Vass, D.: Profit and Gift in the Digital Economy. Cambridge University Press, Cambridge, New York (2016)
10. European Commission (2009–2018). https://ec.europa.eu/eurostat/web/main/data/statistics-a-z
11. Federal State Statistics Service (2018). https://gks.ru/bgd/regl/b18_14p/Main.htm. Accessed 4 Sep 2021
12. Fuchs, C.: Marx in the Age of Digital Capitalism. Brill, Boston (2016)
13. Gates, B.: The Road Ahead. Viking, New York (1995)
14. Gaver, A., Cusumano, M.: Industry platforms and ecosystem innovation. J. Prod. Innov. Manag. **31**(3), 417–433 (2014)
15. Glushkov, V.M.: Cybernetics, Computer Technology, Informatics. Selected Tr. in three volumes. Vol. 1. Mathematical Problems in Cybernetics. Vol. 2. Computer—The Technical Base of Cybernetics. Vol. 3. Cybernetics and Its Application in the National Economy). Naukova dumka, Kiev (1990)
16. Illing, G.: Industrial Organization and the Digital Economy. The MIT Press, London (2006)
17. Digital Economy and Society Index (2018). https://eufordigital.eu/ru/library/digital-economy-and-society-index-desi-2018/.Accessed 28 Sep 2021
18. Indicators of the digital economy: Statistical collection. NRU HSE, Moscow (2019)
19. Inshakov, O.V., Fesyun A.V.: Industry as a form of approval, expansion and domination of economic mode in the economy. Sci. J. Volgograd State Univ. Glob. Econ. Syst. **2**(25), 44–53 (2014)
20. Johansson, B.: The Emerging Digital Economy: Entrepreneurship, Clusters, and Policy (Advances in Spatial Science). Springer, Berlin, Heidelberg (2006)
21. Kehal, H.: Digital Economy: Imacts, Influences and Challenges. IGI Global (2004)
22. Kelly, K.: New Rules for the New Economy: 10 Radical Strategies for a Connected World. Viking, New York (1998)
23. Koh, L.V., Koh, Yu.V.: Analysis of existing approaches to measuring the digital economy. Scientific and technical bulletin of SPbSPU. Econ. Sci. **4**, 78–89 (2019)
24. Lyon, D.: The Information Society: Issues and Illusions. Polity, Cambridge (1988)
25. Machlup, F.: The Production and Distribution of Knowledge in The United States. Princeton University Press, Princeton (1962)
26. Malecki, E.J.: The Digital Economy: Business Organization, Production Processes and Regional Developments. Routledge, NY (2008)
27. Masuda, Y.: The International Society as Post-industrial Society. World Future Society (1980)

28. McKinsey, M.: The Internet of Things: Mapping the Value Beyond the Hype (2015). https://www.mckinsey.com/~/media/mckinsey/business%20functions/mckinsey%20digital/our%20insights/the%20internet%20of%20things%20the%20value%20of%20digitizing%20the%20physical%20world/the-internet-of-things-mapping-the-value-beyond-the-hype.ashx. Accessed 17 Aug 2021
29. McLuhan, M.: The Gutenberg Galaxy: The Making of Typographic Man. University of Toronto Press, Toronto (1962)
30. Mosco, V.: Pushbutton Fantasies: Critical Perspectives on Videotex and Information Technology. Ablex, Norwood, NJ (1982)
31. Naisbitt, J.: Megatrends: Ten New Directions Transforming Our Lives. Warner Books, New York (1982)
32. Negroponte, N.: Being Digital. Knopf, NY (1995)
33. Porat, M.: The Information Economy: Definition and Measurement. United States Department of Commerce, Washington, DC (1977)
34. Rogatnykh, E.B.: The impact of digitalization on the development of the modern world economy. Econ. Manag.: Prob. Solut. **11**, 64–70 (2017)
35. Safronchuk, M.V.: The impact of digital transformation on business and the business environment. Digit. Econ. **2**, 38–44 (2018)
36. Schiller, D.: Digital Capitalism: Networking the Global Market System. MIT Press, Cambridge, MA (1999)
37. Schwab, K.: The Fourth Industrial Revolution. Eksmo Publishing, Moscow (2016)
38. Stonier, T.: Information and the Internal Structure of the Universe: An Exploration into Information Physics. Springer (1990)
39. Strelec, I.A.: The impact of new technologies on the economic behaviour of consumers and firms. SShA i Kanada: Ekonomika, Politika, Kul'tura **8**, 63–72 (2008)
40. Tapscott, D.: The Digital Economy: Promise and Peril in the Age of Networked Intelligence. McGraw-Hill, New York (1997)
41. Twenty-first survey of the leaders of the largest companies in the world (2018). https://www.pwc.ru/ru/publications/e_CEO-Survey2018_rus.pdf Accessed:2 Sep 2021
42. Welfens, P.: Digital Economic Dynamics: Innovations, Networks and Regulations. Springer, Berlin, Heidelberg (2007)
43. Zemtsov S.P.: Potential robotization and the economy of ignorance in the regions of Russia. In: XIX April International Scientific Conference on the Problems of Economic and Social Development. NRU HSE, Moscow (2018). https://conf.hse.ru/2018/program/. Accessed 28 Sep 2021

Chapter 11
Innovative Development Mechanism as a Factor of Inclusive Growth

Elena G. Russkova, Larisa V. Ponomareva, Sergey N. Sokolov, and Vasily A. Yakhtin

Abstract The purpose of the study is to establish the influence of the innovative development mechanism on the formation of the domestic economic growth points. To complete the study objectives, existing scientific and practical methods, official statistical data, and the developments of domestic and foreign researchers were used. Correlation and regression tools provided an opportunity to identify the impact of certain types of innovative technologies on the indicators of the population well-being, infrastructure and employment. The connection between the digitalization degree and indicator values of employment, infrastructure and population well-being has been confirmed. The priorities of state programs to support industries, technologies, territories that ensure the transition to an inclusive growth scenario in order to achieve maximum socio-economic efficiency and meet national interests are substantiated. The special significance of establishing the influence of the innovative development mechanism on the formation of the domestic economic growth points is due to the operational revision of established approaches to ensuring the state national interests caused by a rapid decline in key indicators of socio-economic development, unfavourable macroeconomic, as well as the sanitary and epidemiological situation.

E. G. Russkova (✉) · L. V. Ponomareva · S. N. Sokolov · V. A. Yakhtin
Volgograd State University, Volgograd, Russia
e-mail: russkovaeg@volsu.ru

L. V. Ponomareva
e-mail: larisa.cokyr@volsu.ru

S. N. Sokolov
e-mail: snsokolov@volsu.ru

V. A. Yakhtin
e-mail: ea-191_442755@volsu.ru

E. I. Inshakova and A. O. Inshakova (eds.), *New Technology for Inclusive and Sustainable Growth*, Smart Innovation, Systems and Technologies 287,
https://doi.org/10.1007/978-981-16-9804-0_11

11.1 Introduction

For centuries, ensuring the sustainable growth of the state economic system has been complex, but also a necessary basis for achieving competitiveness in the world community. Sustainable development is still a key national interest and a target for most modern countries. However, the current stage of the world economy functioning implies a significant content rethinking of the "sustainable economic growth" category, which is due to an increase in the aggregate needs of society, a significant depletion of the resource base, as well as rapid population growth. In addition, the currently dominant technological order is gradually moving towards the loss of its potential in terms of increasing additional efficiency, which undoubtedly leads to stagnation and a decrease in the indicator values of socio-economic development. The formed conditions predetermine the search for new ways to resume sustainable development, as well as the formation of economic growth points. Creating an innovative development mechanism based on rethinking the role of production factors in increasing competitive advantages could be one of the ways.

11.2 Methodology

Many foreign and domestic researches, as well as international economic organizations, paid deep attention to the problems of economic growth stagnation at the turn of the first two decades of the twenty-first century. Thus, the study of the economic growth issue in modern conditions was conducted by the Eurasian Economic Commission (EEC), which is a permanent supranational regulatory body of the Eurasian Economic Union. The special report of the EEC-UNCTAD "Inclusive Growth of the Eurasian Economic Union Member States: Assessments and Opportunities" [1] examined the connection between economic growth and the level of social inequality. The EEC-UNCTAD experts concluded that there was a tendency in the global economy to define the worsening inequality as payment for modernization and acceleration of economic growth. After achieving the planned result—a developed economic system of the state, the inequality level significantly decreased. However, in the future, such an assertion was refuted empirically: inequality began to grow in developed countries, while in developing countries the paths of economic development ranged from high inequality and low growth rates (mainly Latin American countries) to high growth rates and low inequality (some East Asian countries). As a result, a new mindset in economic theory has emerged, asserting that equality and growth are interdependent categories, and there has also been a shift in emphasis from the study exclusively applied to inequality in the distribution of economic results (income and wealth) to the study of opportunity inequality [2]. It was during this period that the term "inclusive growth" became more actively used in the scientific community.

Nowadays, the "inclusive development" is one of the central issues in the formation of long-term development strategies at the international level. The EEC-UNCTAD experts, referring to the statistics of the international association "Oxfam" [3], explain this by the following reason: from 1980 to 2016, 1% of the global income distribution accounted for 27% of the growth of the real income in the world economy. At the same time, the poorest 50% accounted for only 12% of the total real income growth. Similarly, the wealth of the world billionaires increased by $900 billion in 2018, or by $2.5 billion per day, while the wealth of the poorer half of humanity (3.8 billion people) fell by 11%. These figures show that the benefits of economic growth are unevenly distributed among people in our societies, and deep inequality is a systemic sign of hyper-globalization. That is why, to achieve integrated sustainable development, the regulatory structures are faced with the task of changing the existing unfavourable trend [4]. This will enable redistribution of the development balance in such a way as to make the global economy productive again, based on full employment with adequate wages; a fair society that seeks to reduce socio-economic gaps both within and between countries, as well as protecting vulnerable groups of the population.

The conditions and opportunities for the growth of the world economy in modern conditions have been repeatedly discussed by the World Economic Forum. As in the case of the EEC, the WEF confirms a strong correlation between the "inclusiveness" degree of the economy and its effectiveness. Thus, "The Inclusive Growth and Development Report 2017" [1] notes the following: the national economy differs from business, where net profit is of the utmost importance. History and science have repeatedly demonstrated that there is a connection between growth and social equality in the national economy. This connection can be either positive or negative. The degree of connection is determined by a diverse combination of structural and institutional aspects of economic policy that go far beyond the two areas most often featured in discussions about inequality: education and the redistribution of economic outcomes. Some world communities that have succeeded in creating a strong middle class, reducing poverty and social marginalization, as a rule, have created effective economic institutions and political incentives in many areas of the economy, and over time have implemented sound macroeconomic policies and reforms aimed at improving efficiency. In this regard, seven key components are identified that determine the degree of social participation in the processes of increment, as well as the distribution of the economic development results (Table 11.1).

In addition to the EEC and the WEF, the Organization for Economic Cooperation and Development (OECD) has repeatedly pointed out the possible causes of the global economic downturn, as well as ways to resume growth in its reports. In the reports "The Productivity-Inclusiveness Nexus" [6], "Making Innovation Benefit All: Policies for Inclusive Growth" [7], the OECD has also linked low economic growth rates and worsening social inequality. Since the poorer part of the population has significantly fewer funds to obtain quality education and the necessary competencies, they are much less likely to be employed for high-paying jobs. Consequently, the human potential in the economy is not fully exploited, which negatively affects the growth of labour productivity in the long term. According to OECD experts, one way

Table 11.1 Components and sub-components of inclusive economic growth

Nos	Components of inclusive economic growth	Sub-components of inclusive economic growth
1	Education and skills	Availability
		Quality
		Equality
2	Key services and infrastructure	Basic and digital infrastructure
		Health-related services and infrastructure
3	Corruption level and rent	Business and political ethics
		Rent concentration
4	Financial investments in the real sector of the economy	Inclusiveness of the financial system
		Participation in business investment
5	Asset creation and entrepreneurship	The number of small businesses
		Real estate ownership and financial assets
6	Employment and labour remuneration	Effective employment
		Wages and other income
7	Budget transfers	Tax legislation
		Social security

Source Compiled by the author based on [5]

to overcome this trend is the development of information and communication technologies (ICT) and their implementation in most social processes. Innovative technologies are a universally recognized factor in accelerating economic development, but in modern conditions, the effect of their influence is much more extensive: such technologies directly affect the level of well-being and social prospects [8], which is an essential prerequisite for the formation of economic growth points. On the one hand, ICT contribute to improving the well-being of the population by increasing the volume of more affordable products, as well as higher quality services. A case in point is the spread of mass open online courses, educational and methodological literature, which provide people with the opportunity to develop the necessary skills via the Internet, often without having to pay a fee, as well as adjust to a given time frame. On the other hand, ICT have had a fundamental impact on rethinking the essence of the functioning of the markets, changing the competition dynamics, the degree of labour requirements and the ability of people to create value.

Also, the issue of ensuring sustainable economic development in modern conditions is of great relevance in the Russian Federation [9]. Since 2007, the GDP growth rate has been rapidly declining, reaching 1.3% in 2019 [10]. A similar situation occurred with the real disposable incomes of the population: the value of an increase of 12.1% in 2007 reached the value of 1% in 2019 [5]. In addition, the current unfavourable sanitary and epidemiological situation has predetermined the operational revision of established approaches to ensuring the national state interests and the search for new ways to form growth points. Accordingly, "National action plan to ensure the restoration of employment and incomes of the population, economic growth and long-term structural changes in the economy" was presented by the Government of the Russian Federation on September 23, 2020 [11]. The key goal of the National Action Plan is to achieve a sustainable trajectory of economic and income growth, ensuring the implementation of national economic development goals based on the use of new technologies, including digitalization, new opportunities in the labour and education market, rapid and high-quality housing construction, export orientation and active import substitution, as well as ensuring a high resilience degree of the economy and the public health system to possible shocks in the future. As can be noted, the document is largely focused on social support of citizens, which is the basis for further long-term development. The key measures include targeted cash payments to socially vulnerable groups of the population, assistance in retraining and re-education, ensuring accessibility of primary health care, drug provision and vaccination for the population. In addition, a key feature of the National Plan is the emphasis on the introduction and use of innovative, including digital technologies as a key factor in achieving the goals. Thus, digital technologies will become a reference point in all sectors of the economy: industry, healthcare, education, as well as construction. Starting from the provision of electronic interaction between medical organizations and their patients to the development of artificial intelligence technologies, the state intends to achieve progressive satisfaction of many key national interests.

Summarizing the above-mentioned, it becomes possible to state that the modern world community defines "inclusivity" and "innovativeness" as the drivers of the further development of the world economy, which imply the involvement of an increasing number of people in the processes of creation, distribution, consumption mediated by information and communication technologies. Digital innovations will help to increase well-being either directly by improving the quality of human lives, or indirectly by providing educational opportunities and the necessary competencies with significantly lower costs. Ultimately, this will offer opportunities to increase total income, enhance competitive advantages, as well as ensure the protection of national interests.

11.3 Results

In order to establish priority state areas of support for economic sectors that ensure the transition to an inclusive growth scenario based on information and communication technologies in the Russian Federation, we conducted a correlation and regression analysis of the innovative technologies impact on human capital, infrastructure and employment. For this purpose, a selection of dependent and influencing variables was made in three specified directions. We present the studied dependent variable, as well as influencing factors concerning the first direction—the relationship of information and communication technologies and rates of employment (Table 11.2).

Table 11.2 Study of the relationship between information and communication technologies and rates of population employment

Dependent variable	Factors of economy digitalization affecting the dependent variable [7]	Factor index
The unemployment rate of the population aged 15 years and older in the constituent entities of the Russian Federation, on average for three months [11] (Y)	The number of subscribers of fixed broadband Internet access per 100 people of the population	X1
	Fixed-line telephone density (including payphones) per 100 people of the population	X2
	The volume of investments in fixed assets aimed at the acquisition of information, computer, and telecommunication (ICT) equipment, million roubles	X3
	The share of people employed in the ICT sector in the total number of employed population	X4
	The share of organizations that had special software tools for managing sales of goods (works, services), in the total number of surveyed organizations	X5
	The share of employees who used Internet access facilities provided by the organization at least once a week, in the total number of the payroll of organizations	X6
	Availability of personal computers, used for educational purposes, in professional educational institutions in the constituent entities of the Russian Federation	X7

Source Compiled by the authors based on [12, 13]

The basis for the general population formation was the data for each of the 85 constituent entities of the Russian Federation in 2019.

As a result of the analysis, the following correlation matrix was formed (Table 11.3).

Based on the presented matrix, the variables X1, X2, X5 have the greatest correlation with the studied indicator: the values of the coefficients exceed the value |0.5|, therefore the link between Y and X1, X2, X5 is discernible. The values of the presented correlation coefficients are negative, which indicates the presence of feedback. Such facts can be interpreted as follows: the greater the spread of the Internet and telephone networks on the territory of the constituent entity of the Russian Federation, the unemployment rate indicator is potentially less important since the

Table 11.3 Correlation matrix No. 1

	Y	X1	X2	X3	X4	X5	X6	X7
Y	1							
X1	−0.656	1						
X2	−0.616	0.688	1					
X3	−0.199	0.343	0.473	1				
X4	−0.441	0.551	0.562	0.367	1			
X5	−0.572	0.588	0.458	0.265	0.395	1		
X6	−0.447	0.539	0.638	0.765	0.503	0.491	1	
X7	−0.409	0.416	0.607	0.421	0.332	0.291	0.347	1

Source Compiled by the authors

Internet provides opportunities to organize remote work and create a significantly larger number of vacancies in the labour market.

We shall now proceed to consider the regression model (Table 11.4).

The data obtained demonstrate that 7 selected factors of digitalization have an impact on the unemployment rate of the population aged 15 years and older (Y) by 55.5%. In this case, such a value of the determination coefficient (R^2) is logically

Table 11.4 Regression model No. 1

Regression statistics	
Multiple R	0.745
R-square	0.555
Normalized R-square	0.515
Standard error	2.280
Observations	85
F	13.719
F crit	2.131
Student's t-criterion	1.991

	Coefficient	Standard error	t-stat	P-Value	Lower 95%	Upper 95%
Y-intersection	16.118	1.280	12.588	0.000	13.568	18.667
X1	−0.148	0.061	−2.420	0.018	−0.270	−0.026
X2	−0.123	0.070	−1.759	0.082	−0.262	0.016
X3	0.00003	0.00014	2.094	0.040	0.000	0.000
X4	−0.146	0.525	−0.277	0.782	−1.191	0.900
X5	−0.134	0.057	−2.364	0.021	−0.247	−0.021
X6	−0.416	0.361	−1.152	0.253	−1.134	0.303
X7	−0.077	0.065	−1.181	0.241	−0.206	0.053

Source Compiled by the authors

justified: in addition to the factors of digitalization, the unemployment rate of the population is influenced by a large set of factors that are not taken into account in the framework of the created model. However, the resulting value ($R^2 > 0.5$) may indicate an acceptable quality of the model. The "Null Hypothesis" is rejected because the value of Fischer F-criterion exceeds the critical value of the distribution F (13.719 > 2.131), therefore, there is a linear relationship between Y and at least one of the factors X. The most significant variables in this model are X1, X3 and X5. This is confirmed by the following: the P-value of these variables is less than the error probability (0.05). In addition, the value of t-stat. variables X1, X3, X5 satisfy the condition $|t - \text{stat.}| > t$—Students's t-Criterion (1.991), which also indicates the greatest significance of these variables within the framework of the model under consideration. The lower and upper bounds of the confidence interval have the same sign, which indicates the significance of the regression coefficients of variables X1, X3, X5. Such data can be interpreted as follows: there is a connection between the unemployment rate and the economy digitalization degree of the constituent entity of the Russian Federation. The most significant factors are the number of subscribers of broadband Internet access, the volume of investments in fixed assets aimed at acquiring information, computer and telecommunications equipment, as well as the number of organizations that had special software tools for managing sales of goods (works, services). The influence of factors is logically explicable: the spread of the Internet, as well as the latest information platforms, allows involving more people in the processes of production, distribution and consumption of goods, works and services by creating new mechanisms of interaction, thereby increasing the level of employment and reducing unemployment.

Thereafter, we proceed to deal with the relationship between the digitalization indicators and the average per capita monetary income of the population (Table 11.5).

The basis for the general population formation was the data for each of the 85 constituent entities of the Russian Federation in 2019.

As in the previous case, a correlation matrix was formed as a result of the analysis (Table 11.6).

Based on the matrix, only factors X1 and X2 have a discernible link with the variable under study. The values of the correlation coefficients for these two factors are positive, which indicates the presence of a direct relationship. The results obtained predetermine the following interpretation: the increase in the average per capita income of the population is mediated by the accessibility of the Internet among households of the constituent entity of the Russian Federation. In addition, there is a relationship between the price of Internet services and the level of population income. In this case, such a connection is likely due to the amount of effective demand, which, in turn, forms a higher price for the service.

We shall now proceed to consider the regression model (Table 11.7).

The data obtained reflect the following: 7 explanatory factors of digitalization have an impact on Per capita monetary income of the population (Y) by 61.5%. As in the previous case, such a value of the determination coefficient (R^2) is logically justified by the influence of a wide range of factors on the Y indicator, most of which are not considered within the framework of the created model. The resulting

Table 11.5 The study of the relationship between the digitalization indicators and the average per capita monetary income of the population

Dependent variable	Factors of economy digitalization affecting the dependent variable	Factor index
Average per capita monetary income of the population in the constituent entities of the Russian Federation, roubles per month [14] (Y)	Percentage of households with Internet access, in the total number of households	X1
	Subscription fee for Internet access, per month	X2
	The number of fixed broadband Internet access subscribers per 100 people of the population	X3
	Payment for providing access to the local telephone network regardless of the subscriber line type of the fixed telephone network	X4
	The volume of investments in fixed assets aimed at the acquisition of information, computer and telecommunication (ICT) equipment	X5
	The share of employees who used personal computers at least once a week, in the total number of the payroll of organizations	X6
	The share of organizations that allocated technical means for mobile Internet access to their employees in the total number of surveyed organizations	X7

Source Compiled by the authors based on [12, 14]

Table 11.6 Correlation matrix No. 2

	Y	X1	X2	X3	X4	X5	X6	X7
Y	1							
X1	0.525	1						
X2	0.582	0.453	1					
X3	0.290	−0.031	−0.163	1				
X4	0.313	0.257	0.428	−0.146	1			
X5	0.349	0.201	0.020	0.343	0.054	1		
X6	−0.024	−0.035	−0.265	0.430	−0.113	0.367	1	
X7	0.285	0.028	−0.132	0.595	0.039	0.383	0.275	1

Source Compiled by the authors

Table 11.7 Regression model No. 2

Regression statistics						
Multiple R	0.784					
R-square	0.615					
Normalized R-square	0.580					
Standard error	9606.208					
Observations	85					
F	17.598					
F crit	2.131					
Student's t-criterion	1.991					
	Coefficient	Standard error	t-stat	P-Value	Lower 95%	Upper 95%
Y-intersection	−38020.73	17112.606	−2.222	0.029	−72096.287	−3945.182
X1	554.221	171.108	3.239	0.002	213.502	894.939
X2	26.020	4.946	5.260	0.000	16.171	35.870
X3	696.757	223.698	3.115	0.003	251.318	1142.196
X4	0.609	0.797	0.764	0.447	−0.978	2.197
X5	0.089	0.042	2.109	0.038	0.005	0.172
X6	−328.348	251.001	−1.308	0.195	−828.154	171.459
X7	272.509	204.635	1.332	0.187	−134.970	679.989

Source Compiled by the authors

value ($R^2 > 0.5$) may also indicate the acceptable quality of the model. The "Null Hypothesis" is rejected because the value of Fischer F-criterion exceeds the critical value of the distribution F (17.598 > 2.131), therefore, there is a linear relationship between Y and at least one of the factors X. The most important in this case are the explanatory variables X1, X2, X3, X5. P-values of the specified variables are less than the error probability (0.05). T-stat values X1, X2, X3, X5 modulo exceed the Student's t-criterion (1.991). The lower and upper bounds of the confidence interval X1, X2, X3, X5 have the same sign, which also indicates the greater importance of variables in the constructed model. The considered statistics are subject to the following interpretation: the value of the average per capita income of the population is interrelated with the level of the economy digitalization of the constituent entity of the Russian Federation. ICT, in particular the Internet, predetermine significant competitive advantages for the economic entities using them by reducing costs and increasing the efficiency of economic processes, allowing them to generate significantly more income. Consequently, the lack of sufficient investments in information and communication infrastructure and technological development predetermine the magnitude of social inequality, as well as lagging in terms of economic growth in general.

Now consider the results of the analysis of the relationship in the last of the designated directions—the impact of economy digitalization on the level of infrastructure development. As in the previous cases, we present the dependent and explanatory variables (Table 11.8).

The basis for the general population formation was the data for each of the 85 constituent entities of the Russian Federation in 2019.

The correlation matrix is as follows (Table 11.9).

Therefore, only factors X4 and X5 have a discernible link with the studied variable. The correlation coefficients for these variables are positive, therefore, the relationship with the Y indicator is direct. Interpretation of the results obtained: the degree of infrastructure development is interrelated with the volume of investments in fixed assets aimed at acquiring information, computer and telecommunication (ICT) equipment, as well as the prevalence of the Internet on the territory of the subject of the Russian Federation. Network infrastructure is increasingly used for ordering goods, performing works and services, consequently, it becomes a necessary attribute of increasing the economic activity efficiency.

Consider regression statistics (Table 11.10).

Table 11.8 Study of the relationship between the digitalization of the economy and the level of infrastructure development

Dependent variables	Factors of economy digitalization	Factor index
Integral Infrastructure Development Index [15] (Y)	Adoption of mobile telephone (cellular) networks per 100 people of the population	X1
	The share of organizations using Internet access with a speed of at least 2 Mbit/s in the total number of organizations, %	X2
	The share of innovation activity costs in the total volume of shipped goods, performed works, provided services	X3
	The volume of investments in fixed assets aimed at the acquisition of information, computer, and telecommunication (ICT) equipment, million roubles	X4
	The share of the population who used the Internet to order goods and (or) services in the total population, %	X5
	The share of educational institutions of higher professional education connected to the Internet in the total number of surveyed institutions of higher professional education with the speed of 2 Mbit/s and faster, %	X6
	The share of organizations using broadband Internet access in the total number of organizations	X7

Source Compiled by the authors based on [12, 15]

Table 11.9 Correlation matrix No. 3

	Y	X1	X2	X3	X4	X5	X6	X7
Y	1							
X1	0.406	1						
X2	0.423	0.292	1					
X3	0.229	0.094	0.188	1				
X4	0.555	0.312	0.266	0.176	1			
X5	0.526	0.196	0.275	0.141	0.261	1		
X6	0.004	0.104	0.225	0.157	0.018	−0.039	1	
X7	0.465	0.180	0.475	0.187	0.151	0.321	0.060	1

Source Compiled by the authors

Table 11.10 Regression model No. 3

Regression statistics						
Multiple R					0.762	
R-square					0.581	
Normalized R-square					0.543	
Standard error					0.343	
Observations					85	
F					15.243	
F crit					2.131	
Student's t-criterion					1.991	
	Coefficient	Standard error	t-stat	P-Value	Lower 95%	Upper 95%
Y-intersection	2.866	0.576	4.974	0.000	1.719	4.014
X1	0.002	0.001	2.114	0.038	0.000	0.003
X2	0.005	0.005	0.940	0.350	−0.005	0.015
X3	0.018	0.024	0.735	0.465	−0.031	0.066
X4	0.000	0.000	4.433	0.000	0.000	0.000
X5	0.013	0.004	3.559	0.001	0.006	0.020
X6	−0.002	0.003	−0.666	0.507	−0.007	0.004
X7	0.020	0.007	2.774	0.007	0.006	0.034

Source Compiled by the authors

The calculations made it possible to establish that 7 selected explanatory factors of digitalization have an impact on the Infrastructure Development Index (Y) by 58.1%. Similarly to the calculations performed earlier, the relatively low value of the indicator R^2 is due to the absence of an influencing factors list, including those not related to digitalization, in the constructed model. The condition $R^2 > 0.5$ is met, therefore, it is possible to determine the quality of the model as acceptable. The "Null

Hypothesis" is rejected because the value of Fischer F-criterion exceeds the critical value of the distribution F (15.243 > 2.131), therefore, there is a linear relationship between Y and at least one of the factors X. The most important in this model are 4 factors—X1, X4, X5 and X7. For these variables, all the above conditions are met: P-values are less than the error probability (0.05); t-stat. values modulo exceed the Student's t-criterion (1.991); the lower and upper bounds of the confidence interval have the same sign. All of the above allows us to draw the following conclusion: the level of the modern infrastructure development of the constituent entity of the Russian Federation directly depends on the digitalization factors influence. In the constituent entity of the Russian Federation, in which much more investment is made in ICT equipment and the development of the network, there is the greatest provision with structures mediating economic activity in general. Internet accessibility provides new opportunities for interaction, contributing to the involvement of an increasing number of people in the processes of creating, distributing, and consuming the manufactured product, as well as increasing the efficiency of these processes. Therefore, a developed digital infrastructure is currently becoming one of the necessary conditions for achieving inclusive economic growth.

11.4 Conclusion

The results of the study lead to the following conclusions.

The current stage of the world economy functioning implies a significant content rethinking of the "sustainable economic growth" category, which is due to an increase in the aggregate needs of society, a significant depletion of the resource base, as well as rapid population growth.

The modern world community defines "inclusivity" and "innovativeness" as the drivers of the further development of the world economy, which imply the involvement of an increasing number of people in the processes of creation, distribution, consumption mediated by information and communication technologies.

Based on the results of the correlation and regression analysis, the relationship between the digitalization degree and the magnitude of the indicators of employment, infrastructure, and well-being of the population was confirmed. Accordingly, it becomes possible to establish that the choice of digitalization by the Russian state as one of the key factors of inclusive growth within the framework of considered directions is well-founded. The introduction of information and communication technologies into most socio-economic relationships will cover a much wider range of social strata, thereby generating significantly more income, while simultaneously ensuring national growth and improving the quality of life.

References

1. The Inclusive Growth and Development Report 2017. World Economic Forum. (2017). http://www3.weforum.org/docs/WEF_Forum_IncGrwth_2017.pdf. Accessed 21 Oct 2021
2. Vandemoortele, M., Bird, K., Toit, A., Liu, M., Sen, K., Soares, F.S.: Building Blocks for Equitable Growth: Lessons from the BRICS. Overseas Development Institute (2013)
3. Public Good or Private Wealth? Oxfam Report, New Zealand (2019). https://www.oxfam.org.nz/news-media/public-good-or-private-wealth/. Accessed 15 Oct 2021
4. Fernandez, R., Isakova, A., Luna, F., Rambousek, B.: Gender equality and inclusive growth. IMF Working Papers **2021**(59) (2021). https://doi.org/10.5089/9781513571164.001
5. Real disposable income of the population, growth rates in % compared to the corresponding period of the previous year. Interstate Statistical Committee of the Commonwealth of Independent States (2021). http://www.cisstat.com/rus/dinamika/rus-t.htm. Accessed 18 Oct 2021
6. OECD: The Productivity-Inclusiveness Nexus. OECD Publishing, Paris (2018). https://doi.org/10.1787/9789264292932-en. Accessed 15 Oct 2021
7. Making Innovation Benefit All: Policies for Inclusive Growth. OECD Publishing, Paris (2017). https://www.oecd.org/innovation/inno/making-innovation-benefit-all.pdf. Accessed 15 Oct 2021
8. Inshakov, O.V.: Strategy and tactics of nanoindustry development in the Russian Federation. Econ. Reg. **3**, 246–250 (2011). https://doi.org/10.17059/2011-3-29
9. Sharafutdinov, R.I., Izmailova, D.O., Akhmetshin, E.M.: Studies in national key performance indicators of inclusive growth and development of the regions of the Russian Federation. Theor. Appl. Econ. **3**, 118–134. https://doi.org/10.25136/2409-8647.2018.3.27061
10. Gross domestic product, growth rate in % compared to the corresponding period of the previous year (in constant prices): Interstate Statistical Committee of the Commonwealth of Independent States (2021). http://www.cisstat.com/rus/dinamika/rus-t.htm. Accessed 12 Nov 2021
11. The National Action Plan to ensure the restoration of employment and incomes of the population, the economy and long-term structural changes in the economy Data portal of legal information "Consultant Plus" (2021). http://www.consultant.ru/document/cons_doc_LAW_333667/. Accessed 21 Oct 2021
12. Monitoring the development of the information society of the Russian Federation. The Federal State Statistics Service (2021). https://rosstat.gov.ru/folder/14478. Accessed: 10 Nov 2021
13. The unemployment rate of the population in the constituent entities of the Russian Federation aged 15 years and older (according to sample surveys of the labor force) (2021). The Federal State Statistics Service. https://www.gks.ru/free_doc/new_site/population/trud/tab_trud6.htm. Accessed 16 Nov 2021
14. Average per capita monetary income of the population in the constituent entities of the Russian Federation (new methodology): The Federal State Statistics Service (2021). https://rosstat.gov.ru/storage/mediabank/YxWdUS2t/urov_11kv-nm.xlsx. Accessed 12 Nov 2021
15. Infrastructure Development Index of Russia 2019: InfaOne Research (2019). https://infraone.ru/sites/default/files/analitika/2019/index_razvitiia_infrastruktury_rossii_2019_infraone_research.pdf?index_id. Accessed 16 Oct 2021

Part III
Key Enabling Technologies: Contribution to Inclusive Development of Modern Society

Chapter 12
Nanotechnology: Contribution to Inclusive Growth in Russia

Irina V. Zaporotskova, Natalya P. Boroznina, and Sergey V. Boroznin

Abstract Creating progressive technological solutions for the development of industry is one of the key areas of social and economic policy of the state. It is possible to provide a comprehensive solution to this problem using the tools of the nano-industry—an interdisciplinary industry consisting of high-tech equipment, a new level of metrology and information technologies, organizational forms of relations in industry, high personnel potential, as well as nanotechnology, in the total aggregate, allowing to obtain innovative products. Nanolevel technologies developed by Volgograd State University employees are described. In particular, we will talk about the methodology of nanomarking using electron microscopes, nanotechnologies in medicine (ultra-thin drug coating of medical stents), the technology of cleaning liquids by using carbon nanotubes as a filtering material, strengthening asphalt concrete with specific nanomaterials (Zaporotskova and Siplivy in Patent of the Russian Federation No. 2515007:S1, 2014), the development of a unique nanopowder based on carbon nanotubes to improve the performance of fuel and lubricants (Zaporotskova, I.V., Arkharova, I.V. (2016). Modeling of promising technologies based on the introduction of carbon nanotubes. Certificate of state registration of database No. 2016620622, 17 May 201). Their practical application contributes to solving urgent problems in the development of healthcare, energy sector, road infrastructure, etc., as a significant factor in ensuring inclusive growth in Russia.

I. V. Zaporotskova (✉) · N. P. Boroznina · S. V. Boroznin
Volgograd State University, Volgograd, Russia

N. P. Boroznina
e-mail: boroznina.natalya@volsu.ru

S. V. Boroznin
e-mail: boroznin@volsu.ru

E. I. Inshakova and A. O. Inshakova (eds.), *New Technology for Inclusive and Sustainable Growth*, Smart Innovation, Systems and Technologies 287,
https://doi.org/10.1007/978-981-16-9804-0_12

12.1 Introduction

The development of high-tech production in all industries is one of the key socio-economic priorities of the state. In the modern world, the most effective mechanism for such development is the nano-industry—an interdisciplinary industry consisting of high-tech equipment, a new level of metrology and information technologies, organizational forms of industrial relations, high personnel potential, as well as nanotechnology, in a total set, allowing to obtain innovative products. Such development and production of a high-tech product becomes possible during the transition from macro parameters to nanosized dimensions [1–4].

This chapter describes nanolevel technologies developed by employees of the Institute of Priority Technologies of Volgograd State University. In particular, we will talk about the methodology of nanomarking using electron microscopes, nanotechnologies in medicine (ultra-thin drug coating of medical stents), the technology of cleaning liquids by using carbon nanotubes as a filtering material, strengthening asphalt concrete with specific nanomaterials [5], the development of a unique nanopowder based on carbon nanotubes to improve the performance of fuel and lubricants [6]. Their practical application contributes to solving urgent problems in the development of healthcare, energy, road infrastructure, etc., which is an important factor in ensuring inclusive growth in Russia.

The chapter also considers the peculiarities of the educational process for the training of highly qualified personnel for the nano-industry, implemented at the Volgograd State University in the direction of "Nanotechnologies and Nanomaterials" as part of ensuring the priorities of the strategic development of the Russian Federation.

Inclusive growth in the Russian Federation should be accompanied by a transition to an equally high standard of living for all segments of the population and social groups, which is impossible without a qualitative change in existing technologies. It is in this area that the products of the nano-industry are a key object designed to increase the effectiveness of all the main sectors of society—from law enforcement and information to health care. Today, the evolution of the country in the framework of its inclusive development is impossible without close connection with high technologies not only in the IT sphere, but also in the field of nanotechnology. Understanding the objective need for such transformations, at Volgograd State University, the development of nanotechnologies that are interesting for the real sector of the economy and allow you to switch to a qualitatively new standard of living was initiated at the initiative of professors I.V. Zaporotskova and O.V. Inshakov many years earlier than at the federal level. Their massive introduction will ensure successful inclusive growth of the region and the country as a whole [7].

In the context of the new Russian statehood, the history of nanotechnology begins in 2007, with the appeal of V.V. Putin—the President of the Russian Federation—to the Federal Assembly and the allocation of nanotechnology as a priority area for the development of science and technology. This was reflected in industry by the establishment of the Russian Nanotechnology Corporation, renamed RUSNANO JSC in

2011. Currently, this industry is booming, federal target programs have been organized to stimulate the development of nanotechnologies in the Russian Federation. It was in 2007 that the specialty "Nanomaterials" was opened at Volgograd State University, designed to ensure the successful training of highly qualified specialists for this industry. In 2012, certified specialists began introducing high technologies in the Volgograd region at key enterprises, such as JSC Kaustik, Krasny Oktyabr, Omsktekhcarbon, Volgograd Neftemash, as well as at a number of research institutes engaged in technology development and research in priority areas, including those related to nanotechnology. Four years later, together with the renaming of the state corporation, there was a change in the name of the direction of training highly qualified specialists from Nanotechnology to Nanoengineering to update the training profile of future graduates. Success in this area of training is made possible by a comprehensive approach to providing high-quality education. This is the involvement of students in scientific activities due to the participation in the projects of the scientific school of the leading specialist in the region in the field of nanotechnology and nanomaterial science—Professor I.V. Zaporotskova; and a unique instrument base, analogues of which are not available in other universities of the regional center; and providing the educational process with methodological materials, textbooks, monographs and training manuals developed at the department and adapted to the specific disciplines and tasks facing future specialists for their involvement in the nanosystem industry. Under the leadership of the rector of the Volgograd State University, O.V. Inshakov and the head of the direction of Zaporotskova I.V., laboratories were created equipped with advanced nanotechnological equipment, which still remain the most staffed in the region. Most of the educational burden is aimed at providing students with practical skills in working with equipment for their more successful integration into production in order to create a large-scale basis for innovative products in all spheres of life for the inclusive growth of the country's economy.

For a better approach to innovative technologies, the university established a specialized Scientific and Educational Center (SEC) "Nanotechnologies and Nanomaterials" in 2007. The main tasks of the SEC were determined by the fact that in the region it was necessary to ensure the introduction of nanotechnologies into industrial production. Therefore, the main priority of the SEC's activities was targeted interaction with the real sector of the region's economy to solve the problems facing enterprises and improve the skills of their specialists in the field of nanotechnology. Also, SEC employees should conduct both practical and theoretical studies of unique new materials to create breakthrough solutions for the needs of industry and basic science in the field of nanotechnology and nanomaterial science. Another area of activity is positioned participation in various competitions and grants aimed at the social and economic development of the region and the state. Among the indicators of achievement of tasks we can mention the participation of SEC employees in the Presidential program for the training of engineering personnel, thanks to which a number of industrial enterprises in the region were able to improve the qualifications of their employees in the nano-industry. Also, the relevance of research conducted within the framework of the SEC for the region is confirmed by the active support of the team from grants from the Administration of the Volgograd Region (AVO) and joint grants from AVO with the Russian Foundation for Fundamental Research. As a result of active grant activity, SEC employees raised funds in the amount of more than 20 million rubles from regional and federal sources to conduct successful research on nanomaterials and their properties. The projects produced new data on the properties of nanomaterials, thanks to basic research, which formed the basis of a wide range of innovative technologies related to the use of new substances for the needs of industry.

12.2 Nanomarkings

The challenges of modern society are forcing the use of increasingly advanced methods of protecting information or protected objects. Moreover, it sometimes happens that the surfaces of these objects perceive different types of effects differently depending on their chemical nature and surface topology. To overcome the existing difficulties, the technology of applying protective nanoscale images (nanomarkings) using scanning probe microscopy tools was proposed. This technology is reflected in Russian Patent No. 2365989 "Method of application of nanomarkings on products" [8].

Obtaining a unique degree of protection that is 3 times higher than all known methods of applying protective markings was made possible by a complex approach in which electron microscopy technologies such as scanning tunnel microscopy, various nanolithography methods, scanning probe, and atomic force microscopy are combined. Due to its novelty and high degree of protection, the resulting development has high competitiveness.

In addition to the technology described above, there are currently a number of other ways to apply markings on the surface: mechanical, laser, electrophysical, chemical and photochemical, ultrasonic, thermal, and many others [9, 10]. Their main difference from nanomarking is that the fixation of applied images occurs using traditional optical devices. That is, the minimum individual portions of such images can reach dimensions of at least 0.1 mm.

The latency of the marking increases its protective properties. In this case, an additional protection against counterfeiting is its location on the protected object and a method of visualizing its content.

If we go to the critical dimensions of nanomarkings, then the effective effect on the surface with the help of electron microscopes, for example, scanning tunnel or atomic power, allows to apply nanolithography up to 0.1 nm in size. Practical installations for nanomarkings are shown in Fig. 12.1.

They were created on the basis of microscopes "NanoEducator" and "Solver-Pro" (manufactured by NT-MDT, Zelenograd). To demonstrate the possibilities of the proposed technology, scientists from the Volgograd State University applied nanomarking in the form of the coat of arms of the university on the surface of a polymer material. The result of the experiment is shown in Fig. 12.2.

Fig. 12.1 Technology complexes for creating nanomarkings based on microscopes "NanoEducator" and "SolverPro"

Known optical devices are powerless in attempts to detect or counterfeit applied nanomarkings. Without the use of scanning microscopes, it is impossible to fully obtain all information about the applied image and the features of nanomarking. A separate difficulty is the attempt to detect nanomarking without special knowledge, since 1 mm^2 contains more than a hundred sites on which it can be localized.

The field of application of innovative technology is extremely wide and diverse. Of course, the most obvious consumers are the "power" departments of the Russian Federation, such as the Ministry of Internal Affairs, the Ministry of Justice, and the Ministry of Emergencies. In this case, nanomarkings will be necessary to designate weapons, protected or classified materials. Nanomarkings can also be used in the customs committee, where there is a need to take into account special marks on antiques, various works of art, and other valuable objects exported outside the Russian Federation. Another potential consumer is industrial enterprises that can not only protect their products from counterfeiting in such a unique way, but also use nanomarking as information cards (read QR code) or a system for designating nodes in particularly accurate processes where the application of macro-symbols will lead to defects in the operation of devices (metrological equipment, high-precision devices).

12.3 Cleaning Liquids Using Carbon Nanomaterials

One of the most important points for ensuring the quality of life of the population in modern conditions is the use of high-quality products, including liquid. However, modern methods of cleaning food and technical liquids still require refinement to obtain a truly pure product. And here, nanotechnologies can also be used.

As part of the work of the Scientific Educational Center "Nanotechnologies and Nanomaterials," a technology for finishing cleaning food and technical liquids (including alcohol-containing ones) was also proposed and brought to the consumer due to the use of unique carbon nanomaterials with high sorption capacity, for example, fullerenes and nanotubes [10].

Materials known in the art for filters [11, 12] lose the proposed one of the most important parameters. This is a specific surface area that in carbon nanotubes is several times the same parameter for known sorbents. At the same time, carbon nanotubes are able to realize selective adsorption due to the fact that not only their external, but also internal surface acts as an active one. Also in favor of the proposed material is the fact that at present one of the most effective filters is coal, which contains a carbon sorbent. But even taking into account the identical chemical nature, these filters have sorption activity, tens of times less than carbon nanotubes. Therefore, the developed technology consists in the introduction of a carbon nanotube filter into the technological cycle at the finishing stage, which can retain even microamounts of harmful impurities such as methanol and luminal oils (strong toxicants). The recovery of these impurities is currently a time-consuming and financially expensive task for industrial production. And the introduction of a filter model with carbon nanotubes will significantly reduce the cost of the described part.

Thus, the main advantages of the proposed cleaning technology can be distinguished. First of all, this technology at a relatively low cost will significantly improve the quality of the product, and therefore the production efficiency. The nanomaterial used is a modification of carbon, that is, no additional studies are required related to the study of the side effects of the new filter material. To carry out selective adsorption, the manufacturer will no longer need additional technological solutions in the process. The nanomaterial used in the production of the filter can be reused, which gives the proposed model an additional economic benefit.

The promising consumers of the developed innovative product are primarily enterprises related to the chemical and food industries, as well as the liquor industry. In addition to them, such filters will be interesting to developers of the pharmaceutical, electronic, optical industries. The absence of significant costs is also due to the fact that the introduction of the proposed technologies does not require a revision of production lines and changes in the stages of industrial production. This technology was implemented in a filter for the purification of alcohol-containing products, which was successfully tested at the PermAlko enterprise (Perm) (Fig. 12.3).

Thus, the introduction of the technology for cleaning liquids in industry developed at the Volgograd State University will lead to an improvement in the state of health of citizens through the use of high-quality water, the development of chemical and food industries, which will be one of the integral elements of the inclusive growth of the state.

Fig. 12.2 Example of nanomarking applied to the surface of a polyvinyl chloride sample

12.4 Nanotechnology in Medicine

Biomedical nanotechnologies represent the most promising direction for the development of the nano-industry, however, this direction is also fraught with a number of difficulties. New materials can find application in virtually all medical spheres [13]. First of all, with the help of nanotechnologies, it is possible to design completely new drugs, make breakthrough discoveries in prosthetics using robotic nanobioconstructures, as well as contribute to the creation of new biocompatible materials, etc., these developments are now in their initial state, as they cannot be put into practice without detailed and comprehensive experiments and theoretical modeling, despite the breadth and relevance of the tasks set by the needs of society. As mentioned above, Presidential Decree No. 642 of December 1, 2016 "On the Strategy for the Scientific and Technological Development of the Russian Federation" formulated one of the areas, namely, the transition to personalized medicine, high-tech healthcare, and health-saving technologies, including through the rational use of medicines. Scientists from Volgograd State University also contributed to the solution of this problem and proposed a number of developments that contribute to the transition to high-tech healthcare.

12.4.1 Stomatology

Currently, there are a number of disadvantages inherent in impression materials, including those from which dentistry fillings are made. Therefore, a scientific group engaged in solving current problems of nanotechnology and nanomaterial science at the Volgograd State University proposed a technology for creating polymer composite materials using carbon nanotubes. The obtained polymer nanomaterials had one of the main advantages of sealing blanks—ductility, however, thanks to the introduction of carbon nanotubes, they acquired additional strength.

12.4.2 Medicinal Nanocoating for Stents

The provision of medical care becomes more rapid and effective when introducing new technologies [14, 15]. One of the main negative factors in the transition to high technology is the high cost of consumables used in operations—from catheters to stents, without which medical intervention and assistance to the patient is sometimes simply impossible. The high cost of this type of material is explained by the fact that most of them are imported and, despite the support of the state and numerous state programs, still does not become possible for implementation everywhere. At the same time, the cost of used products and materials in domestic production becomes significantly lower.

One of the worst problems of modern society is the high mortality rate from constriction and blockage of vessels entering the circulatory system. The solution to this problem is to carry out special medical actions on stenting (stent implantation), which is a low-traumatic method and contributes to the avoidance of surgery.

However, these technologies are also not devoid of disadvantages: when a stent is used for a long time, they are blocked by various blood clots or crystalline bodies, resulting in the need for repeated intervention or restatement. Therefore, it is necessary to increase the service life of this type of medical stents. An effective solution will be to apply nanocovering to the surface of the stent with a drug, which will provide a prolonged effect of this product.

As part of the scientific work carried out at Volgograd State University, it was proposed to use a special nanofilament consisting of a polymer material and a drug, which will be a homogeneous nanofilm covering the surface of coronary (metal) and bile (polymer) stents. This solution can be applied to various biomedical needs and will significantly reduce the impact of negative factors that reduce the efficiency of using stents. Another modification of the proposed technology is the introduction of carbon nanoparticles into the obtained nanoparticle to more accurately distribute the concentrations of substances included in the nanofilm and will contribute to a given drug intake into the patient's body along with an increase in the operation time of the stent.

12.5 Ultra-Strong Asphalt Concrete

Traditionally, an urgent task for all regions of Russia, including the Volgograd region, is the creation of modern expanded road surfaces, which is impossible without conducting comprehensive scientific research for the needs of the road industry in order to transform it in accordance with the challenges of time. Some innovative solutions should be aimed at optimizing transport flows and developing high-safety systems on the roads, but the development of innovative materials used in the construction of expensive ones, as well as the use of breakthrough technologies in their repair and modernization, will make a great contribution to improving the road situation. That is, a high-quality road surface is a basis, followed by a clear improvement in the road situation as a whole. But it is impossible to create it without new materials and technologies. The most likely and cost-effective solution will be the use of nanomaterials that allow you to obtain macro objects with projected physico-chemical properties. Therefore, a method for strengthening the asphalt road surface using carbon nanomaterials has been developed at VolSU.

One of the key requirements for asphalt concrete is that its properties must meet the conditions of the surrounding road environment and change with them. A measure of these properties is the compressive strength measured at given temperatures and the rate of application of the load. Also significant for the operation of asphalt concrete is their water and wear resistance.

It is possible to improve the listed properties of the asphalt road surface by adding carbon nanotubes to the component of the mixture, whose strengthening properties are well known. It is proposed to introduce carbon material into molten bitumen, which will make it possible to obtain a stronger and more wear-resistant road surface.

The suggested assumption was tested during a series of experiments by scientists from Volgograd State University. Asphalt concrete samples were tested for compression, and it was found that even a small introduction of nanotubes (up to 0.0001% by weight) leads to strengthening of the starting material.

Therefore, reinforcement of asphalt concrete with carbon nanotubes can be proposed as one of the effective methods for improving the performance of asphalt concrete.

12.6 Improving the Performance of Petrol, Oil and Lubricants

Improving the properties of fuels and lubricants is an important task when introducing improved engines. It is necessary to improve the resistance of oils to oxidation due to the fact that this will solve one of the most important operational tasks facing engine manufacturers, namely, reducing the frequency of oil replacement in them.

Already at the present stage, only doped oils are used in industry, that is, fuels and lubricants containing various mineral and synthetic additives to the initial motor oil in their structure. This is done to give the oil the desired properties. The content of additives and their amount is now responsible for the effectiveness of certain types of oils, ensuring their compliance with operational properties and increased service life. The advent of new combinations of these components also provides an expanding range of fuels and lubricants. The unique sorption properties of carbon nanotubes allowed scientists from the Volgograd State University to put them forward as one of the main components, which will increase the operational properties and physicochemical characteristics of existing base oils.

The use of lubrication is necessary to reduce the breakability of parts due to wear and increase their service life. Oil performs a number of important functions, such as increasing corrosion resistance, reducing the temperature of parts in constant friction, denser piston travel in cylinders, as well as eliminating coke and metal microparticles from working surfaces. The main requirement for metal and rubber bonds of engines is to increase their service life, which is achieved by high tribological characteristics of lubricants, which can be increased when carbon nanoparticles are introduced into them. That is, the use of carbon nanotubes as additives will just solve the main technological problem facing manufacturers of fuel and lubricants, and as a result, engine manufacturers.

High sorption activity of carbon nanotubes will lead to higher resistance to oxidation of oils, which will affect the growth of their stability and preservation of viscosity, which affect performance characteristics of fuel and lubricant material as a whole [16]. At Volgograd State University, experiments were carried out to add carbon nanotubes to Lukoil Super 10w-40 motor oil, which showed that the resulting composite material has changed parameters of acid number and alkalinity, as well as viscosity and acid number.

Both after aging and in the initial state, the addition of carbon nanotubes leads to a decrease in acidity and acid number, which means a slowdown in the oxidizing process in oil. The change in the alkaline parameter means that the resulting composite mixture neutralizes the sulfur products during the operation.

Thus, it can be concluded that the addition of an additive in the form of carbon nanotubes to motor oil avoids corrosion of the main components both during the operation of the engine or motor and during its preservation, that is, it is possible to unequivocally talk about improving the functional characteristics of fuel and lubricants by introducing such an additive.

The result of the research was the innovative product "NanoOIL" (Fig. 12.4), the implementation of which is currently being negotiated with OAO LUKOIL. Also, this development interested a number of foreign partners of the university.

Fig. 12.3 Carbon nanomaterial filter for cleaning liquids from harmful impurities

Fig. 12.4 Innovative product nano-oil

12.7 Conclusion

The inclusive development of the Russian Federation is impossible without comprehensive measures for socio-economic growth, which can ensure the active use of nano-industrial products. Nanotechnological products will make significant changes in all areas of society and industry, from medicine and environmental protection to chemistry, biotechnology, and materials science. Scientific potential, technological capabilities, and grant support create a positive background for the Russian Federation to provide leadership positions in the field of breakthrough technologies, create unique products for foreign and domestic markets, and therefore overcome socio-economic challenges that impede its inclusive development.

But the existing sphere at the present stage has a number of risk factors. The most significant of these are the substantial financial resources necessary for the mass introduction of new materials and technologies into the industry; a technology race involving advanced countries; the need for production in highly skilled personnel and the creation of conditions for their retention in domestic production. Also, the mass production of nanotechnological products is hindered by the easy availability of domestic raw materials for the creation of final products; problems with the import substitution of high-tech products and, as a result, the lack of modern scientific equipment necessary for the successful integration of nanotechnologies into industrial technological processes. The emergence of the Russian Federation as a technological leader and its successful inclusive development are impossible without accelerating the pace of transition to high-tech production using nano-industry and its products.

Therefore, in the Volgograd region, to solve these urgent problems for the entire state, scientists at Volgograd State University are making a large-scale and comprehensive introduction of nanotechnologies, ranging from numerous proposals for high-tech products and technologies to training personnel with unique professional competencies. This was possible due to the systematic development of this scientific direction for three decades, the emergence of a unique scientific and pedagogical composition and a scientific school, a scientific and technological base that has no analogues in other universities in the region, as well as accompanying their methodological support activities.

Acknowledgements The chapter was prepared with the financial support of the scholarship of the President of the Russian Federation No. 798.2019.1 and a grant from the President of the Russian Federation No. MK-1758.2020.8.

References

1. Inshakov, O.V.: Strategy and tactics of nanoindustry development in the Russian Federation. Econ. Region **3**, 246–250 (2011). https://doi.org/10.17059/2011-3-29
2. Karabasov, Y.S. (ed.). New Materials. MISIS, Moscow
3. Alferov, J.I., Kopyev, P.S., Suris, R.A., et al.: Nanomaterials and nanotechnologies. Nano- and microsyst. Technol. **8**, 3–13 (2003)
4. Alfimov, S.M., Bykov, B.A., Grebennikov, E.P., et al.: Development in Russia of works in the field of nanotechnology. Nano- and Microsyst. Technol. **8**, 2–8 (2004)
5. Zaporotskova, I.V., Siplivy, B.N.: Method of strengthening asphalt pavement with carbon nanomaterials. Patent of the Russian Federation No. 2515007 S1 (2014)

6. Zaporotskova, I.V., Arkharova, I.V.: Modeling of promising technologies based on the introduction of carbon nanotubes. Certificate of state registration of database No. 2016620622, 17 May 2016 (2016)
7. Zaporotskova, I.V.: Nanotechnologies and nanomaterials: scientific, economic and political realities of the new century. Bulletin of Volgograd State University. Series 3: Economics. Ecology **1**, 18–29 (2015)
8. Zaporotskova, I.V., Kislova, T.V.: Method of application of nanomarkings on articles. Patent of the Russian Federation No. 2365989, 17 August 2009
9. Perriere, J., Millon, E., Fogarassy, E. (eds.): Recent Advances in Laser Processing of Materials. Elsevier Science, Oxford (2006)
10. Usanov, D. A., Suchkov, S. G., Zaporotskova, I. V., Skripal, A. V., Kuzmina, R. I., Panina, T.G.: Method of purifying water-ethanol mixtures from isopropyl alcohol. Patent No. 2359918, 17 June 2009
11. Watanabe, T.A.: Nano-plating II Metallurgical Approach to Electrochemical Theory and its Applications to Technology. Elsevier Science, Oxford (2020)
12. Lei, Y., Tian, Z., Sun, H., Liu, F., Zhu, Z., Liang, W., Li, A.: Low-resistance thiophene-based conjugated microporous polymer nanotube filters for efficient particulate matter capture and oil/water separation. ACS Appl. Mater. Interfaces **13**(4), 5823–5833 (2021). https://doi.org/10.1021/acsami.0c20484
13. Higson, S.: Biosensors for Medical Applications, 1st edn. Woodhead Publishing, Cambridge (2012)
14. Zeng, A., Yao, X., Gui, Y., Li, Y., Jones, K., Yu, L.: Inhibiting surface crystallization and improving dissolution of amorphous loratadine by dextran sulfate nanocoating. J. Pharm. Sci. **108**(7), 2391–2396 (2019). https://doi.org/10.1016/j.xphs.2019.02.018
15. Li, Y., Yu, J., Hu, S., Chen, Z., Sacchetti, M., Sun, C., Yu, L.: Polymer nanocoating of amorphous drugs for improving stability, dissolution, powder flow, and tabletability: the case of chitosan-coated indomethacin. Mol. Pharm. **16**(3), 1305–1311 (2019). https://doi.org/10.1021/acs.molpharmaceut.8b01237
16. Liu, Y., Zhao, C., Meng, M., Yanga, X., Biana, W., Wu, H., Zhang, C., Wang, J., Tsai, C.-J., Yang, R.T.: Adsorption and desorption of gaseous naphthalene on carbonaceous sorbents: insights into advantageous pore sizes and morphologies. J. Clean. Prod. **314**(10), 127905 (2021). https://doi.org/10.1016/j.jclepro.2021.127905

Chapter 13
Neural Network Prediction of Economic Structural Changes in the Context of Industry 4.0

Elena A. Petrova, Alla E. Kalinina, and Petr V. Bondarenko

Abstract Current studies consider structural changes not as a consequence, but as one of the sources of economic growth. However, this connection is not absolute and the changes intensity depends on many factors, one of which is technological development, which is associated primarily with the formation of Industry 4.0. This chapter presents the results of the structural changes analysis of the European Union countries economies, the United States and Russia between 2007 and 2017, calculated by Gatev's integral structural change coefficient, in accordance with the growth rate indicators of the information technology sector. The highest annual structural changes were observed in 2009 in foreign countries. Since 2010 the structure of value added has remained relatively stable, but there was a slight increase in 2015 in the United States. The Russian Federation shows two peak values in 2009 and 2011 and a smaller high value in 2015. A neural network is used as a tool for predicting structural changes. The prediction results show a direct dependence between structural changes in the economy and the IT industry indicators, and the dependencies configuration allows us to conclude that the processes in this sector of economy are stabilized and unlikely to lead to structural changes in the near future.

E. A. Petrova (✉) · A. E. Kalinina · P. V. Bondarenko
Volgograd State University, Volgograd, Russia
e-mail: ea_petrova@volsu.ru

A. E. Kalinina
e-mail: rector@volsu.ru

P. V. Bondarenko
e-mail: petrbondarenko@volsu.ru

E. I. Inshakova and A. O. Inshakova (eds.), *New Technology for Inclusive and Sustainable Growth*, Smart Innovation, Systems and Technologies 287,
https://doi.org/10.1007/978-981-16-9804-0_13

13.1 Introduction

Modern foreign and Russian studies consider structural changes as one of the most important components of economic growth. Structural changes are understood as the redistribution of economic activity between the three traditional sectors of the economy (agriculture, industry, and the service sector in Clark's three-sector model) [1].

This process usually involves an increase in labor productivity in all sectors of the economy, surplus labor lay-offs and high-tech industry development, which in turn supports industrialization, urbanization and high-productivity service sector development. Sustainable economic growth is based on the reallocation of resources from traditional agriculture and other low-productivity activities to more productive sectors in both urban and rural areas.

Thereby, structural transformation (or structural changes) is a priority in economic policy for countries at all development levels. There are five main features: a steadily declining share of agriculture in economic production and employment; a growing share of urban economic activity in industry and modern services; an increasingly high share of manufactured goods in production and exports; the migration of agricultural workers to urban areas; and the demographic transition, which usually involves a sharp increase in population before a new equilibrium is achieved.

However, modern economic development changes the classical division into three sectors, differentiating the service sector [2, pp. 90–91]. Since the 1980s, information technologies have acquired a new role in the production process, changing its content, technologies, and requirements for the employees qualification. A new stage of technological progress is associated not only with the development of information technologies, but also with the mutual interaction of nano-, bio-, information-, cognitive—(hereinafter—NBIC)—technologies, that also affects the dynamics of structural changes [3]. Therefore, economic growth prediction should take into account these changed economic conditions, which implies supplementing the prediction models with parameters that take into account the dynamics, direction, and intensity of the implementation of NBIC technologies.

Globalization of production is equally important for structural changes due to many factors, including profound changes in geopolitical relations between the countries, the widespread growth of digital transformation in all sectors of the economy, reduced transportation costs, the development of computer-assisted production technologies and the proliferation of bilateral and multilateral trade agreements. These processes made it possible to decentralize supply chains into independent but coherent global networks and allowed multinational companies to place their separate production facilities around the world. Creative product design, search for materials and components, and manufacturing of products can now be done less expensive and more efficiently from virtually any region of the planet, while end products and services are customized to meet the needs of customers in national markets.

Modern analysis of industry transformation was initiated by Fisher (1935, 1939) and Clark [1] and dealt with sector shifts related to the labor resources [4]. However, in the later years, the study of the long-term transformation of the sectoral structure by Kuznets [5] established the factors of structural transformation that measure the dynamics of such indices as indicators of production and consumption.

The structural transformation production indicator analyzes sectoral changes in the employment rate and value added in gross domestic product (GDP) as the economy grows. Thus, it is believed that structural changes are possible when an increase in GDP per capita is associated with a decrease in both the employment rate and the nominal value-added rate in agriculture, as well as an increase in both the employment rate and the nominal value-added rate in industry and service industries.

Another important area of research is the work of Baumol [6], who showed that there is a negative feedback between economic growth and the expansion of the tertiary sector characterized by low productivity, the so-called "Baumol's cost disease". Further studies of this process show both its confirmation [7–10], and negative results of its prediction [11, 12]. Studies of structural changes in the economies of Asian countries and in Japan [13–15] focus on the industry structure and mainly examine the causality of productivity at the branch industry level as well as at the level of an individual enterprises.

The structural transformation of the economy, both in highly developed countries and countries with a lower level of development, is primarily associated with the peculiarities of country economic systems evolution, the integration into the world economy, as well as the role in the world labor division. Over the past decades, the highly developed countries of Western Europe and the United States have been actively improving their innovation systems, including the introduction and development of converging technologies, which also led to a change in both the economical structure and the contributions of individual sectors to the total value added. An important part in these processes was played by the implemented state programs for the digital economy and IT development.

The main state programs for the digital economy development in Europe and the United States are presented in Table 13.1.

Therefore, the objective of the study was to trace how the economic structure is changing in response to large-scale changes, as well as to carry out a comparative

Table 13.1 State programs for the digital economy and IT sector development in the world

Country	State programs for the digital economy and IT sector development
Germany	Industry 4.0., Smart networking
France	Alliance Industrie de future
United Kingdom	UK digital strategy
USA	Digital economy
Japan	New strategy in information and communications technology
China	Strategic plan for reindustrialization using new information and communication technologies

Source Compiled by the authors

analysis based on measurements of similar processes in the Russian Federation and Western Europe and the United States.

Another important aspect of this research was the modeling of the studied processes using neural networks.

13.2 Methodology

The information database of the study consists of the official data of the United Nations Statistics Division, analytical reports of the International Telecommunication Union (ITU) on information and communication technologies between 2007 and 2018 [16], the statistical publication of the Federal State Statistics Service (Rosstat) "Regions of Russia. Socio-economic indicators", "Russia in figures", "Regions of Russia. The main characteristics of the subjects of the Russian Federation" for the period between 2005 and 2018 [17–19].

Gatev's structural change coefficient together with the indicators of the growth rates of the information technology sector indicators is used as the main tool that will allow measuring structural changes in countries economies:

Gatev's integral coefficient:

$$K_t^{\Gamma} = \sqrt{\frac{\sum_{i=1}^{n}(x_{it} - x_{it-1})^2}{\sum_{i=1}^{n} x_{it}^2 + \sum_{i=1}^{n} x_{it-1}^2}} \quad (13.1)$$

where x_{it}—is the share of the *i*-industry in the gross indicator in the period *t*, $x_{it-1} \geq 0$; $x_{it} \geq 0$;

n—number of structure elements.

To analyze the processes under study, the authors calculated Gatev's integral structural change coefficient based on gross value added in the Russian Federation, USA, France, Germany, Italy, United Kingdom of Great Britain, and Northern Ireland between 2007 and 2017 (Fig. 13.1).

Figure 13.1 represents the calculations of Gatev's annual structural change index based on the gross value added in the Russian Federation, France, Germany, Italy, and the USA. The annual change in structural changes shows that the highest change was observed in 2009 in foreign countries, and since 2010, the structure of value added has remained relatively stable, but in the United States there was a slight increase in 2015.

The Russian Federation had two peaks in 2009 and 2011 and a small high value in 2015. However, the change rate is much higher than the rate in Western Europe and the United States. This is explained by the fact that the Russian Federation began to actively introduce information and communication technologies much later. For example, the "Strategy for Development of Information Society in the Russian Federation until 2015" was approved on February 7, 2008, by the President of the

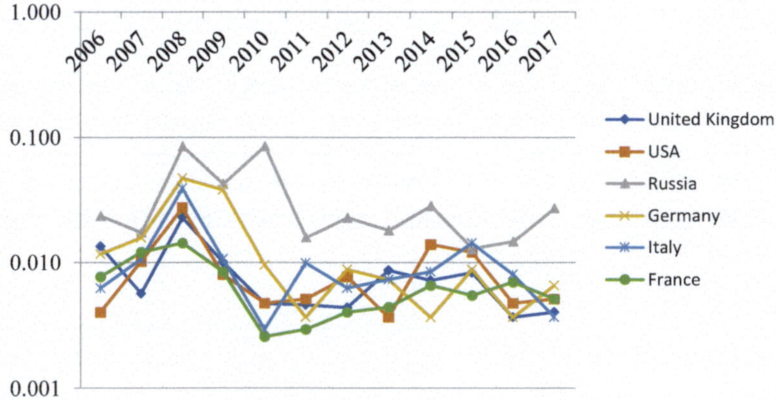

Fig. 13.1 Gatev's structural change coefficient based on gross value added in the Russian Federation, France, USA, Germany, Italy and the United Kingdom of Great Britain and Northern Ireland. *Source* Compiled by the authors based on [16]

Russian Federation, the Federal Target Program "Electronic Russia (2002–2010)" was approved in 2002 (the program was amended and supplemented in 2004, 2009 and the last changes were made in 2010). Recent years have shown the stabilization in changes both in the Russian Federation and abroad.

Diagrams in Figs. 13.2 and 13.3 reflect the comparison of Gatev's index by the sectoral structure of gross value added and the growth rates of the main IT industry indicators in France and Germany in 2007–2018.

The largest changes in the IT sector indicators values fall between 2007 and 2010, then they appear stabilized. However, some indicators (in France it is the percentage of people using the Internet, and in Germany it is the number of cellular subscribers

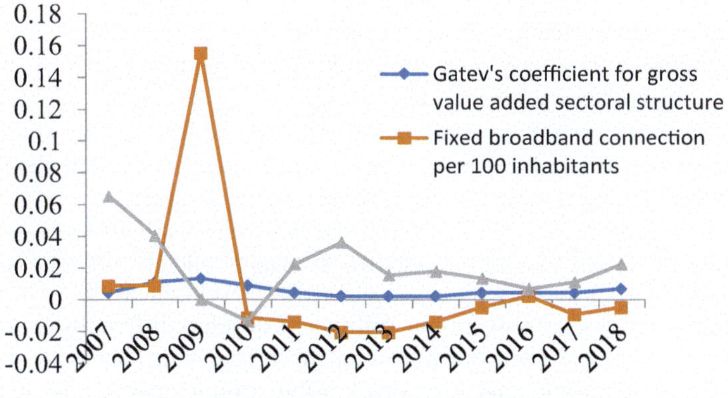

Fig. 13.2 Dynamics of the structural change coefficients and the growth rates of information and communication technologies indicators in France between 2007 and 2018. *Source* Compiled by the authors based on [20]

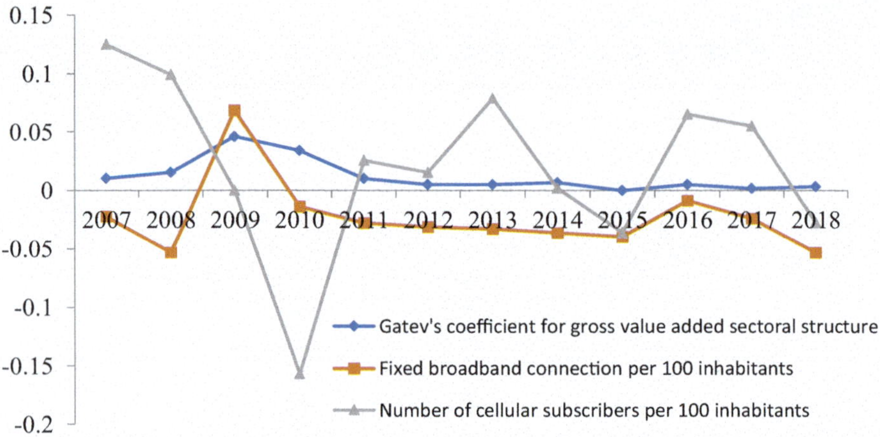

Fig. 13.3 Dynamics of the structural change coefficients and the growth rates of information and communication technologies indicators in Germany between 2007 and 2018. *Source* Compiled by the authors based on [20]

per 100 inhabitants) have negative growth rates in 2010 and 2015. Structural changes in this period are stabilized and there are no sharp spikes in the structure of gross value added.

In the long term, these indices and IT sector indicators show a clear trend towards slight growth in France. There is a wider range of values in the IT industry in Germany. Since 2016 there is an increase in the growth rate of the indicator "Percentage of people using the Internet", but the indicators "Number of cellular subscribers per 100 inhabitants" and "Fixed broadband connection per 100 inhabitants" tend to decrease, while Gatev's structural change Index remains stable.

Similar processes are observed in the Russian Federation, but with a slight time delay; the structural change peaks occurred in 2009 and 2011. The IT industry in the Russian national economy shows an upward trend in the indicators growth, in particular, the largest increase is observed in the cost of ICT (Fig. 13.4), while the structural change index remains stable.

Therefore, the same processes are being observed both in foreign countries and in the Russian Federation. It is not possible to unambiguously assert the existence of a link between IT sectors and structural changes in value added.

The next stage of the study is devoted to the construction of an economic growth prediction model, taking into account structural changes and growth rates of the main IT sector indicators of the Russian economy.

The studied processes, both in the Russian Federation and in the countries of Western Europe and the USA, have a complex nonlinear dependence. Therefore, neural networks are used as a modeling tool, because they solve with good accuracy the problems of classification, prediction, and search for associations in the data array, especially if the data is heterogeneous and there is a wide range of values.

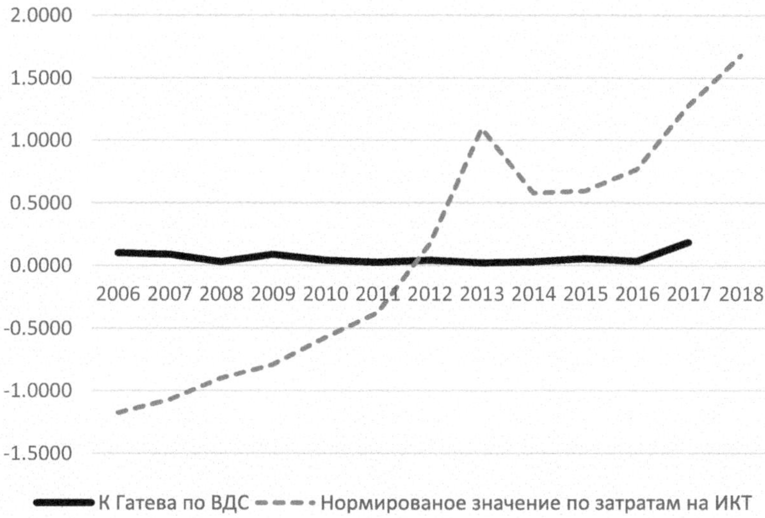

Fig. 13.4 Dynamics of Gatev's structural change coefficient for the sectoral structure of gross value added and for the normalized ICT costs value in the Russian Federation. *Source* Compiled by the authors based on [18, 19]

The Deductor Studio cross-platform solution was chosen as a neural network modeling tool. The Deductor Studio allows to implement a complex scenario for collecting data from a source, to perform preprocessing and data cleaning, as well as to build a neural network with the specified parameters of the activation function and to set the number of hidden layers in the neural networks.

The authors selected 11 indicators (presented in Table 13.2) to design a neural network taking into account the correlation analysis of indicators describing structural changes in the economy of the Russian Federation and the information and communication technologies integration.

The data set for the study was a set of 1,246 observations for all regions of the Russian Federation between 2006 and 2019.

The designed neural network is based on the Deductor Studio cross-platform solution. It is a network with two hidden layers and a hyperbolic tangent activation function.

The authors implemented an error backpropagation algorithm within the framework of this neural network. The gradient descent method was chosen as the network training method. This method allows for the weights to be corrected after all examples of the training set are presented.

All indicators were normalized and brought to a standard form in order to be used by a neural network. Using linear normalization the values of the studied indicators in the data set were presented in the interval (−1, 1). A logarithmic transformation of the value in the interval (0, 1) was used to calculate Gatev's coefficient. Randomly determined 95% of the original data set was used for the training set, the rest 5% was used for the test set.

Table 13.2 Structural changes indicators in the Russian economy and the information and communication technologies integration

Indicator	Designation
Gatev's structural change coefficient	Y
Use of personal computers	X1
Server use	X2
Use of local networks	X3
Use of global networks	X4
Hardware costs	Z1
Software costs	Z2
Communication service costs	Z3
Costs for the ICT use and development staff training	Z4
The cost for the third-party ICT specialists and services (except of communication and training services)	Z5
Other costs	Z6

Source Compiled by the authors

An error of 0.05 between the values at the output and the input and reaching 5000 epochs were chosen as the parameters for the completion.

13.3 Results

As a result of the neural network training 91.99% of examples were recognized in the learning set, and 90.05% in the test set (Fig. 13.5).

Figure 13.5 shows a scatter diagram that is nonlinear due to complex nonlinear dependence in the studied data.

According to the diagram, most of the measurements are within the margin of error, and the chosen neural network configuration fits for this study. The number of unrecognized observations is about 10%. This can be explained by the fact that there is a strong differentiation between the regions of the Russian Federation for the selected indicators and the regions are not homogeneous in the data set.

The authors used the "What-if" tool in Deductor Studio to predict the processes under study. This toolkit allows not only to build diagrams from the parameters under study, but also to conduct research on the effect of changing the value of one indicator on the value of the output parameter.

Of greatest interest are the following dependency diagrams (Figs. 13.6, 13.7, 13.8, 13.9).

The presented graphs illustrate the direct dependence between structural changes in the economy and the studied indicators, such as the use of global networks, the

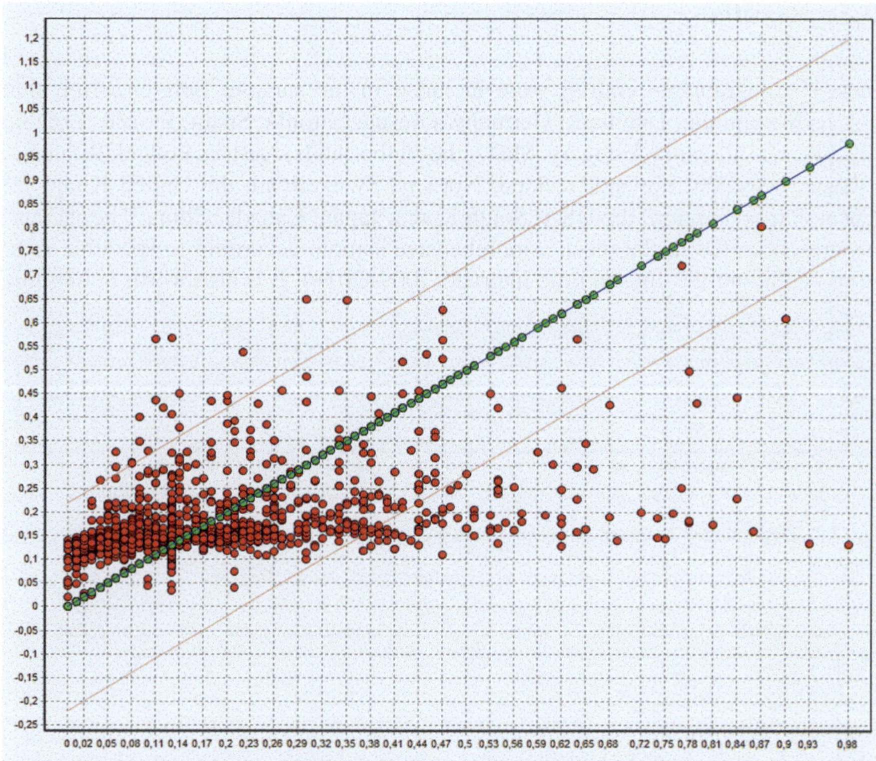

Fig. 13.5 Scatter diagram with a preset error threshold of 0.05

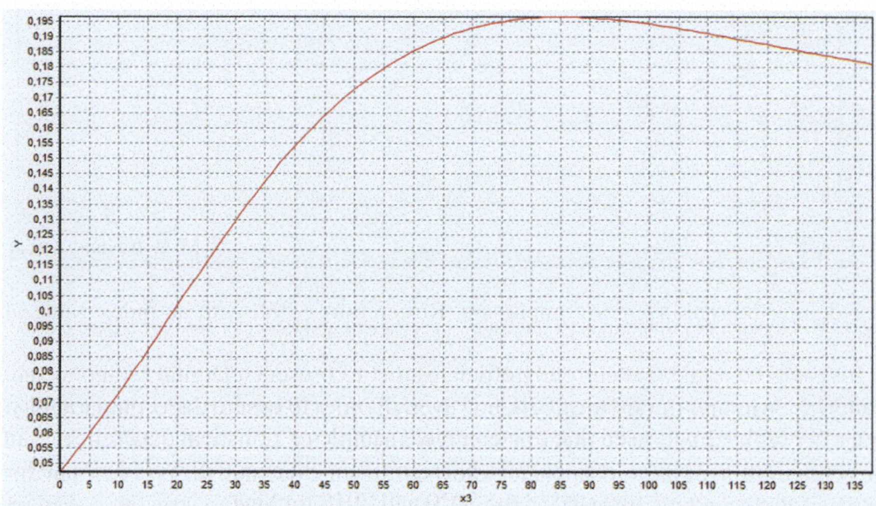

Fig. 13.6 Dependency diagram between Gatev's coefficient and indicator for the use of local networks

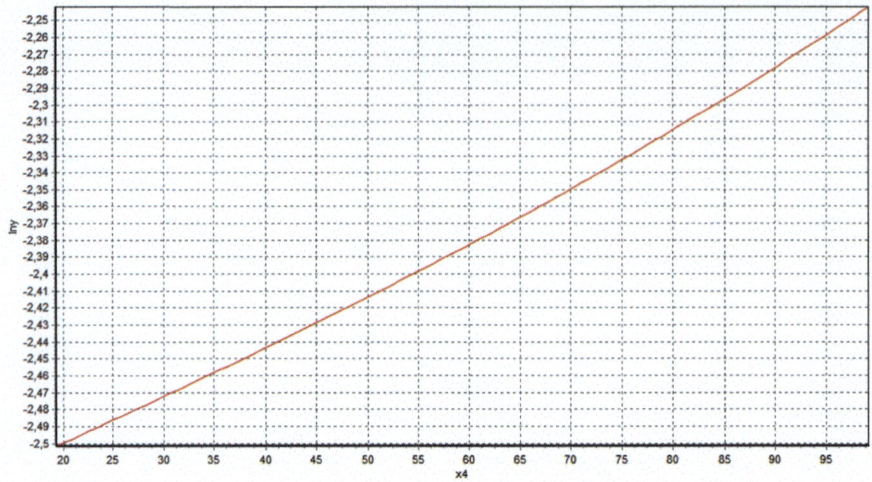

Fig. 13.7 Dependency diagram between Gatev's coefficient and indicator for the use of global networks

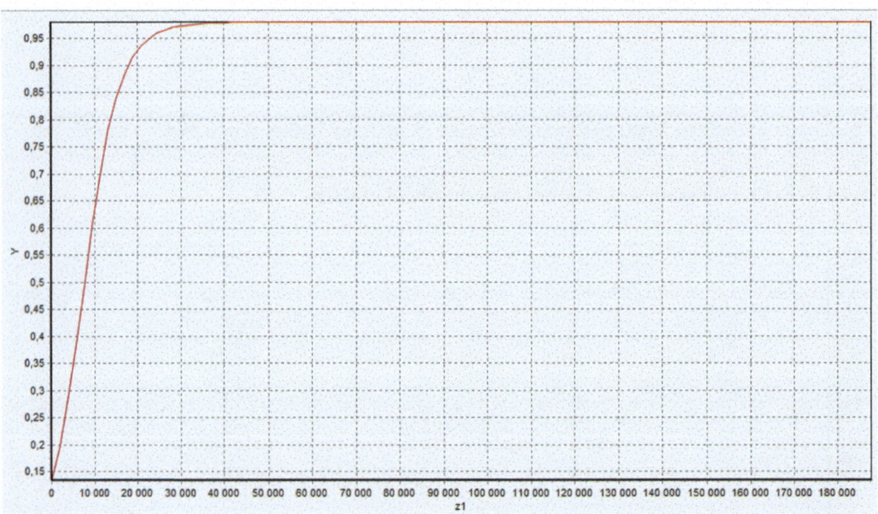

Fig. 13.8 Dependency diagram between Gatev's coefficient and indicator for the hardware cost

hardware cost, and the software costs. The prediction results show a direct dependence between structural changes in the economy and the IT industry indicators, and the configuration of the dependencies allows us to conclude that the processes in this sector of the economy are stabilizing and are unlikely to lead to structural changes in the Russian economy in the near future.

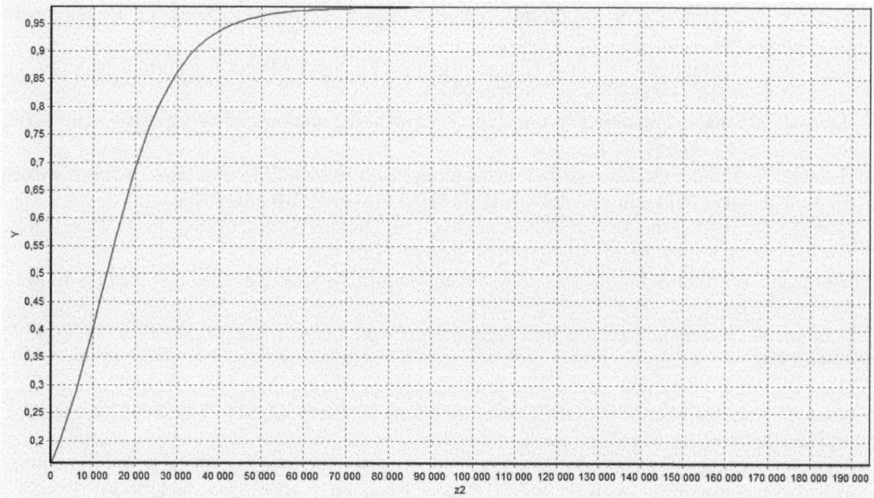

Fig. 13.9 Dependency diagram between Gatev's coefficient and indicator for the software costs

13.4 Conclusion

Therefore, the analysis of the processes taking place in the IT sector and in the economy in general in foreign countries allows us to conclude that the highest structural changes occurred between 2007 and 2009. In the Russian Federation, these changes are delayed by 2–3 years.

The IT sector shows the stabilization of the main indicators, which is unlikely to lead to structural changes in the Russian economy in the near future.

The predictive model, built on the basis of a neural network, shows the nonlinear dependence between structural changes and the development of Industry 4.0. In particular, the nonlinear dependence is observed between the structural change coefficient and the use of local networks.

This type of dependence behavior is explained by the high degree of influence of this indicator on the studied processes and is the most significant in the implementation of programs for the introduction of information and communication technologies to the Russian economy.

References

1. Clurk, C.: The Conditions of Economic Progress. Macmillan, London (1957)
2. Galbraith James, K.: Created Unequal. The Crisis in American Pay, N.Y. 378 (1998)
3. Inshakov, O.V.: Strategy and tactics of nanoindustry development in the Russian Federation. Econ. Region **3**, 246–250 (2011). https://doi.org/10.17059/2011-3-29

4. Syrquin, M.: Patterns of structural change. Chapter 7. In: Handbook of Development Economics, vol. 1, pp. 203–273 (1988)
5. Kuznets, S.: Economic Growth of Nations. Harvard University Press, Cambridge MA (1971). https://doi.org/10.4159/harvard.9780674493490
6. Baumol, W.: macroeconomics of unbalanced growth: the anatomy of urban crisis. Am. Econ. Rev. **57**(3), 415–426 (1967)
7. Hartwig, J.: Testing the Baumol-Nordhaus model with EU KLEMS data. Rev. Income Wealth **57**(3), 471–489 (2011). https://doi.org/10.1111/j.1475-4991.2010.00409.x
8. Hartwig, J.: Testing the growth effects of structural change. Struct. Chang. Econ. Dyn. **23**(1), 11–24 (2012). https://doi.org/10.1016/j.strueco.2011.09.001
9. Nordhaus, W.: Baumol's diseases: a macroeconomics perspective. B.E. J. Macroecon. **8**(1), 1–37 (2008)
10. Peneder, M.: Industrial structure and aggregate growth. Struct. Chang. Econ. Dyn. **14**(4), 427–448 (2003)
11. Dietrich, A.: Does growth cause structural change, or is it the other way around? A dynamic panel data analysis for seven OECD countries. Empir. Econ. **43**, 915–944 (2012). https://doi.org/10.1007/s00181-011-0510-z
12. Maroto-Sanchez, A., Cuadrado-Roura, J.R.: Is growth of services an obstacle to productivity growth? A comparative analysis. Struct. Chang. Econ. Dyn. **20**(4), 254–265 (2009)
13. Fukao, K., Miyagawa, T. (eds.): Productivity and Japanese Economic Growth: Empirical Analysis at Industrial and Firm Level. Using JIP Database. Tokyo University Press, Tokyo (2008)
14. Ito, K., Lechevalier, S.: The evolution of the productivity dispersion of firms: a reevaluation of its determinants in the case of Japan. Rev. World Econ. **145**(3), 405–429 (2009). https://doi.org/10.1007/s10290-009-0027-0
15. Morikawa, M.: Productivity Analysis in Service Industries: An Empirical Analysis Using Microdata. Nihon hyorosha, Tokyo (2014)
16. Department of Economic and Social Affairs. Statistic Division. https://unstats.un.org/unsd/snaama/Basic (2006–2017). Accessed 21 Sept 2020
17. Federal State Statistics Service: Regions of Russia. Main characteristics of the constituent entities of the Russian Federation (2019). https://gks.ru/bgd/regl/b19_14s/Main.htm. Accessed 20 Feb 2020
18. Federal State Statistics Service: Regions of Russia: socio-economic indicators (2019). https://gks.ru/storage/mediabank/Region_Pokaz_2019.pdf. Accessed 20 Feb 2020
19. Federal State Statistics Service: Russia in figures (2019). https://gks.ru/storage/mediabank/rus19.pdf. Accessed 20 Feb 2020
20. International Telecommunication Union (ITU) Statistics. https://www.itu.int/en/ITU-D/Statistics/Pages/stat/default.aspx (2007–2018). Accessed 20 Feb 2020

Chapter 14
Artificial Intelligence Technologies for Business Continuity Protection in Industry 4.0

Michael I. Zabezhailo and Yu. Yu. Trunin

Abstract Computer data analysis, which is inherent to the inclusive digital economy, is considered one of the crucial components for Industry 4.0. The pivotal role of computer techniques is determined by their universal nature, regardless of the geography of their implementation. Big Data analysis is critical for decision-making support systems to ensure the protection and stability of business continuity in a rapidly evolving economic environment. This chapter discusses technological capabilities to tackle the vital need in Industry 4.0—the personalization of products, services, and decisions. Personalization receives attention due to the increasingly intensive development of various global network communities and collaborations. From a technological perspective, it is crucial to effectively identify homogeneous groups of network users and balance the needs anticipated by an individual and group. Modern high-precision evidence-based medicine requires computer technology for decision-making support and utilizes with efficacy the capabilities of computer-based techniques to differentiate cases of pseudo-progression in human brain tumors. The approach proposed can be expanded to other relevant for Industry 4.0 subject areas.

14.1 Introduction

Emerged with the advent of Industry 4.0 [1], the term digital economy has become a full-fledged concept embraced by scholars, researchers, etc. To date, the term denotes an economy based on the implementation of computer technologies and

M. I. Zabezhailo
Federal Research Center "Computer Science and Control", Russian Academy of Sciences (RAS), Moscow Institute of Physics and Technology (MIPT—National Research University), Moscow, Russia
e-mail: m.zabezhailo@yandex.ru

Yu. Yu. Trunin (✉)
Department of Radiology and Radiosurgery, N.N. Burdenko National Medical Research Center of Neurosurgery, Ministry of Health of the Russian Federation, Moscow, Russia
e-mail: ytrunin@nsi.ru

E. I. Inshakova and A. O. Inshakova (eds.), *New Technology for Inclusive and Sustainable Growth*, Smart Innovation, Systems and Technologies 287,
https://doi.org/10.1007/978-981-16-9804-0_14

techniques in economic activities. Big Data and Artificial Intelligence (AI) transform the environment, traditional concepts (smart home, unmanned vehicle systems, smart contracts, and blockchains (distributed ledgers), economic activities, and industries into a qualitatively new state (smart agriculture, digital healthcare, etc.). At the same time, these activities rely on data exchange and decision support systems as vital architectural components required to respond efficiently to possible challenges.

The operational features that are crucial for the reliable operation of Industry 4.0 computer technologies include.

(1) processing large amounts of data taking into account not only the Big effect but also the Open effect, which consists in regular updating of the current database with new information,
(2) dealing with the time constraints imposed on the Data Analysis (DA) and Decision-Making time, the so-called process-real time.

The development and implementation of Industry 4.0 technologies have shown that in the conditions under consideration, Big Data analysis can be carried out effectively and reliably only by using Artificial Intelligence techniques (where this term represents a general name for the relevant field of research and development). Thus, a problem-oriented approach using Intelligent Data Analysis (IDA), Big Data analysis performed by computer systems that simulate and enhance the cognitive capabilities of an expert researcher, is an effective means to respond to these kinds of challenges.

The need to develop approaches, techniques, and means to ensure reliable operation of the infrastructure components and aid to protect and maintain business continuity in the complex technologies employed in Industry 4.0. appears to be a critical component for appropriate decision-making in the industry.

When developing techniques for decision-making systems, it is vital to take into account the following requirements [2–7, etc.]:

- supported by IDA techniques, guidelines, conclusions, and recommendations for Decision-Makers (DM) must be transparent that is meaningfully interpreted and clearly explained in meaningful terms and concepts adopted in the analyzed subject area because the decision-maker takes full responsibility for the consequences that arise as a result of their implementation;
- conclusions and recommendations provided by decision support systems (DSS) should be sustainable when updated with new evidence on the control object; causal factors in the DA and DSS, which determine the presence of the studied effect or phenomenon, need to be identified. Consistent refinement of ideas about such causal influences, formed in the process of constantly updated empirical data on the behavior of the control object, allows us to reliably and effectively maintain the business continuity mode of the relevant technologies and end solutions in Industry 4.0;
- Big Data processed under process-real time constraints requires special efforts to control and optimize the inspection of possible alternatives when searching for

appropriate management decisions (more formally speaking, targeted control of computations to identify a reliable alternative).

Previous studies have shown that when we deal with relatively small initial (see [3–7, etc.] for mathematical problems of diagnostic type), some computationally hard—provably intractable combinatorial problems—arise [6, etc.]. In this case, a critical role belongs to various methods to provide control case study inspection when searching for a reliable alternative (the development of effective heuristics for finding particular solutions, development of some approximate methods, etc.).

The techniques discussed seem promising for diagnostics and support in large technical systems such as data centers and infrastructure telecommunication complexes. These techniques are efficient for fraud protection in the financial sector and decision-making support under force majeure in the industrial sector. The latter includes crisis management, support for the survivability (business continuity) of the control object under the conditions of failures, and other incidents that disrupt its operation.

14.2 The Scientific Problem

From a formal point of view, the described above industrial applications tackle specific diagnostic tasks [3–7, etc.]. In this case, the analyzed data represent descriptions of samples with the presence (or, conversely, the absence) of the phenomenon or effect under study—the so-called precedents. The initial database, regularly updated with descriptions of new precedents of the controlled object behavior, can be summarized in the table Precedents × Features, where a concrete value is assigned to each of the observed features of a specific precedent. The purpose of the IDA is to identify the causal factors, which determine the presence/absence of the analyzed effect/phenomenon. Based on the collected empirical data, reconstruction of causal factors enables the effectiveness and reliability of decisions based on the conclusions and recommendations derived by Intelligent Data Analysis due to a targeted search of the identified causal factors.

In Industry 4.0, the management of complex industrial facilities involves risk assessment for further development of the current situation—the state of the control object that can evolve in various possible ways. Simultaneously, it is vital to consider several requirements and restrictions that vary from the desire to enhance profits to the need to maintain business continuity by ensuring the operational stability of high-tech business structures by repairing various technical failures and targeting disruptive impacts. In this case, decision-making should rely on specific control systems that use built-in models to control an object's behavior in the changing environment. Here, we deal with solutions of two types. The first type uses stationary models of large objects and the cooperation of objects ported on large geographically distributed computing systems (see, for example, [8, 9] and others). The second type includes various mobile and onboard systems taking into consideration their limitations in functionality and performance.

The computer models for the control systems described above are as follows. (1) Mathematical models that now have the status of classical use system of balance type relations, e.g., models based on systems of equations, in particular, describe different effects in physics, economic cross-production or cross-industrial balances, etc. (2) Interpolation-extrapolation models implement learning procedures based on precedents, machine learning, data mining, etc.

To date, the use of balance models prevails in cases with a sufficiently small set of influence factors (possible causes), e.g., effects and phenomena characteristic description in natural science, such as physics, and technology, such as technical diagnostics, in particular. The efficient use of mathematical Models in Big Data analysis of multi-causal effects and phenomena is limited significantly in analytical and numerical ways. Let us take, in particular, the objectives of sociological studies that should consider the typology of society, identification of rational grounds for the electoral decisions in various social groups in society, etc., or the purposes of modern high-tech medical diagnostics. It is precisely these restrictions that have become the most significant driver for the development and improvement of interpolation-extrapolation models, which use training samples of precedent descriptions in dynamically changing conditions for the subject under study with its characteristic effects and phenomena. The COVID-19 pandemic represents a situation, in which the virus is subject to constant mutations. The need exists to analyze its impacts on people and the environment (pets, etc.) promptly.

The mechanism for this class models functioning is the interpolation of the training set of precedents with the presence (and/or absence) of the studied phenomenon/effect by empirical dependencies of one kind or another, followed by checking the expansion of the dependencies found in this way to the newly analyzed precedents to predict the presence (or, vice versa, the absence) of the studied target effects/phenomena.

To date, a topic of general interest is the personalization of products and services that is Industry 4.0 systemic feature. The individual settings and the need to classify consumers into homogeneous groups give rise to the issue of finding a balance between individual and group needs. We mean the following: can an individual consumer need simultaneously be a commercially significant need of a group of consumers that, when extracting the costs, can bring a profit? Where should we draw economically meaningful boundaries between homogeneous groups with retained, to a certain extent, characteristics? So, researchers set the task of quickly identifying economically significant homogeneous groups of consumers for Industry 4.0 products and services. Various network communities and collaborations [10–12, etc.] have become a topical research objective. It became clear that at the technical—procedural, algorithmic—level, identification of the boundaries between homogeneous groups of consumers in terms of their properties such as needs, and interests, has several specific features. First, it is necessary to promptly analyze large amounts of initial data (Big Data) to identify sought-for patterns under the specified time constraints for data analysis and decision-making. It is required to process many alternatives (candidates for a group in question) in constrained time intervals because of the specifics of the problem solved. An example of a successful solution to the issue under discussion is the well-known story of voting in the UK and USA associated with the Cambridge Analytica company [13].

Apart from deciding whether a specific object belongs to a group of similar objects, it is vital to explain why we assign it to this group as it aids to suggest a course of action, or select a treatment scheme based on medical diagnosis, or develop a repair plan based on the results of technical diagnostics, or take measures to counteract cyber attacks, etc. A lack of consideration of the causal factors can lead to undesirable effects—the inability to resist threats efficiently.

Today, it is evident that this class of problem solutions in industrial applications requires artificial intelligence computer systems. This type of problem solution relies on IDA and machine learning for supporting decision-makers, and professional collaborations (including operational formation and reorganization in response to regularly occurring changes in the functional environment) as efficient techniques for Industry 4.0 [10–12, etc.], given time limits for data analysis and decision-making.

14.3 The Scientific and Practical Relevance of the Discussed Problem

Modern high-tech medicine as a research and development area uses Industry 4.0 technology to provide effective treatment. The use of modern techniques and technologies for medical decision-making support is a universal component of the inclusive growth and development of national digital economies. Using these tools allows creating favorable conditions for improving the quality of life and ensuring the equality of opportunities for all social groups of the population, regardless of geographical locations. Evidence-based medicine aims at personalization, proactivity, and prevention of medical interventions in conditions when it is necessary to analyze Big Data within restricted time. The requirements to ensure sufficient evidence for decision-making and, finally, enable decision-makers to be responsible for the consequences of the decisions made present strong arguments in favor of choosing medicine as a model for illustrating the approaches discussed in this chapter.

In high-tech medicine, the data required for successful diagnosis and subsequent therapy comprise a complete personalized record of a patient. It includes the anamnesis data and follow-up—immune status, heredity, genomic and metabolic data, etc., exhaustive lists of causal factors when analyzed effects and pathologies occur, the dynamics of changes in the observed indicators over time. Here, computer techniques for data analysis will lead to improved diagnosis and assessment of medical therapeutic effects and possible complications, side effects, etc.

An example of a mathematical technique adequate to the peculiarities of the performed computer DA (see [3–6] and others) is the procedural construction of the so-called Characteristic Functions (CF), which enables developing empirical theories based on regularly updated experimental data. Each of these theories is a consistent set of formulae that are empirical dependencies that interpolate on the available training samples of precedents—the Database of Facts (Fact Base). The theory describes all the facts available—the precedents from the Fact Base—as the logical consequences of the assumptions that form this evidence-based theory. These statements are accepted as valid because they are incontestable on the current samples of precedents from the interpolated Fact Base. At the same time, the Fact Base updated with descriptions of new precedents requires a re-evaluation of the acceptability of the assumptions that form the current evidence-based theory that is a dynamically changing structure, reflecting the evolution of our knowledge about the subject area under study.

Within the framework of these theories, verification of the uncontested conclusions can be completed using the argumentation scheme for the assessment of the acceptability of conclusions. The arguments PRO, empirical dependencies responsible for the target effect/phenomenon presence, are analyzed against the current Fact Base. The PRO arguments oppose the CONTRA arguments, facts, and dependencies characterizing the precedents of the target effect absence. Thus, medicine, a subject area regularly updated with new evidence and knowledge, is subject to formalization based on the principles and techniques of open theories. The latter represent expandable collections of empirical evidence (facts) and knowledge representation by local empiric dependencies.

14.4 Competitive Advantages of the Approach

By definition, a Characteristic Function (CF) is a logical condition that binds especially salient elements of case descriptions—examples and counterexamples of a diagnosed effect/phenomenon from the current Fact Base (FB), which takes:

- the value that is true on all facts (examples) of the current FB, characterized by the presence of the analyzed target property (diagnosed phenomenon),
- the value that is false on all facts (counterexamples) of the current FB of the CF—in the absence of the analyzed target property.

Papers [3, 5, 6] give a fairly detailed consideration of the procedures for generating CFs for a given FB. The mathematical properties of the CFs procedural toolbox in their most general form are as follows:

1. in the general case, the exponentially fast growth of the size of the classes of generated empirical dependencies with a linear increase in the number of elements in the original training sample of precedents;
2. as a consequence, there is a lack of opportunities in many significant applications to identify all conflicting pairs of empirical dependencies by the so-called "brute-force" method—by exhaustive enumeration of all the candidates, which leads to a controversy in the diagnosis;

3. and an efficient, fast (polynomially complex) solution to some key combinatorial problems.

In a formalized form, this can be summarized in the following statements:

- the problem of checking the causal representativeness of the current training sample of FB precedents (see [5, 6, etc.] on the non-emptiness of the set of CFs formed on its base) is effectively solvable: there is an algorithm of polynomial computational complexity that generates its solution;
- the problem of enumeration of all CFs formed on a specific FB is enumeratively complete [5, 6];
- the problem of enumeration of all elements in the set CF(FB ∪ FB) formed on the current FB upgraded by descriptions of new precedents ΔFB is enumeratively complete [5, 6] (with the effective solvability of the problem of non-emptiness—causal representativeness—checking of such a set CF(FB ∪ ΔFB);
- the problem to define the number of elements in the set of characteristic functions that are true on a newly diagnosed patient—extrapolated to his description—is enumeratively complete (with the effective solution for the solvability of the problem of non-emptiness—causal representativeness—of a set of such CFs).

14.5 Results

14.5.1 Intelligent Data Analysis and Medical Diagnosis Based on Empirical Data

Using the example of identification and prediction of the post-treatment radiation effect, the so-called pseudoprogression of brain tumors (PsP) (see, for example, [14–20] and others), we will show how the presented IDA technique works in medical diagnostics. Pseudoprogression is temporary changes in the tumor and perifocal tissues surveyed on magnetic resonance imaging. It is challenging to differentiate post-radiation treatment effects from tumor recurrence by MRI alone. It is important to understand the radiological and clinical presentation that distinguishes these two entities to guide management. A radiation-induced increase in tumor size and contrast-enhancing lesion usually occurs shortly after radiation therapy and resolves spontaneously without treatment in majority of cases. The observed changes that characterize pseudo-progression are as follows:

- increased accumulation of contrast enhancement with or without the border expansion (T1 c/i),
- increased swelling edema the tumor and perifocal tissues (T2/Flair),
- an increase in tumor size due to a cyst and/or a solid component.

From the point of view of data analysis and medical decision support, pseudo-progression is specific primarily by the lack of accurate data that can reliably determine:

- the time of PsP development,
- the duration of PsP occurrence,
- diagnostic standards for PsP that differentiate PsP from tumor progression,
- recommendations for the treatment of patients with PsP, prognostic factors for PsP development.

The study of the PsP effects with the aid of IDA focuses on the three main objectives that include the diagnosis of the PsP effect, the diagnosis that differentiates PsP from tumor progression, and the identification of the so-called markers of pseudo-progression that allow for an uncontested diagnostic conclusion about the presence of PsP in a new patient.

The initial data (FB) of the study have descriptions of 410 patients of the N.N.Burdenko National Medical Research Center of Neurosurgery collected over about 15 years. The Database of Facts comprises data on 67 precedents of the PsP effect (described in the terminology we have adopted above) and 343 precedents of its absence (counterexamples). Each of the precedents is described by 150 parameters, quantified by values (boolean, numerical, etc.), but some parameters lack data in some patients. Although FB is regularly upgraded, due to obvious reasons, it occurs at a slow pace, with as few as several dozen descriptions of new patients annually.

Thus, in terms of mathematical models and methods of computer data analysis, we have to deal with a training sample of a limited size. At the same time, when processing these data, we need to operate with real Big Data effects. It is clear that even if we consider only the Boolean quantification of the data on precedents (when a description attribute significant for a PsP is either involved (has an impact) or not), we should deal with more than 2^{150} combinations of values for the parameters under consideration. We have to identify meaningful combinations of the impact factors out of all the descriptions of the current FB precedents empirical dependences, which determine (see above) the occurrence of the target effect—PsP of the tumor.

Some knowledge about the nature of pseudo-progression occurrence in the situation under consideration can be obtained using traditional statistical tools for testing medical empirical data. Literature (see, in particular, [17–19] and others) analyzing the risks of PsP development found that a statistically significant indicator that characterizes the empirical dependence correlation is the presence of a cyst in the tumor, which has an infratentorial localization in patients over 11 years of age in the FB under study. However, the approach that uses statistical tests to analyze the sets of correlations between the parameter values in the descriptions of the current FB cases does not solve vital questions concerning the accuracy of diagnostic conclusions and personalization of recommendations based on possible correlations of the observed parameter values. It does not provide meaningful explanations and informal interpretations of diagnostic decisions based on the identified correlations.

Papers [4, 5, 7] present a detailed description of the IDA results; in particular, it enabled us to divide the examples and counterexamples (descriptions of precedents characterizing the presence and absence of the PsP effect), based on an explicit form of the Characteristic Function, by meaningfully interpreted conditions, and give sufficiently detailed (multifactorial) descriptions of markers of the development of pseudo-progressions. One of the empirically established results is the structure of the Characteristic Function that is common for all that is formed on a given FB, in which two groups of the impact factors are present:

- the presence of those causal influences that force the emergence of PsP,
- the presence of some PRO arguments that provide sufficient grounds for accepting the results of the performed IDA, and
- the absence of those impact factors that indicate any deviations from the development of events typical for the PsP effect (the absence of actual CONTRA arguments in assessing the sufficiency of grounds to consider the generated IDA results acceptable).

14.5.2 The Proposed Approach and COVID-19 Pandemic-Induced Medical Services Transformation

In the context of the discussed features of Industry 4.0, we have to agree that the personalization of decisions and recommendations has fundamental significance. The current situation with COVID-19 viruses is one of the strongest arguments in favor of prioritizing personalization. Indeed, vaccination as a universal (proposed for everyone) means of tackling a pandemic has both clear advantages and disadvantages. Combating COVID-19 has taught us to perceive its effects on people as a combined multi-stage pathology. At the first stage, the virus accommodates in the human body, then, at the second stage, the immune system responds to the virus, and at last, at the third stage, we have to deal with various complications, the consequences of the first and the second stages. It is undeniable that the multidimensionality of the second and third stages (the variety of personalized characteristics of the course of these stages of the disease in different patients) is the prominent argument in favor of the resource futility of attempts to develop a universal remedy (medicinal product) for the second and third stages. There is too much variety in the individual responses of various patients to the effects of COVID-19 viruses.

Given this variability, the fabrication of a highly targeted drug that would consider individual responses to the infection will require enormous resources and time to conduct appropriate laboratory and clinical tests. At the same time, the formation of a typology of patients at the first stage of the infection, and the elaboration of treatment schemes that target homogeneous groups of patients is one of the few (if not the only one) alternatives for the COVID-19 personalized treatment. Identification of typologies, which would start with express diagnostics at the early stages of the disease, based on the constantly growing evidence (factual base) of clinical practice (both positive and negative), is an obvious tactic for improving therapeutic measures in the current situation. Moreover, we note that the described above clinical practice does not need to counteract the destructive activities of various COVID dissidents and all kinds of anti-vaxxers).

Procedurally, using the methodology and the relevant IT tools, the identification and maintenance of the proposed typologies that take into account the dynamics of changes in the epidemic, is a practice that has been elaborated in detail (see, for example, the case with Cambridge Analytica [13], etc.). Speaking about the latest approaches and mathematical models developed to identify typologies of the type under discussion, one can consider the experiences of using artificial intelligence and data mining (see, in particular, [21] and others).

Various research collaborations could play a critically important role in the identification of the discussed typologies. Cross-national and cross-border integration of regularly updated empirical data from research teams can allow us to fine-tune the personalized characteristics of each group of patients in the typologies identified. An example of a setting is the ability to manage the balance between the size of the group and the information content of its description, which is critically important for the effectiveness of the relevant recommendations and treatment measures, taking into account the limitations on the resources available when fighting emerging pathologies. In this context, the growing importance of network collaborations accumulated in the framework of Industry 4.0 is indisputable [10–12].

The proposed typologies identification requires not only to classify the studied precedents but also to highlight the causal grounds for the classification. The informativeness of the description of causal influences that allow for a meaningful explanation and interpretation of the emerging typologies provides opportunities to use the characteristic features of the formed classes of similarity of the analyzed precedents when tackling the related pathologies and threats. Medicine, technical diagnostics, the fraud protection in the financial sector, cyber security, etc. are some of the many areas that can use types of causal relations as a basis for combat measures that, in the first place, target the causes of the identified threats.

The approach we have considered using the example of combating COVID-19, which discusses the personalized characteristics of objects of analysis, support of collaborations, and typologies formation, can also be expanded on several other areas characteristic of Industry 4.0. Along with the previously mentioned medical and technical diagnostics, these areas can include providing business continuity of large infrastructure solutions for Industry 4.0, the fraud protection in banking and finance (including identification and countering/protection insider activities, etc.), analysis of public opinion (typology of society, allocation of rational grounds for the electoral choice of various social groups, etc.), and others.

14.6 Conclusion

Research programs undertaken in industrialized countries [22, etc.], which focus on support and development of the technology initiatives of Industry 4.0, draw our attention to the development of mathematical techniques and computer systems for the automated formation of evidence-based theories of the type discussed above. The AI technique allows us to effectively manage large amounts of empirical data from various sources: data fixed by different indicators and other measuring equipment, production and operation monitoring data, etc. AI dealing with regularly updated databases and promptly utilizing the outcomes in decision-making support systems ensures an adequate level of situational awareness for decision-makers and predicts further development of the current situation.

Here, we assign the fundamental role to:

(a) methods and technologies of computer-oriented representation of knowledge about the analyzed subject area in the form of dynamically changing logical theories (consistent sets of formulas—empirical dependencies) of a special type, the so-called partial empiric (evidence-based) theories. At the same time, such partial (empirical, regularly reorganized\reformed with the arrival of new empirical data) theories are used as a computer-oriented technique for providing support for decision makers' situational awareness;

(b) methods and technologies utilized for predicting the development of the current situation when analyzed by the decision-maker (including the identification of potential risks that may arise from making decisions based on suggested alternatives) by expanding the existing empirical dependencies,

(c) AI systems that provide just in time automated generation and maintenance/modification of evidence-based theories that target a wide range of subjects varying from heads of local business units to analysts in large centers for data analysis and decision support.

References

1. World Economic Forum. ICT for Economic Growth: A Dynamic Ecosystem Driving the Global Recovery. Annual Meeting Report—World Economic Forum Publishing. https://www.ifap.ru/pr/2009/n090910b.pdf (2009). Accessed 10 Nov 2021
2. Grusho, A.A., Zabezhailo, M.I., Piskovski, V.O., Timonina E.E.: Industry 4.0: opportunities and risks in the context of information security problems. Autom. Doc. Math. Linguist. **54**(2), 55–63 (2020)
3. Grusho, A., Zabezhailo, M., Timonina, E.: On causal representativeness of training samples of precedents in diagnostic type tasks. Inform. Primen. **14**(1), 80–86 (2020)
4. Zabezhailo, M.I., Trunin, YuYu.: On the problem of medical diagnostic evidence: intelligent analysis of empirical data on patients in samples of limited size. Autom. Doc. Math. Linguist. **53**(6), 322–328 (2019)
5. Zabezhailo, M.I., Trunin, Y.Y.: On the reliability of medical diagnosis based on empirical data. Sci. Technol. Inform. Process. **48**(5), 1–8 (2021)
6. Zabezhailo, M.I.: To the computational complexity of diagnostic predictions designed by means of characteristic functions. Autom. Doc. Math. Linguist. **54**(6), 298–305 (2020)
7. Zabezhailo, M.I., Trunin, Y.Y.: On the importance of empirical contradiction for reliability estimation of intelligent data analysis results. Autom. Doc. Math. Linguist. **55**(3), 94–100 (2021)
8. National Center for Supercomputing Applications. University of Illinois. http://www.ncsa.illinois.edu/ (2021). Accessed 10 Nov 2021
9. Argonne National Laboratory. https://www.anl.gov/ (2021). Accessed 10 Nov 2021

10. Inshakov, O.V., Yakovlev, A.R.: Modification as a form of the firm's goods evolution. Bulletin of Volgograd State University. Series 3: Economics. Ecology, **2**(15), 115–123 (2009)
11. Inshakov, O.V.: Collaboration as a form of knowledge-based economy organization. Econ. Region **3**, 38–45 (2013)
12. Inshakov, O.V., Voloshina, A.Y.: Megacollaboration as a Form of Creative Cooperation in Economy of Knowledge. VolSU Publ, Volgograd (2013)
13. The Cambridge analytica files. In: The Guardian. https://www.theguardian.com/news/series/cambridge-analytica-files (2021). Accessed 13 Oct 2021
14. Nudnov, N.V., Zheludkova, O.G., Mnatsakanova, I.V., Sidorova, E.V., Podoksenova, T.V., Shevtsov, A.I.: Pseudo-progression in a patient with anaplastic ependymoma after radiation therapy. Med. Vizualiz. **2**, 18–24 (2018)
15. Hygino da Cruz, L.C., Rodriguez, I., Domingues, R.C., Gasparetto, E.L., Sorrensen, A.G.: Pseudoprogression and pseudoresponse: imaging challenges in the assessment of posttreatment glioma. Am. J. Neuroradiol **32**(11), 1978–1985 (2011)
16. Parvez, K., Parvez, A., Zadeh, G.: The diagnosis and treatment of pseudoprogression, radiation necrosis and brain tumor recurrence. Int. J. Mol. Sci **15**(7), 11832–11846 (2014)
17. Trunin, Yu.Yu., Golanov, A.V., Kostyuchenko, V.V., Galkin, M.V., Khukhlaeva, E.A., Konovalov, A.N.: Pseudo-progression of benign glioma on the example of piloid astrocytoma of the midbrain. Clin. Observ. Oncol. J.: Luchevaya Diagn., Luchevaya Terap. **1**(1), 94–97 (2018)
18. Trunin, Y.Y., Golanov, A.V., Kostyuchenko, V.V., Galkin, M.V., Khukhlaeva, E.A., Konovalov, A.N.: Increased volume of piloid astrocytoma of the midbrain: relapse or pseudo progression? Clin. Observ. Onkol. Zh.: Opukholi Golovy Shei **6**(1), 68–75 (2016)
19. Trunin, Y., Golanov, A.V., Kostjuchenko, V.V., Galkin, M.V., Konovalov, A.N.: Pilocytic astrocytoma enlargement following irradiation: Relapse or pseudoprogression? Cureus. https://www.cureus.com/articles/3962-pilocytic-astrocytoma-enlargement-following-irradiation-relapse-or-pseudoprogression (2017). Accessed 23 Oct 2021
20. Ellingson, B.M., Wen, P.Y., Clouhesy, T.F.: Modified criteria for radiographic response assessment in glioblastoma clinical trials. Neurotherap **14**, 307–320 (2017)
21. Mikheyenkova, M.A.: Cognitive knowledge discovery in social sciences. In: Golenkov, V., Krasnoproshin, V., Golovko, V., Azarov, E. (eds.) Open Semantic Technologies for Intelligent System, OSTIS 2020. Communications in Computer and Information Science, vol. 1282, pp. 48–63. Springer, Cham (2020)
22. NSCAI Final Report (January 2021, Draft). https://www.nscai.gov/wp-content/uploads/2021/01/NSCAI-Draft-Final-Report-1.19.21.pdf (2021). Accessed 23 Oct 2021

Chapter 15
Artificial Intelligence Algorithms in Diagnosis of Breast Cancer

Alexander G. Losev and Andrey V. Svetlov

Abstract The field of this research is the development of intelligent systems based on the use of modern methods of artificial intelligence. Such systems are able to provide a diagnosis with high accuracy and also to be an advice-giver for a diagnostician. At the present stage of development of medical diagnostic systems, it is assumed that it can analyze a screening data, and it is often an extremely complex process that requires a long preparation and many years of experience for a physician. Intelligent advisory systems should provide a rationale for the proposed diagnostic decision, and a description of the data analysis results in a form that healthcare professionals can understand. The main method of our research is microwave radiothermometry. This is a biophysical method for non-invasive examination of biological tissues, based on the measurement of internal and surface temperatures by the intensity of their electromagnetic radiation in the microwave and infrared ranges. The study uses specific methods for intelligent analysis of thermometric data and the results of computer modeling of spatial and temporal temperature distributions inside mammary glands. The proposed approach significantly increases the sensitivity and specificity of the method.

15.1 Introduction

The development of society is marked by technological waves and industrial revolutions. Most researchers agree that the present time period refers to the sixth technological wave. Its main feature is development of robotics, biotechnology, nanotechnology, artificial intelligence systems, etc. [1, 2]. At the same time, "the introduction

A. G. Losev (✉) · A. V. Svetlov
Volgograd State University, Volgograd, Russia
e-mail: alexander.losev@volsu.ru

A. V. Svetlov
e-mail: andrew.svetlov@volsu.ru

E. I. Inshakova and A. O. Inshakova (eds.), *New Technology for Inclusive and Sustainable Growth*, Smart Innovation, Systems and Technologies 287,
https://doi.org/10.1007/978-981-16-9804-0_15

of nano-, bio-, information and cognitive technologies, cloud computing and artificial intelligence in all spheres of society" [3] is a characteristic feature of the fourth industrial revolution.

We also note one of the most important social and state problems, which is to increase the efficiency of early diagnosis and treatment of diseases of the mammary glands. Statistics from recent decades shows that breast cancer is the most common cancer among women. At the same time, rather serious scientific results in this area, obtained over the past decades, significant changes in approaches to treatment, the use of high-tech medical technology, all of it did not lead to a solution to the problem. In 2020, according to the World Health Organization, 2.3 million cases of breast cancer were detected. More than half a million people die from this disease in the world every year.

According to a number of experts, at present, one of the most promising methods of mammological screening and early diagnosis of breast cancer is the method of microwave radiothermometry. Microwave radiothermometry is a screening method based on measuring the spatial distribution of the temperature field in the human body by the intensity of its own electromagnetic radiation, analyzing and interpreting the obtained data [4]. In recent years, this method of medical examination has become widespread in a number of branches of medicine. The experience gained during this time has identified one of the serious obstacles to its widespread use for early diagnosis. Namely, the analysis of thermometric data turned out to be an extremely difficult task, in fact, available only to a doctor of very high qualifications with a sufficiently large experience. However, it is not a unique phenomenon, but rather one of the most urgent problems of mathematical modeling, artificial intelligence and medicine.

One of the most important and complex tasks of modern medicine was generated by the following phenomenon. The use of modern high-tech equipment has led to the fact that the main problems in making a diagnostic decision arise not due to a lack of information but because of the complexity of its processing and analysis. The main way to solve this problem is the creation of systems for the interpretation and analysis of medical data based on artificial intelligence algorithms. In recent years, a number of studies have appeared showing the effectiveness of this approach [5–11]. At the same time, the main interest here is in development of intelligent systems with a mechanism for explaining and justifying the proposed solutions in a form understandable to a diagnostician [12]. Such systems, using the latest developments in the field of machine learning, will provide decision support for specialists in the tasks of forming a diagnostic solution, predicting possible options for the development of diseases, etc.

The aim of this study is a creating the foundations of an intelligent system for diagnosing breast cancer based on the method of microwave radiothermometry. The proposed approach is based on the combined use of methods for intelligent analysis of thermometric data and the results of computer modeling of spatial and temporal temperature distributions inside mammary glands. This technique significantly increases the sensitivity and specificity of the method.

15.2 Methodology

As mentioned above, in the last two decades, the microwave radiothermometry method has shown very high efficiency in the diagnosis of a number of diseases, including breast cancer. In 1975, the American researcher A. Barrett proved the possibility of using this approach for the diagnosis of breast cancer [13]. However, microwave radiothermometry was widely used in clinical practice only at the beginning of the twenty-first century [7].

Note that the temperature field of the human body is characterized by significant spatial variability, and its fluctuations are of great importance for the detection of pathological processes.

Generally speaking, thermometry as a method of medical diagnostics has been known for a long time. The most important stage in its development was the emergence of infrared thermometry. It is based on measuring the infrared radiation of the human body and has been quite actively used in the past few decades. The most serious disadvantage of this examination method is the fact that infrared thermometry can only determine the temperature of the skin. As a result, the accuracy of the spatial thermometric description of the human body leaves room for improvement and this is why in a large number of cases it is rather difficult to detect signs of pathological processes. For this reason, this method of examination has a very ambiguous assessment in the medical community.

The method of microwave radiothermometry makes it possible to obtain a much more adequate spatial thermometric description of human body. At the first stage, this examination method was used exclusively for the diagnosis of breast cancer. Among other reasons, this is due to the fact that in the area where the malignant neoplasm is located, a local increase in the internal temperature occurs. Thus, the method of microwave radiothermometry is able to provide high accuracy in solving the problem of early detection of malignant neoplasms.

Note that modern technologies for creating radiothermometers have provided a fairly accurate measurement of the temperature fields inside various human organs. It makes possible to develop an advisory diagnostic system based on machine learning algorithms. However, for a diagnostician without special and sufficiently long training the thermometric information analysis is quite difficult, what significantly reduces the potential for using this examination method in screening. This problem can be completely solved by developing effective algorithms for data analysis based on artificial intelligence algorithms. Thus, in order to effectively use the unique capabilities of microwave radiothermometry in early diagnosis, we need to create an intelligent diagnostic system with high accuracy, specificity, and sensitivity in identifying patients at risk of breast cancer.

15.3 Results

At the first stage of our research, we developed several mathematical descriptive models for describing the diagnostic state of patients of various classes. For this, it is necessary to detect a dependence between the presence of breast cancer and the spatial dynamics of its temperature fields. It is important to take into account the ambient temperature, patient's age, breast size, weight index, and a number of other parameters of the history and preliminary examinations [7, 10]. And we should note that in most cases we do not use the temperature values at individual points of mammary glands for descriptive models of a patient's diagnostic state. It turned out that some mathematical functionals (norms, seminorms, etc.) more accurately describe the aspects of the behavior of temperature fields in patients of various diagnostic classes. In what follows, we refer to them as thermometric characteristics.

The method of setting up spaces of thermometric characteristics (also known as highly informative thermometric features) is one of the most vital aspects of this research. These spaces are used in algorithms for evaluating a diagnostic solution, in a module of justifying the proposed diagnostic solution of an advisory intelligent system, as well as in algorithms for analyzing thermometric data.

The obtaining the elements of the space of diagnostic signs based on the study and analysis of areas of change in functional characteristics describing the corresponding effects of the temperature fields behavior in the mammary glands of patients within various diagnostic classes. To develop a thermometric descriptive model of the patient's diagnostic condition we have analyzed the existing medical knowledge about emriogenesis, development and structure of mammary glands, various heuristics used by specialists during examinations. The key elements of this model are determining the presence and location of areas of mammary glands with an abnormally high (or abnormally low) temperature, the presence and degree of mirror symmetry of temperature fields, the level of their variability, dispersion characteristics, the rate of temperature change in significant directions (internal gradient, axial gradient), the rate of change of the internal gradient in the radial direction, etc. Further, on the basis of the developed mathematical models of human thermal and radiation fields, statistical analysis of the results of examinations carried out in medical institutions over the past two decades, we develop a mathematical descriptive model of the patient's diagnostic state. Next, on the basis of computational experiments and intellectual analysis of training samples of thermometric data, we will form a fairly complete set of mathematical constructions describing the corresponding effects.

This approach has shown significantly greater efficiency in making a diagnostic decision based on the operation of machine learning classification algorithms and in justifying the resulting decision. Let us describe this result in more detail.

To evaluate the effectiveness of classification algorithms in machine learning, one can usually use such parameters as accuracy, completeness, F-measures, and so on. In particular, to evaluate diagnostic methods in medicine in the case of binary classification, we should adhere to the tradition and use the following indicators.

Sensitivity, also known as Recall, is calculated using the formula

$$\text{Sens} = \frac{\text{TP}}{\text{TP} + \text{FN}}$$

where TP is the number of patients from the "Cancer" class who were assigned to the class "Cancer" by the classifier, FN is the number of patients from the "Cancer" class who were not assigned to the class "Cancer" by the classifier.

Specificity is calculated by

$$\text{Spec} = \frac{\text{TN}}{\text{TN} + \text{FP}},$$

where TN is the number of patients of the "Healthy" class who were classified by the classifier as "Healthy", FP is the number of patients of the class "Healthy" who were classified as not "Healthy".

As a metric of the classification efficiency one can usually use various variants of F-measures. In this study, based on the results of computational experiments, we use for this purpose the value $\text{G} = \sqrt{\text{Spec} \cdot \text{Sens}}$.

Now we consider the logistic regression as an example of classification algorithm. If we take the temperature values at different points of mammary gland as the feature space, we get the sensitivity 0.75 and the specificity 0.81. If we take the values of thermometric characteristics as the feature space in this classification algorithm, we get the sensitivity 0.89 and the specificity 0.81. This way, the sensitivity has increased by 15 percent. A detailed description of the algorithm and feature spaces you can see, for example, in [8]. Similar results we also obtained using artificial neural networks [7] and some other classification algorithms.

For comparison, we present the results of breast cancer diagnostics, which were achieved in some medical centers in Russia and the United States using the microwave radiothermometry method from 1997 to 2013. According to [4], the diagnostic sensitivity ranged from 0.85 (Medical College, Arkansa, USA) to 0.96 (Radiology Center, Moscow, Russia) and even 0.98 (Burdenko Hospital, Moscow, Russia). The specificity of the diagnosis ranged from 0.57 (Radiology Center, Moscow, Russia) to 0.88 (Burdenko Hospital, Moscow, Russia). It should be noted that these results were obtained by highly qualified specialists with extensive experience in medical practice. Naturally, to make a diagnosis, they used not only the thermometric information but also a fairly large amount of additional data about the patient's condition. Meantime, it is now obvious that the results obtained with artificial intelligence algorithms are comparable to the results obtained by experienced specialists of well-known medical centers. Obviously, if the developed intelligent system is introduced into medical practice, the sensitivity and specificity will be even higher. Moreover, it is quite possible to involve nurses to measure temperatures at the screening stage, and it also can increase the efficiency of examinations.

There is another and rather non-trivial problem to establish the location of the tumor according to temperature data. This problem is much more difficult than detecting the presence of a malignant tumor. However, in [14] we propose an approach to solution of this problem and obtain quite consistent results. To be exact, the accuracy of tumor localization ranges from 55 to 65 percent, depending on the test samples.

To make a correct diagnosis we use the voting method with a set of classification algorithms, and then we analyze and interpret the work of some of them to justify the diagnosis. When we use as classification algorithms the ensemble of an artificial neural network, logistic regression, and decision forest, the sensitivity ranges from 0.88 to 0.92, and the specificity is in range 0.85–0.9. It is important to use for the justification of the proposed set of thermometric characteristics [7, 10].

At the same time, this study develops a fairly new approach to data mining. It is based on the results of computer modeling of the spatial and temporal distribution of breast temperatures [10, 14–16] (Fig. 15.1). Here we proposed fairly consistent approaches to the application of machine learning methods in the analysis of thermometric data obtained with physical, mathematical, and computer models of thermal and radiation fields. We developed quite specific methods for the computer modeling of the temperature fields of patients' mammary glands within various diagnostic classes. In [10] we proved the validity of the resulting set of computer models for temperature fields of mammary glands. The proof bases on computational experiments and using a set of machine learning algorithms (logistic regression, naive Bayesian classifier, support vector machine, decision tree, gradient boosting, K-nearest neighbor method, etc.).

It is interesting to note that our technique involves the use of computer models of temperature fields and it allows us to propose an approach to solving the problem of errors in the raw data. Actually, it is highly relevant problem to minimize the influence of errors in the training samples of artificial intelligence algorithms. Moreover, it will hardly be solved with existing standard methods. This is especially true for medical data, where elements of uncertainty and inconsistency are a distinguishing feature. If one can create sufficiently valid computer models of the temperature fields of various human organs, then the results of these models can be used in the analysis of "natural" thermometric data, i.e., obtained as a result of real medical examinations.

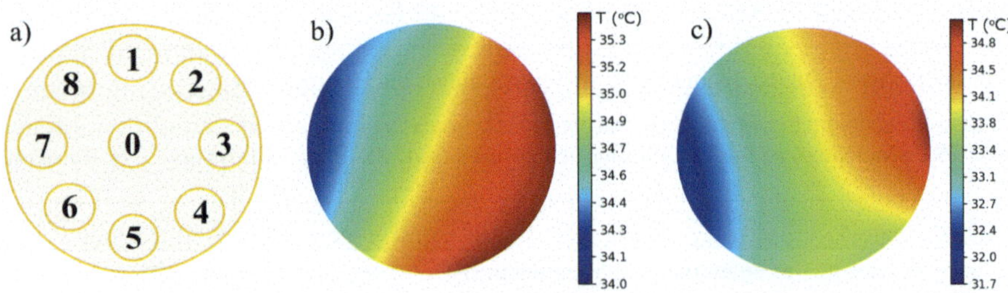

Fig. 15.1 **a** Scheme used in simulation of brightness temperatures for right breast; **b** Internal temperature distribution for a model with a 1 cm radius tumor located at point "3"; **c** Surface temperature distribution for a model with a tumor with a radius of 1 cm located at point "3". *Source* [10]

Another application of appropriate computer models is computational experiments aimed at enhancing the existing survey techniques. In particular, computer models make it possible to obtain temperature values not only at the points of the standard examination technique but also at any other point of the breast. It provides significantly more accurate visualization of the temperature fields of mammary glands. According to analysis of the obtained visualizations it is known that sometimes it is necessary to measure internal and skin temperatures at some specially selected additional points during a medical examination.

15.4 Conclusion

Thus, the results of researches in the past several years show the possibility of creating an effective intelligent system for early diagnosis of breast cancer based on microwave radiothermometry data mining. At the same time, it was shown that the following aspects are crucial for developing an appropriate advisory diagnostic system.

The preprocessing of the temperature data of mammary glands is required to create correct training samples for classification algorithms of the developed diagnostic intelligent system. For this purpose, at first, we build a mathematical descriptive model of a patient's diagnostic state. It includes mathematical functionals that characterize the behavior of the temperature fields of patients within various diagnostic classes. This approach can significantly increase the efficiency of the classification.

To minimize the influence of errors and inaccuracies in medical databases, we recommend to construct training samples using the results of computer modeling of the temperature fields of examined organ. This is why it is urgent to create appropriate mathematical models with a sufficiently high degree of validity.

Also, it is necessary to increase the accuracy of the proposed diagnostic solution with the voting method of the ensemble of classification algorithms. Here it is reasonable to use models of two types as the constituent parts of the ensemble. First, it is possible to use artificial neural networks, which are good for accuracy, but not very good for justifications. Second, it is possible to use expert system models that have a good skill of justification but are not so good for accuracy.

Acknowledgements This work was supported by the RFBR, project no. 19-01-00358, "Mathematical models of radiation fields and analysis of microwave radiothermometry data in the early diagnosis of breast cancer".

References

1. Averina, I.S.: Industrial revolution and technological pattern: essential characteristics, similarities and distinctive features. J. Volgograd State Univ. Econ. **23**(1), 52–63 (2021). https://doi.org/10.15688/ek.jvolsu.2021.1.4
2. Inshakov, O.V., Inshakova, E.I.: Technological platforms in Russian nanoindustry: problems and prospects of development. J. Volgograd State Univ. Econ. **19**(3), 7–20 (2017). https://doi.org/10.15688/jvolsu3.2017.3.1
3. Shvab, K., Devis, N.: Technologies of the Fourth Industrial Revolutions. Eksmo Publications, Moscow (2018)
4. Vesnin, S., Turnbull, A.K., Dixon, J.M., Goryanin, I.: Modern microwave thermometry for breast cancer. J. Mol. Imaging Dynam. **7**(2), 1000136 (2017). https://doi.org/10.4172/2155-9937.1000136
5. Araujo, T., Aresta, G., Castro, E., Rouco, J., Aguiar, P., Eloy, C., Polonia, A., Campilho, A.: Classification of breast cancer histology images using convolutional neural networks. PLoS ONE **12**(6), e0177544 (2017). https://doi.org/10.1371/journal.pone.0177544
6. Ghoneim, O., Soliman, G., Galal, A., Mahgoub, H.: Breast cancer histological image classification using ensemble convolutional neural network and triplet loss. IOSR J. Comput. Eng. Ser II **II**, 30–42 (2021). https://doi.org/10.9790/0661-2304023042
7. Levshinskii, V., Galazis, C., Ovchinnikov, L., Vesnin, S., Losev, A., Goryanin, I.: Application of data mining and machine learning in microwave radiometry (MWR). In: Communications in Computer and Information Science, 1211 CCIS, pp. 265–288 (2020). https://doi.org/10.1007/978-3-030-46970-2_13
8. Levshinskii, V.V.: Mathematical models for analyzing and interpreting microwave radiometry data in medical diagnosis. J. Comput. Eng. Math. **8**(1), 3–14 (2021). https://doi.org/10.14529/jcem210101
9. Mohanty A.K., Senapati M.R., Lenka S.K.: An improved data mining technique for classification and detection of breast cancer from mammograms. Neural Comput. Appl. **22** (SUPPL.1), 303–310 (2013). https://doi.org/10.1007/s00521-012-0834-4
10. Polyakov, M.V., Popov, I.E., Losev, A.G., Khoperskov, A.V.: Application of computer simulation results and machine learning in the analysis of microwave radiothermometry data. Math. Phys. Comput. Simul. **24**(2), 27–37 (2021). https://doi.org/10.15688/mpcm.jvolsu.2021.2.3
11. Yassin, N.I.R., Omran, S., El Houby, E.M.F., Allam, H.: Machine learning techniques for breast cancer computer aided diagnosis using different image modalities: a systematic review (Review). Comput. Methods Programs Biomed. **156**, 25–45 (2018). https://doi.org/10.1016/j.cmpb.2017.12.012
12. Kobrinskii, B.A.: Advisory intelligent medical systems: classification, principles of construction, efficiency. Inf. Technol. Phys. **2**, 38–47 (2008)
13. Barrett, A.H., Myers, P.C.: Subcutaneous temperature: a method of noninvasive sensing. Science **190**, 669–671 (1975). https://doi.org/10.1126/science.1188361
14. Glazunov, V.A., Polyakov, M.V.: Testing the algorithm of tumor localization in breast cancer on the results of temperature field simulation. J. Phys. Conf. Ser. **1794**, 012006 (2021). https://doi.org/10.1088/1742-6596/1794/1/012006
15. Polyakov, M.V., Khoperskov, A.V., Zamechnic, T.V.: Numerical modeling of the internal temperature in the mammary gland. Lecture Notes in Computer Science, 10594 LNCS, pp. 128–135 (2017). https://doi.org/10.1007/978-3-319-69182-4_14
16. Polyakov, M., Levshinskii, V., Khoperskov, A.: Modeling of brightness temperature in biological tissue. J. Phys: Conf. Ser. **1368**(4), 042057 (2019). https://doi.org/10.1088/1742-6596/1368/4/042057

Chapter 16
Automating the Audit Process of Management Systems Through Artificial Intelligence Methods

Valentin Dzedik, Valentina Moiseeva, and Alex Ezrakhovich

Abstract The chapter explores the impact of the COVID-19 pandemic on the extensive transformation of management processes audit, the prerequisites for this transformation and the development of models and methods to automate the management processes audit through artificial intelligence methods. The key feature of the transformation of audit processes is achieving inclusive and sustainable growth. Inclusive growth is considered not only in terms of the process itself but also in regard to its final result. Automation not only accelerates audit procedures, reduces time- and labor-intensive, mitigates the epidemiological risks to participants but also applies the findings to a wider range of processes. The impact of improvements is getting extensive for more people involved in various production operations. A detailed study has led to the development of the model for the application of artificial intelligence methods for automating the process audit of management systems. The creation of the model was prompted by the need to improve the capabilities to save both the sustainability of adaptable systems and processes and stable economic growth. The chapter identifies the most important tasks in developing an effective automated audit model, as well as the functionality of each element of the automated audit model and its implementation prospects. Also were analysed risks emerging from the implementation of this approach.

V. Dzedik · V. Moiseeva (✉)
Volgograd State University, Volgograd, Russia
e-mail: v.chistobaeva@volsu.ru

V. Dzedik
e-mail: prorector1@volsu.ru

A. Ezrakhovich
AEConformity Pty Ltd, Sydney, NSW, Australia
e-mail: alex@aeconformity.com

E. I. Inshakova and A. O. Inshakova (eds.), *New Technology for Inclusive and Sustainable Growth*, Smart Innovation, Systems and Technologies 287,
https://doi.org/10.1007/978-981-16-9804-0_16

16.1 Introduction

The analysis of the basis for applying management systems audit methods currently needs to highlight three main components: the state of management systems audit methods in the pre-pandemic period, research on-site of audit process automation and modern artificial intelligence methods in the perspective of applying them to audit management systems.

16.2 Methodology

The theoretical and methodological basis to improving quality management systems is laid in the [1–5] writings. The issues of the influence of informatization on the functioning of management systems of different levels were studied by [6, 7] and others.

The concept of management systems is specified by a number of international and national standards, including [8–20] etc. Worldwide, millions of organisations are using these standards to improve performance, efficiency and competitiveness. These standards require such an essential management system feedback and monitoring method as an internal audit. The external audits of management systems, for example in pre-contractual, 2nd party or 3rd party certification activities, follow the requirements specified in the relevant documents, such as [21, 22]. All these documents focus on-site audits, where a team of auditors carries out the field audit at the premises of audited entity, for example at an organization or a particular unit.

As far back as the pre-pandemic period there have been efforts to reduce the share on-site work in management system audits through, for example, remote audits using information and communication technology [23, 24]. The main reason for using remote audits was to reduce costs. However, remote auditing methods do not provide the only alternative on-site audits. Another approach to streamlining the audit process is its automation.

The rapid spread of coronavirus infection has led not only to huge economic losses but also to the transformation of various business processes. A large-scale study conducted by the CQI—Charted Quality Institute (UK) and Qualsys, a quality management software developer company, resulted in formulating the main trend of quality management system audits under COVID-19: an automated quality system plays the key role in solving many problems [25].

Almost all information-analytical tools can be applied in some way to automate the audit process. Such tools include database management systems, search engines, cloud technology, etc. However, applying these tools is only possible in a fragmented way and does not significantly reduce the proportion on site audits.

The application of artificial intelligence algorithms can lead to a major breakthrough in the automation of the audit process. Artificial Intelligence is "… the property of automatic systems to perform certain functions of human intelligence …

to select and make optimal decisions based on prior experience and rational analysis of external influences" [26]. The concept of "machine learning" is also used as a relative of "artificial intelligence". There are many types of artificial intelligence algorithms, and the most popular are the following:

A. K-Nearest Neighbour Algorithm

It is the identification of the class affiliation of the studied object through exploring a predetermined neighbourhood in the multidimensional space of objects with a predetermined affiliation to a certain class (Fig. 16.1).

B. Neural Network Algorithms

It relies on mathematical modelling of the structure and connections between artificial neurons transformed by processed data sets (Fig. 16.2).

C. Support Vector Machine Algorithms

It uses the transformation of numerical vectors into a multidimensional space and the subsequent search for the hyperplane that most effectively separates different objects to be assigned to a particular class of unidentified objects (Fig. 16.3).

D. Random Forest Algorithm

It allows creating a predetermined set of "decision trees" based on the structure of the data samples, which acts as a sequence of actions to compare the available data with the criterion values (Fig. 16.4).

However, there were some studies of applying artificial intelligence algorithms to analyse management systems [26], but the main focus was on analysing data for management processes.

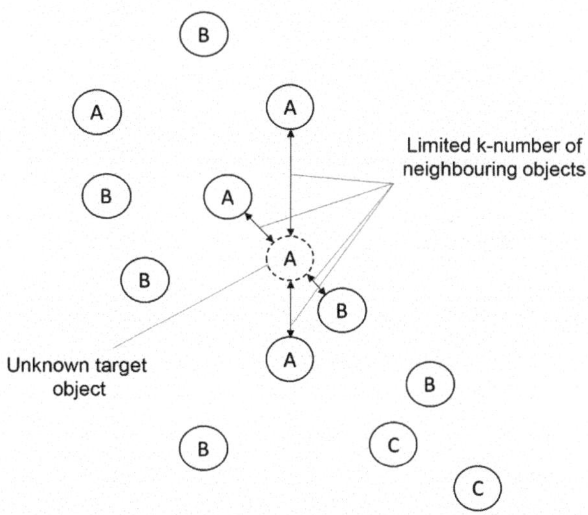

Fig. 16.1 Example of a K-Nearest Neighbour scheme. *Source* Developed and compiled by the authors

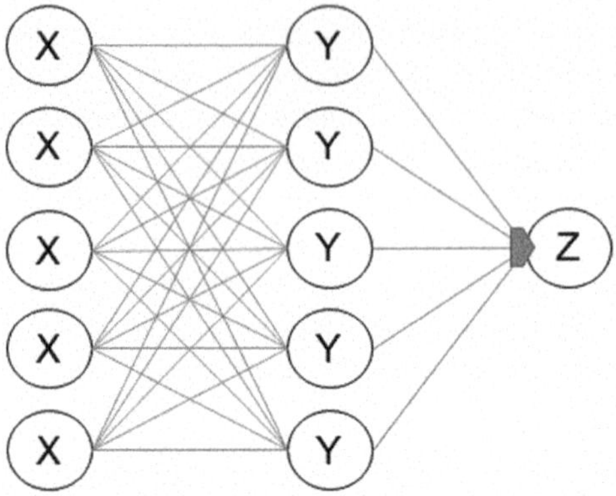

Fig. 16.2 Example of the Neural Network Algorithms simplified scheme. *Source* Developed and compiled by the authors

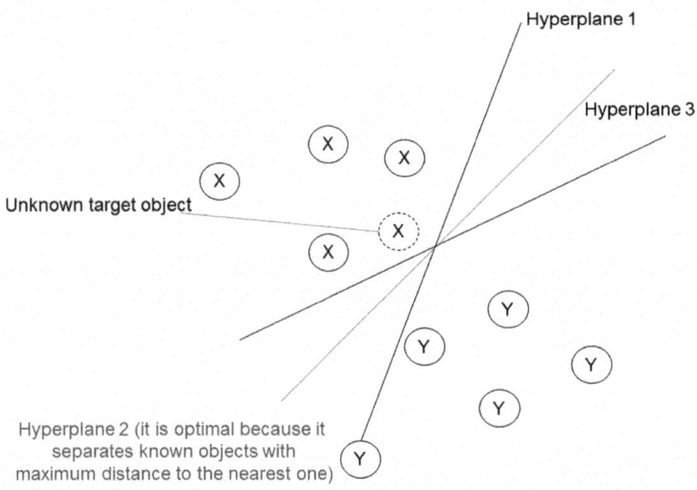

Fig. 16.3 Example of the Support Vector Machine Algorithms simplified scheme. *Source* Developed and compiled by the authors

16.3 Results

16.3.1 Model for Applying Artificial Intelligence Algorithms to Automate the Process Audit of Management Systems

To achieve the goal of streamlining the audit process through its automation [27] it is necessary to structurally link the following elements:

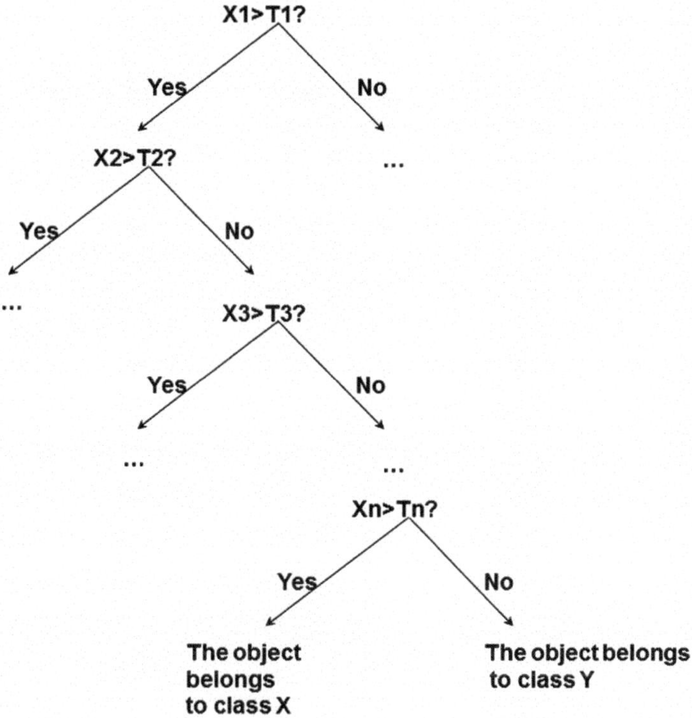

Fig. 16.4 Example of the simplified scheme of Decision Tree class algorithm. *Source* Developed and compiled by the authors

- Intelligence is important to create a core system, based on artificial intelligence methods, which is capable of comparing objective evidence of compliance with established criteria, e.g. the requirements of one of the previously mentioned international standards and of making a compliance/non-compliance decision through rapid real-time reconfiguration.
- Mobility is essential to provide the auditee with a quick and easy interface to the analytical system.
- Activity is the ability of the system to independently communicate with the auditee, to actively dialogue with the auditee pursuing the audit objectives.
- Adaptability is the ability of the analytic system to adequately identify and respond appropriately to non-standard words and phrases of the auditee during the dialogue and correctly relate them to the evidence of conformity/non-conformity [26].

To realize these objectives, the authors developed a model of an intelligent system to perform automated audits of management systems (hereinafter "systems") (Fig. 16.5).

The model consists of the following elements:

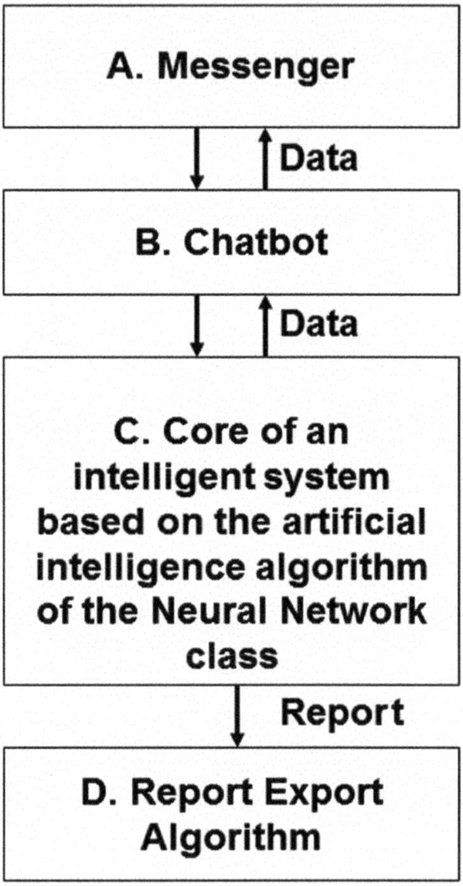

Fig. 16.5 Model of intelligent system for automated management system Audits. *Source* Developed and compiled by the authors

A. Messenger

The most important challenge in developing an effective automated audit model is to ensure its high mobility. Traditional approaches involving the deployment of specialized software on the auditee's equipment entail an unreasonably high cost in terms of technical, time, human, and financial resources. There are also significant risks of incompatibility between software and hardware. It is suggested that any of the popular messenger programmes installed in almost any modern mobile or desktop computer or mobile phone may serve as an interface between the auditee and the system.

B. Chatbot

A chatbot is created to enable the intelligent system to communicate with the auditee using the basis and means of a specific messenger. To control the chatbot, the system exchanges the relevant data. The chatbot solves the problem of system activity. Chatbot communicates with the audited staff by exchanging text messages and uploading data files from the audited party. He initiates the audit interview by contacting the auditee by sending a message via an appropriate messenger. Next, the chatbot provides the interview process by taking the lead, i.e. the role of the auditor asking questions and reacting appropriately to the answers of the auditee, providing an interface for downloading documents requested from the auditee for further transmission to the system for analysis. There are plans to develop a system capable of providing voice communication during audits in the future. Another task of the chatbot is to ensure the debriefing of the interview and the correct farewell to the auditee.

C. The core of an intelligent system based on Neural Network class algorithms

The analysis showed that the most effective class of artificial intelligence algorithms for solving the problem of automating the audit process of management systems is the Neural Network. Appropriately structured, configured, pre-trained and equipped with additional interfaces, amongst which you can distinguish the interfaces for collecting statistics and administration, it allows you to adequately apply pre-prepared audit scenarios, quickly switching between them and combining them. The system can retrain during each new interview with the verified party and can draw the right conclusions from a wide range of user messages that are not identical to the scenarios originally laid down in the system. It provides for the system administration, including the management of additional training.

This can be extremely useful, for example, when in the course of additional training, the system makes incorrect conclusions due to the inadequate behaviour of the audited staff. It is proposed to equip the system with a modified module for automated compliance analysis of documented information with the established criteria.

D. Report Export Algorithm

It is proposed to equip the system with an additional module for generating and exporting a report containing, according to the requirements of [22]:

- an audit purposes;
- an audit scope, in particular, information about the organization (auditee) and verified functions or processes;
- the information about the audit client;
- the list of participants from the audited organization;
- the name and version of the system;
- the full name of the system administrator(s), if applicable;
- dates and the electronic tools used for the audit activities;
- audit criteria;
- audit conclusions and related evidence;
- audit findings;
- an application for the degree of compliance with the audit criteria;
- any objections of the audited organization to the conclusions made by the system;

- a clarification that audits are inherently selective in nature; thus, there is a risk that the received audit certificates do not provide a complete and accurate picture;
- an audit plan, including schedule;
- a summary of the audit process, including any encountered obstacles that could reduce the reliability of the audit findings, including technical failures;
- the confirmation that the audit purposes were achieved within the audit area following the audit plan;
- everything that was not covered by the audit, but was covered by the audit scope, including all problems with obtaining certificates, resources, or confidentiality with appropriate justification;
- an agreed follow-up plan, if any;
- an application for the confidentiality of the contents;
- any implications for the audit program or subsequent audits.

16.4 Conclusion

Inclusive and sustainable growth is based not only on long-term strong economic growth but also on social integration and providing conditions and opportunities for decent work and working environment for all.

The authors proposed the intelligence system of automated auditing of management systems that can significantly reduce the costs and risks associated with face-to-face audits. These include the following:

- transport and other related expenses (e.g. accommodation, meals, etc.);
- payment in relation to non-productive in travel time;
- risks on-site of environmental protection, associated with the movement of auditors' risks in the on-site of occupational safety and professional security associated with the presence of auditors on the audited organization's site;
- epidemiological risks;
- risks of the transregional and transboundary restrictions related to political, epidemiological and similar events and phenomena;
- risks to impartiality (e.g. corruption risks, etc.).

The model of the intelligence system of automated auditing of management systems was implemented as an integrated software product. It is tested in the course of AEConformity Pty Ltd consulting projects implementation around the world.

The approaches and methods in this chapter were applied by the authors in the development of GOST R 59,424-2021 "Guidelines for remote production condition analysis and management systems auditing".

These approaches and methods are currently used to develop international standards ISO/IEC 23,894/"Information technology. Artificial intelligence. Risk management" and ISO/IEC 42,001 "Information technology. Artificial intelligence. Management system" at the international level.

References

1. Liker, J.K.D.: The Toyota Way: 14 Management Principles from the World's Greatest Manufacturer. McGraw-Hill (2004)
2. Deming, E.W.: Out of the Crisis: The New Paradigm of Managing People. MIT Press, Systems and Processes (1986)
3. Masaaki, I.: Gemba Kaizen: A Commonsense, Low-Cost Approach to Management. McGraw Hill Professional (1997)
4. Aleksakhina, Yu.V., Kuz'mina, T.I., Gorshkova, E.A.: Evolution of documented quality systems. Econ. Manag. XXI Century Dev. Trends **13**, 165–173 (2013)
5. Kachalov, R.M., Sleptsova, Y.A.: Dynamics of risk in the processes of innovative development of enterprises. Sci. J. VolSU. Global Econ. Syst. **4**(27), 72–86 (2014). https://doi.org/10.15688/jvolsu3.2014.4.8
6. Porter, M.: Competitive Strategy Techniques for Analyzing Industries and Competitors. Free Press, New York (2004)
7. Kalinina, A.E.: The development of the information space of the regional labor market and employment. Dissertation for the degree of doctor of economic sciences, specialty 08.00.05—Economy and Management of National Economy. VolSU Publishers, Volgograd (2005)
8. ISO 9001:1987: Quality systems—Model for quality assurance in design/development, production, installation, and servicing. https://www.iso.org/standard/16533.html. Accessed 22 June 2021
9. ISO 14001:2015: Environmental management systems—Requirements with guidance for use. ISO: International Organization for Standardization. https://www.iso.org/standard/60857.html. Accessed 22 June 2021
10. ISO 45001:2018: Occupational health and safety management systems—Requirements with guidance for use. https://www.iso.org/standard/63787.html/. Accessed 23 June 2021
11. IATF 16949:2016: Quality management systems requirements for automotive production and relevant service part organizations. IATF: International Automotive Task Force. https://ru.scribd.com/document/330831757/IATF16949-2016%E8%8B%B1%E6%96%87%E7%89%88/. Accessed 11 June 2021
12. ISO 21001: Educational organizations—Management systems for educational organizations—Requirements with guidance for use. https://www.iso.org/ru/standard/66266.html. Accessed 10 June 2021
13. ISO 50001:2011: Energy management systems—Requirements with guidance for use. https://www.iso.org/standard/51297.html. Accessed 22 June 2021
14. ISO/IEC 27001:2013: Information Technology—Security Techniques—Information security management systems—Requirements. ISO: International Organization for Standardization. https://www.iso.org/standard/54534.html. Accessed 29 June 2021
15. ISO/TS 29001:2010: Petroleum, petrochemical and natural gas industries—Sector-specific quality management systems—requirements for product and service supply organizations. ISO: International Organization for Standardization. https://www.iso.org/standard/55499.html. Accessed 22 June 2021
16. SAE. AS9100C: http://docplayer.net/12074923-Aerospace-standard-quality-management-systems-requirements-for-aviation-space-and-defense-organizations-rationale.html. Accessed 22 June 2021
17. STO Gazprom 9001–2012. Quality management systems. Requirements. VIII, 43. Moscow (2014)
18. GOST R 56853–2016. Conformity assessment. Requirements for authorities conducting audits and certification of management systems. Part 3. Competency requirements for auditing and certification of quality management systems. 12. Standartinform, Moscow (2016)
19. GOST R ISO/MEK 17021–2012. Conformity assessment. Requirements for bodies providing audit and certification of management systems. 44. Standartinform, Moscow (2013)

20. GOST RV 0015-002–2012. Systems for the development and production of military equipment. Quality management systems. General requirements. Introduction. 2012–06–05. IV, 38. Standartinform, Moscow (2012)
21. ISO 19011:2018. Guidelines for auditing management systems. ISO: International Organization for Standardization. https://www.iso.org/standard/70017.html (2018). Accessed 22 June 2021
22. ISO/IEC 17021:2011: Conformity assessment—Requirements for bodies providing audit and certification of management systems. ISO: International Organization for Standardization. https://www.iso.org/standard/56676.html. Accessed 22 June 2021
23. IAF MD 4:2018. IAF Mandatory Document for the use of Information and Communication Technology (ICT) for auditing/assessment purposes. IAF: International Accreditation Forum. https://www.iaf.nu/upFiles/IAF%20MD4%20Issue%202%2003072018.pdf (2018). Accessed 22 June 2021
24. Vladimircev, A.V., Dzedik, V.A., Zagorulko, Yu.V., Ryazanova, E.E.: Remote audits of management systems: prospects, risks, advantages. Stand. Qual. **9**(999), 60–63 (2020)
25. COVID-19: The impact on quality professionals. https://quality.eqms.co.uk/hubfs/Branding/Covid-19_Quality_Industry_Impact_FINAL_web%5B1%5D.pdf. Accessed 22 Aug 2021
26. Kobzar, A.I., Tikmenov, V.N.: Artificial intelligence: legends and myths, dreams and realities. Electron. Inf. Syst. **2**(2), 5–22 (2014)
27. Dzedik, V., Ezrakhovich, A.: Analysis of quality management systems with the use of machine learning methods. Qual. Access Success **19**(164), 40–42 (2018)

Chapter 17
Data Centers: Market Trends and Contribution to the World Economy Development During the COVID-19 Pandemic

Elena I. Inshakova and Roman M. Kachalov

Abstract The chapter investigates trends in the global data center market in the context of data centers as drivers of inclusive economic and social growth, enhancing business and social communication in the world economy, which has become distinct during the pandemic lockdowns. Regardless of challenges induced by the COVID-19 pandemic, the data center market is experiencing an acceleration in demand for data center services, cloud services, in the first place. The study identifies the following main trends in the global data center market: consolidation in the data center industry arising from M&As, that grow in number and increase in transaction values; a growth in hyper-scale data centers; formation by large-scale companies their global data center networks by transnationalization of activities to optimize their functions and reduce operating costs. The Russian data center market is classified as emerging, with its growth rates outstripping the global market in 2020. The contributors reveal several current trends in its development that, on the one hand, are consistent with global trends, but, on the other, have national specifics, involving the dominance of the capital cities in the regional structure of the data center market, with the predicted stability of this trend at least in the short term.

17.1 Introduction

Data centers, spaces that house computer systems used for storing, processing, analyzing, providing, and managing Big Data [1], emerged as a response to an impetus on behalf of businesses to reduce costs and, accordingly, enhance profits [2]. Data centers effectively manage the challenge by generating revenues for both data centers providers and consumers of their services. Data centers enable connectivity

E. I. Inshakova (✉)
Volgograd State University, Volgograd, Russia
e-mail: inshakovaei@volsu.ru

R. M. Kachalov
Central Economics and Mathematics Institute, Russian Academy of Sciences, Moscow, Russia
e-mail: kachalov1ya@yandex.ru

E. I. Inshakova and A. O. Inshakova (eds.), *New Technology for Inclusive and Sustainable Growth*, Smart Innovation, Systems and Technologies 287,
https://doi.org/10.1007/978-981-16-9804-0_17

between employees, partners, and customers, provide a platform for new business models, in which the efficient management of large amounts of data and the elasticity of computer technologies are crucial. They generate innovation in information technology selling their services to TNCs, national vertically integrated companies, national and transnational networks, who use their servers as platforms to develop and test advanced solutions to define the scale, stages, and levels of projects and evaluate the economic activity effectiveness.

The construction of data centers reflects the intensive digital transformation of the economy and society in the conditions of the emerging Industry 4.0. as the need for data storage and processing has rocketed in parallel with the amount of data consumed by individuals, businesses, and new technologies. Although the COVID-19 pandemic has severely affected the global economy, its impact on the use of data has been monumental. In the context of the pandemic, data centers have become vital in supporting connectivity in the global economy to meet the urgent needs of the state, business, and society. As a result, in many countries, data centers have been classified as objects of major infrastructure [3] and, consequently, enjoy a policy of mitigated quarantine restrictions.

Pandemic-related social restrictions, which have severely affected business activity in most sectors of the economy, have given impetus to enterprises and organizations to operate in the remote access mode. Companies have been encouraged to shift business online by using digital platforms B2B and B2C online stores, B2B marketplaces to organize interaction between manufacturers and suppliers, exporters and importers, wholesalers, and retailers. It enabled many economic subjects to survive despite a sharp reduction in the number of customers and a drop in demand and allowed consumers to gain access to necessary goods and services. The pandemic has accelerated data usage trends, and companies, especially large-scale ones, use data centers for cloud services that provide easy, affordable access to applications and resources, without the need for internal infrastructure or hardware. These opportunities offered by datacenters enable businesses to transfer digital business operations to the cloud or construct a hybrid IT infrastructure. Consequently, the desire of business entities to ensure business continuity in the challenging epidemiological conditions has become a vital factor in the development of the market for data centers services that level the decline in the IT industry as a whole.

In the context of the pandemic challenges, educational institutions, social infrastructure facilities, and state institutions have to operate in the remote access mode in compliance with pandemic-related social restrictions. The IT infrastructure in these organizations is under growing pressure, so they rely on cloud services provided by data centers as an effective way to enhance the availability and reliability of socially significant functions.

According to the International Data Corporation (IDC) estimates, the COVID-19 pandemic has induced accelerated investments in the digital transformation of organizations. Direct investments in digital transformation are expected to grow at a CAGR of 16.5% between 2022 and 2024, compared to 15.4% between 2019 and 2024, which will comprise more than half of all the ICT investments by the end of 2024 [3].

Thus, the dynamics in the market for data centers services are due to factors like growing demand for storing and processing significant amounts of data, companies' centralization of their IT resources, and the transition to cloud services. However, with the rising costs of construction sites, the number of data centers has reduced, which is confirmed by empiric data on the downtrend in the number of data centers in the world that has decreased from 8.55 million units in 2015 to 7.2 million in 2021 [4].

However, these data are not indicative of a weak data center market. Thus, Research and Markets experts predict the growth of the global Internet data center market by 13.4% annually, from USD 59.3 billion in 2020 up to USD 143.4 billion by 2027. High annual growth rates are projected in China (+17.5%), Canada (+11.8%), Germany (+10.6%) and Japan (+9.5%) [5]. GlobalData analysts expect the data center revenues growth at a compound annual growth rate (CAGR) of 6.7% between 2020 and 2030, from USD 466 billion to USD 948 billion [2].

Data centers have become the cornerstone of the information economy as they produce economic and social effects that ensure the inclusive development of the global economy and relieve the adverse impact of the COVID-19 pandemic on it. That is why the task of identifying current trends and determining the prospects for a market for data centers services in Russia has been embraced by researchers and practitioners of digital transformation of the economy.

17.2 Materials and Methods

The study utilizes conceptual foundations formed on theoretical assumptions that deal with the following issues: foreseeing the onset of the information society [6, 7], strategic management to ensure competitiveness in an era of global changes [8], the development of mechanisms (including institutional) for the transition to a new technological structure and the digital economy and the identification of the economic effects they produce [9, 10], determining the economic characteristics and architecture of modern data centers, the economic and innovative effects they generate [2, 11, 12].

Analytical reports and studies delivered by specialized foreign and national organizations such as International Data Corporation, GlobalData, Synergy Research Group, Data Center Dynamics, Mordor Intelligence, Research And Markets, Knight Frank, Statista, CNews, iKS-Consulting, etc. formed an empirical database for investigating the current dynamics of the global and Russian market for data centers services, identifying the main trends and prospects for their development, considering the impact of the COVID-19 pandemic on the world and national economy.

The research employs the methodology of the evolutionary approach using structural and functional, temporal and spatial, comparative and documentary analysis, along statistical and graphical techniques.

17.3 The Trends in the Global Data Center Market in the Context of Industry 4.0

17.3.1 Consolidation Process in Data Center Industry

A growing trend for consolidation and scale enlargement by mergers and acquisitions (M&As) has led to a decline in the number of data centers, and the data center total space increased. With a fewer number of data centers, their total area worldwide was estimated at 180 mln.m^2 (compared to 167 mln.m^2 in 2017 and 148.6 mln.m^2 in 2013) [13], and this trend is growing. These figures allow us to build an optimistic projection for the global data center market [2, 3, 5], noted above in this study.

Data centers, organizations that benefit from them, consumers of data center services determine the main factors that exert a complex effect on the processes in the world data center market (including M&As) that are as follows:

growing global demand for storing and processing of large volumes of data;
increasing automation level in the work of data centers to reduce the influence of the human factor, and modernization of equipment to significantly enhance the operation of networks;
the transition to the so-called third platform [14] that rests on the following constituents: mobile, big data/analytics, cloud computing, and social technologies, which has a direct impact on data center construction and renovation;
the focus on IT innovations that reduce the number of employees and save time to solve technical problems. For instance, technology developed by Facebook requires only one employee to maintain 25 thousand servers, and the new design of server racks allows engineers to replace hard drives in 2 min as compared to an hour, which is huge time savings in replacement time;
the decline in average server prices due to reduction in component costs as a result of economy and innovation;
the shift of the part of users to the public and hybrid cloud services (a model of joint utilization of private and public clouds) [15];
organization and logistic problems caused by lockdowns, for instance, a slowdown in the construction of new large American and European data centers by Google, Apple, and Facebook in 2020 due to supply disruptions and lack of necessary components (fiber optics, batteries, etc.) [16];
transformations in the enterprise behavior that boosted their demand for cloud services during the COVID-19 pandemic;
the need for significant consumption of electricity and other resources;
the impetus of enterprises to rent server capacities from large cloud service providers instead of buying servers;
the need to ensure conditions for effective operation by reducing operating costs;
implementation of corporate strategies to optimize and improve performance to strengthen competitiveness.

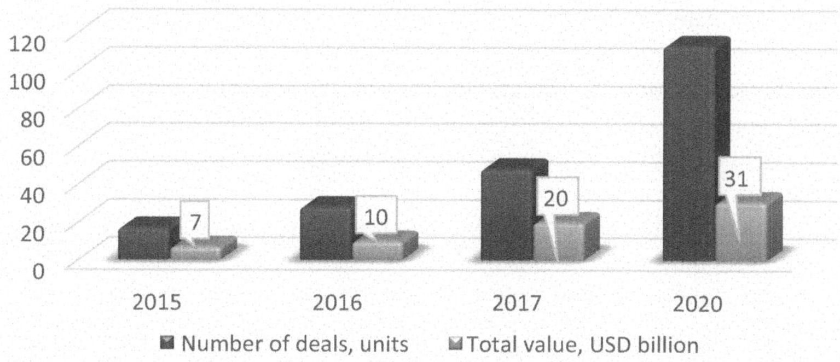

Fig. 17.1 Data centers M&A global activity. *Source* Compiled by the authors based on [17]

In 2020, following a slight slump in the data center market activity in 2018 and 2019 compared to 2017 [17], the data center industry reached a record figure in the number of M&A transactions (Fig. 17.1). Experts point out the emergence of special purpose acquisition companies (SPACs) created to acquire data centers [2].

In 2021, Blackstone Infrastructure Partners completed the acquisition of QTS Realty Trust. It was the industry-record deal (USD 10 billion) exceeding the figures of the previous year. In general, in the period between 2015 and 2020 in the data center industry, 483 mergers and acquisitions were completed for a total of USD 107 billion [18], with 31% of transactions performed by the world-leading colocation providers Digital Realty and Equinix, which have become the largest investors in this market.

Accelerated growth in M&A activity in the data center industry is consistent on the whole with the global trend to the increasing M&A activity in the technology sector in the context of emerging Industry 4.0. The logical result of the industry consolidation is the growing trend in the number of hyper-scale data centers (HDC) (Fig. 17.2), which has more than doubled since mid-2016.

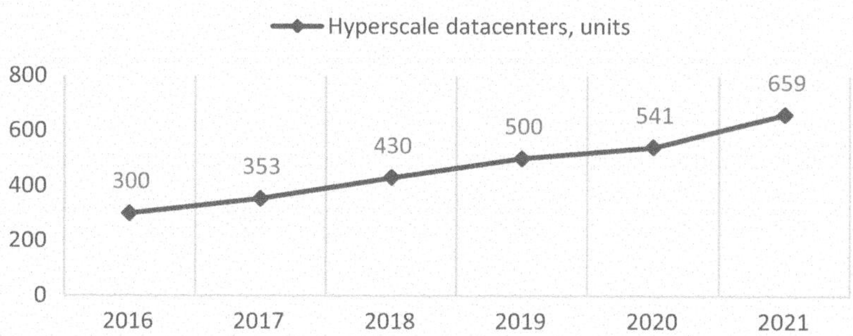

Fig. 17.2 The hyper-scale data centers dynamics, 2016–2021. *Source* Compiled by the authors based on [15, 16, 19]

HDCs can scale and develop massively since the costs of attracting another 1,000 or even 1 million users to their services or applications are insignificant. As HDCs provide unique opportunities to manage costs in purchase and supply transactions, large-scale companies with a turnover of billions of dollars can no longer function effectively in the current conditions without utilizing HDC for data connectivity and data computing.

By July 2020, despite the Covid-19 pandemic, as many as 26 HDC were established worldwide. The number of HDC shows sustainable growth, with the emergence of 16 new data centers on average every quarter over the last three years [19]. Such HDC operators as Amazon, Microsoft, and Google jointly account for over half of the data center space. The HDC providers such as Amazon Web Services, Microsoft Azure, and Google Cloud continue extending their global data center network [20]. According to expectations, the HDC market will be growing at a CAGR of about 3.4% per year between 2021 and 2026 [16]. On a global scale, the HDC market for the year 2020 has the following distribution: out of the total number of HDC, 39% is located in the United States, while China (9%) and Japan (6%) take second and third place, respectively [19].

HDC development is expected to be consistent with other megatrends of the global economy in terms of personnel and competencies, technology and materials, institutional and organizational features, implementation of appropriate modernization projects, and environmentally friendly innovative technologies [12].

17.3.2 Formation of Global Data Center Networks

Recently, there has been an accelerating trend on behalf of the largest IT companies and companies for which IT technologies are outside their core business toward the extension of their data center networks to various countries to optimize their functions and reduce operational costs (Table 17.1).

Data center networks provided by multinational companies Amazon, Microsoft, and IBM cover over 40 countries in at least two of the following [Synergy] mega-regions: North and South America, Europe, Middle East, Africa, and Asia–Pacific. Google and Oracle also have extensive global data centers. Data centers of other major players in the market are located mainly in the US (Facebook, Apple, Twitter, etc.) or China (Baidu, Tencent Holdings). The Chinese mega-company Alibaba is actively developing its global data center network, which has extended to the United States, Hong Kong, Singapore, Japan, and the United Arab Emirates.

TNC IBM (https://www.ibm.com/cloud/data-centers/) consistently extends its global data center network, which comprises more than sixty data centers in eighteen countries, including North and South America, Europe, Asia, and Australia. The network integrates the IBM data centers, data centers acquired from SoftLayer company, newly established centers in China, Hong Kong, Japan, India, Canada, Great Britain, US, Mexico, Spain, France, Italy, Germany, Switzerland, Norway, etc.). IBM plans to extend its data center network to the world's largest geographic and

Table 17.1 Global data center networks of the selected largest companies, 2020–2021

Company name	Headquarters home country	Number of data centers	Number of national markets served
Equinix	USA	202	24
Digital Realty	USA	214	14
China Telecom	China	456	> 10
NTT Communications	Japan	48	17
Telehouse/KDDI	UK/Japan	40	12
Coresite	USA	22	8
Verizone	USA	150	40
Cyxtera Technologies	USA	60	9
China Unicom	China	550	2
Amazon Web Services	USA	116	16
Google Cloud	USA	24	11
Microsoft Corporation	USA	16	8
IBM Cloud	USA	> 60	19
Facebook/Meta	USA	17	5

Source Compiled by the authors based on: [21; official websites of the largest IT-companies]

financial centers, including the Middle East and Africa. However, in 2020, intending to optimize and improve the network efficiency, the company stopped funding the construction of three data centers in Dallas, Houston, and Seattle, the US, and one data center in Melbourne, Australia.

TNC Google (Google Cloud, https://cloud.google.com/infrastructure) has established twenty-four data centers in eleven countries: fourteen located in the states of South Carolina, Iowa, Georgia, Oklahoma, North Carolina, Oregon, and others, one in Chile, South America and one in Canada, two are located in Taiwan and Singapore, Asia. Six data centers are established in Finland, Belgium, Ireland, Denmark, Netherlands, and Europe. Google Cloud will continue extending its network of data centers into another five new countries: Qatar, France, Italy, Spain, and Germany. The company intends to ensure its presence on the African continent by establishing a data center in this part of the world. As TNC embraces the policy of establishing carbon–neutral data center capacity, it is expected that the new data center will be powered by renewable energy.

Hewlett-Packard uses another way of establishing corporate data center networks (https://baxtel.com/data-centers/hewlett-packard). In 2006, the company announced the plan to consolidate eighty-five of its data centers into six huge mega-centers located in three US cities, Atlanta, Houston, and Austin, each housing two data centers. Based on the factors such as space availability, electricity supply, network capacity, and small likelihood of disaster occurrence, these cities were selected as the centers for the network. According to estimates, cost reduction was projected at 25% per year. At present, the current company's savings comprise $ 1 billion annually.

The largest social network Facebook, rebranded as social technology company Meta since October 28, 2021, has been extending its global data center network (https://datacenters.fb.com) since 2010. In 2010, Facebook created two HDC in the USA (in Oregon and North Carolina) and one data center in 2011 in Europe (Sweden). In addition, the company leases space at the data centers in Silicon Valley (Santa Clara, San Jose), San Francisco, and Virginia. In 2021, the global Facebook/Meta network comprises 17 data centers: 13 located in the USA, Ireland, Denmark, Sweden, and Singapore, each having one Facebook data center. Strategies for the construction and operation of large-scale data centers focus mainly on the imperatives of a green economy and an inclusive society: the use of renewable energy sources, the utilization of natural and climatic resources at the data centers location sites to provide natural cooling of machine rooms, using thermal energy released during the operation of data centers' equipment for heating buildings, etc. For instance, the data center in Finland, owned by the Russia-based company Yandex, provides heating and hot water in residential houses for twenty thousand residents of the host town Mäntsälä.

17.3.3 The Development of Data Center Market in Russia: Trends and Regional Prospects

In the Russian Federation, the data center market with its relatively small size can be classified as emerging. Like other world data center markets, it was affected by the pandemic, which inflicted expenses for personnel protection, required implementation of automation tools for remote monitoring of the systems, caused delays in the commissioning of new capacities, etc. However, compared to 2019, in 2020, it grew by 19%, and with the total capacity of data centers, it occupied the fifteenth position in the rank of the countries in Europe, Asia, and the Middle East [22]. At the same time, the pace of its development in 2020 was ahead of the global data market. The potential demand for the services of commercial data centers in the Russian Federation is estimated to increase due to the rapid digital transformation of the economy that involves an exponential growth in data volume, a sharp increase in demand for cloud services. In addition, the prospects for commercial data centers to provide services to enterprises and the public sector enable to enhance the data center market growth [23].

In 2020, the Russian data center market continued to grow (Table 17.2, Fig. 17.3). At the end of the year, the total capacity of the 30 largest commercial data centers increased by 12%. Rostelecom-DPC, ranked first in the country, exceeds the competitors by a significantly greater margins as the company accounts for almost a third of all server racks. In 2020, the company launched a new data center in St. Petersburg and data center Nord 5 in Moscow. The total revenue of Russian data center providers is growing, and profitability per server rack increased by 17% in 2020 compared to 2019 [24].

Table 17.2 Top 8 suppliers of the data center services in Russia, 2020

2020 Ranking	2019 Ranking	Company name	Number of racks	Number of data centers	Revenue growth 2019/2020, %
1	1	Rostelecom/DataLine	13,170	31	36.0
2	2	IXcellerate	3,415	4	108.6
3	6	Selectel	3,028	6	43.4
4	3	DataPro	2,800	2	67.2
5	4	MTS (DPC "Avantazh")	2,240	1	n/a
6	5	Linxdatacenter	2,020	2	5.9
7	–	Rosenergoatom/StoreData, Xelent	1,660	3	–
8	7	3data/CoreDataNet	1,453	15	−14.4

Source Compiled by the authors based on [24]

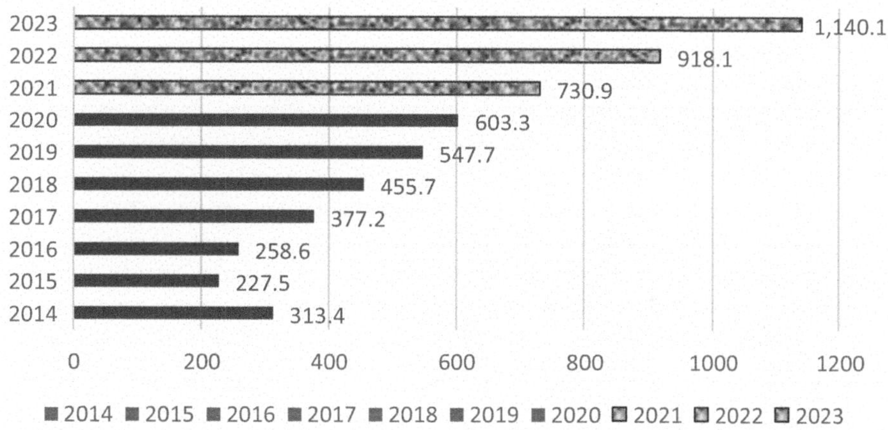

Fig. 17.3 Dynamics of revenue in commercial data centers of the Russian Federation. *Source* Compiled by the authors based on [25]. *Note*: 2020 revenue value – preliminary estimate; 2021–2023 revenue values—forecast

In 2020 and 2021, the main trends in the development of the Russian data center market can be described as follows:

(1) the emergence of new participants, companies for which information technology is outside the core business area (JSC Rosenergoatom, a Russia-based energy company, the operator of nuclear power plants; PNK Group, the industrial property developer in Russia; JSC Goznak, one of the world leaders of secured printing), establish their powerful data centers;
(2) the development of consolidation and scale enlargement processes through mergers and acquisitions. Thus, the emerging data center network developed

by Rosenergoatom (7th in the 2020 rating) comprises the greenfield data center Kalininsky located in the Tver Oblast, and the two data centers Xelent and StoreData data centers acquired in 2021. In April 2021, a consortium of investors (Russian Direct Investment Fund, ER-Telecom Holding, and based in the Middle East Talos Fund I LP) acquired 100% of Svyaz VSD LLC, a Russian branch of the Dutch company Linx data center (6th in the rating). The goal of the acquisition was the cloud platform LinxCloud and two data centers located in Moscow and St. Petersburg [26]. In early July 2021, MTS PJSC (5th in the rating), a leading Russian company providing digital, media, and telecommunications services, announced the acquisition of data center Green Bush DC, one of the largest in the Russian Federation;

(3) Moscow still dominates in the regional structure of the data center market in the Russian Federation (Fig. 17.4): the metropolitan region contains more than 73% of the total number of commercial data centers of the country, while the growth rate of Moscow data centers exceeds the growth of the market as a whole.

At the same time, other regions of the Russian Federation are implementing large projects for the construction and acquisition of data centers. So, several companies are actively building data centers in Innopolis in the Republic of Tatarstan. In 2020, the Tele2 company established 12 data centers in six regions of the Russian Federation (in Moscow, St. Petersburg, Rostov, Novosibirsk, Nizhny Novgorod, and Yekaterinburg). In addition, the above noted Rosenergoatom and ER-Telecom Holding have recently completed data center M&A deals in the St. Petersburg segment of the Russian data center market [25, 26]. However, Fig. 17.4 shows that the regional share in the data center services market can hardly be called significant. In addition, it saw a decrease between 2016 and 2020. Analysts predict this trend to continue, foreseeing that the main economic reason—a higher average profit per server rack and, accordingly, shorter payback periods for projects [25]—will maintain the investors' interest in the Moscow region, at least in the short term.

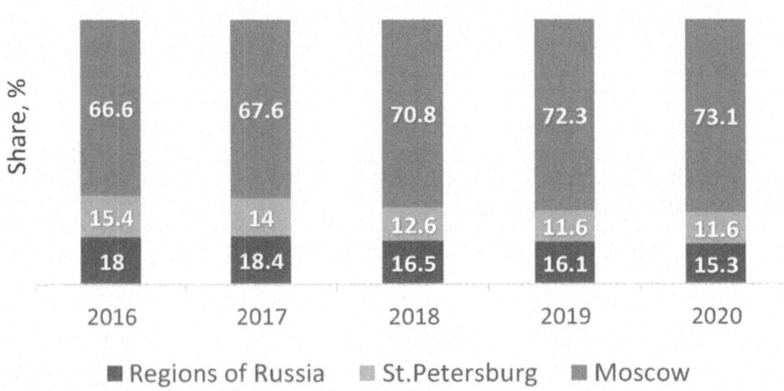

Fig. 17.4 Data center market in the Russian Federation: the regional structure according to the number of racks. *Source* Compiled by the authors based on [25]

17.4 Conclusion

Data centers are a new organizational form of the global economic system transformation following the megatrends of economic development [27], including the intensively developing digital transformation. Data centers that accumulate and store, process and analyze, provide, and manage Big Data are becoming a critically important factor in ensuring sustainable and comprehensive development of the economy and society in the context of Industry 4.0.

Data centers as drivers of inclusive economic and social growth, providing business and social communications in the global economy, have proven their utility during the pandemic lockdowns. Despite the heavy economic impact of the COVID-19 pandemic, the data center market continues to develop due to the explosive demand for data center services, primarily in the cloud segment.

The identified trends in the development of the global data center market are as follows:

- the consolidation processes in the data center industry that involve a rise in the number of M&As, and growth in the M&A transaction value in the period under study;
- an increase in the number of hyper-scale data centers, with the dominant share located in the United States;
- formation by large-scale companies their global data center networks by transnationalization of activities to optimize their functions and reduce operating costs.

The Russian data center market is defined as emerging, with its growth rates in 2020 outstripping the global ones. Among the main trends in the market development, the following factors are identified: the emergence of new participants—large companies for which information technologies are outside their core business; a rise in consolidation and scale enlargement processes through M&As; the dominance of the capital cities Moscow and St. Petersburg in the regional structure of the data center market, where a shorter payback period on the investment projects is ensured, with the predicted stability of this trend at least in the short term.

References

1. Clark, J.R.: The rise of the mega data center. Data Center J. **23**(3), 2–4 (2012)
2. Opiah, A.: Datacenter revenues to hit USD948bn by 2030. https://www.capacitymedia.com/articles/3828002/data-center-revenues-to-hit-948bn-by-2030-says-globaldata (2021). Accessed 15 Sept 2021
3. Carosella, G. Findling, S., Fitzgerald, S., Jimenez, D., Little, G., O'Brien, A., Parker, R., Pucciarelli, J., Yorifuji, Y.: IDC FutureScape: Worldwide digital transformation 2022 predictions (IDC #US47115521). IDC, Needham, MA (2021)
4. Number of datacenters worldwide in 2015, 2017, and 2021. https://www.statista.com/statistics/500458/worldwide-datacenter-and-it-sites/. Accessed 16 Sept 2021

5. Internet data centers—global market trajectory & analytics (2021). Report. Research and Markets; Global Industry Analysts, Inc., Dublin
6. Machlup, F.: The Production and Distribution of Knowledge in the United States. Princeton University Press, Princeton (1973)
7. Castells, M.: The Rise of the Network Society (Information Age, vol. 1). Blackwell Publishing, Oxford (1996)
8. Drucker, P.: Management Challenges for the 21st Century. Routledge, London (2007). https://doi.org/10.4324/9780080942384
9. Swaminathan, A., Wade, J.B.: Institutional environment. In: Augier, M., Teece, D. (eds.) The Palgrave Encyclopedia of Strategic Management. Palgrave Macmillan, London (2016). https://doi.org/10.1057/978-1-349-94848-2_608-1
10. Inshakov, O.V., Fesyun A.V.: Industry as a form of approval, expansion and domination of economic mode. Sci. J. Volgograd State Univ. Global Econ. Syst. **2**(25), 44–53 (2014)
11. Beaty, D.: Designing a mega data center. Data Center J. **23**(3), 14–18 (2012)
12. Johnson, B.: Is the landscape cool enough? DataCentre Magazine, 99–103. https://datacentremagazine.com/issuu?id=173747&startPage=96 (2021). Accessed 17 Nov 2021
13. Worldwide datacenter census and construction 2014–2018 forecast: aging enterprise datacenters and the accelerating service provider buildout (Doc #251830). IDC, Needham, MA (2014)
14. Explore the 3 Chapters of 3rd Platform Evolution. IDC. https://www.idc.com/promo/thirdplatform?tab=innovation#learnmore (2021). Accessed 20 Sept 2021
15. Hyperscale data center count reaches 541 in mid-2020. Synergy Research Group. https://www.srgresearch.com/articles/hyperscale-data-center-count-reaches-541-mid-2020-another-176-pipeline (2020). Accessed: 21 Sept 2021
16. Hyperscale datacenter market—growth, trends, COVID-19 impact, and forecasts (2021–2026). Mordor Intelligence, Hyderabad, India. https://www.mordorintelligence.com/industry-reports/hyperscale-datacenter-market (2020). Accessed 21 Sept 2021
17. 2020 Data center M&A deal value passes the USD30 billion milestone. Synergy Research Group. https://www.srgresearch.com/articles/2020-data-center-m-deal-value-passes-30-billion-milestone (2020). Accessed 21 Sept 2021
18. Scannell, E.: Data center market M&A deals hit new high in 2020. TechTarget network. https://searchdatacenter.techtarget.com/news/252494080/Data-center-market-MA-deals-hit-new-high-in-2020 (2020). Accessed 21 Sept 2021
19. Hyperscale Data Center Count Grows to 659—ByteDance joins the leading group. Synergy Research Group. https://www.prnewswire.com/news-releases/hyperscale-data-center-count-grows-to-659--bytedance-joins-the-leading-group-301376729.html (2020). Accessed 21 Sept 2021
20. Number of hyperscale data centers worldwide from 2015 to 2020. Statista. https://www.statista.com/statistics/633826/worldwide-hyperscale-data-center-numbers/ (2021). Accessed 21 Sept 2021
21. Top 11 data center companies in the world. Software Testing Help. https://www.softwaretestinghelp.com/data-center-companies/ (2021). Accessed 18 Oct 2021
22. The data centre report: EMEA & APAC, Q1 2021. Knight Frank, London, Singapore. https://app.dcbyte.com/knight-frank-data-centres-report/Q1-2021 (2021). Accessed 18 Oct 2021
23. Monitoring SMI. 02 avgusta 2021: Dajdzhest (Media monitoring. 02 August 2021: Digest). ANO Digital Technologies. https://files.data-economy.ru/Digest/2021-08-02-digest.pdf (2021). Accessed 15 Sept 2021
24. Obzor. Centry obrabotki dannyh 2021 (Review: Data Centers 2021). CNews Analytics. https://www.cnews.ru/reviews/rynok_tsod_2021/articles/rynok_kommercheskih_tsod_prodolzhaet (2021). Accessed 16 Sept 2021
25. Rossijskij rynok kommercheskih data-centrov 2020 (Russian market for commercial data centers 2020). iKS-Consulting. http://survey.iksconsulting.ru/page14877965.html (2020). Accessed 18 Sept 2021

26. RFPI i “JeR-Telekom” vykupili rossijskij biznes data-centrov Linxdatacenter (RDIF and ER-Telecom have bought out the Russian business of Linxdatacenter data centers). Kommersant. https://www.kommersant.ru/doc/4781263 (2021). Accessed 18 Sept 2021
27. Inshakov, O.V.: Collaboration as a form of knowledge-based economy organization. Econ. Region **3**, 38–45 (2013)

Part IV
Financial and Investment Technologies in Provision of Inclusive Growth of the Economy and Human Capital Quality

Chapter 18
Drivers and Stoppers of Inclusive Development of Financial Services and Products in Modern Russia

Leyla A. Markevich, Ekaterina A. Shkarupa, and Lyubov V. Grigoryeva

Abstract Currently, the services and products of leading financial and credit organizations demonstrate asymmetry in the context of the target audience, range of services, their cost, quality, and technology for companies and individuals. In such conditions, the formation of a systematic approach to identifying drivers and stoppers of inclusive development of the Russian market of financial services and products is highly relevant. The inclusive development of financial services and products refers to their production and provision with real access for all groups of the population and their consumption that is individually effective and socially useful. The empirical basis of the study was the main indicators characterizing the current state of inclusiveness of financial services and products in modern Russia. The authors identified the main problems of the production, provision, and consumption of financial services and products by Russians and determined the directions of their inclusive development by applying the six-factor meta-production function of O. V. Inshakov (Inshakov in Economics of Contemporary Russia **1**, 11–25, 2003).

18.1 Introduction

The inclusiveness of the population in the consumption of financial services and products is considered as their "real accessibility" to different socio-economic groups of the population: people with different incomes, different places of residence in the context of city-village and region-financial center, different gender, age, health status, level of financial and digital literacy, etc. [1–4]. "Real access" implies convenient and

L. A. Markevich (✉) · E. A. Shkarupa · L. V. Grigoryeva
Volgograd State University, Volgograd, Russia
e-mail: mytarevala@volsu.ru

E. A. Shkarupa
e-mail: shkarupaea@volsu.ru

L. V. Grigoryeva
e-mail: grigorievalv@volsu.ru

E. I. Inshakova and A. O. Inshakova (eds.), *New Technology for Inclusive and Sustainable Growth*, Smart Innovation, Systems and Technologies 287,
https://doi.org/10.1007/978-981-16-9804-0_18

responsible provision of services at a price that is affordable for the consumer and economically justified for the supplier, as a result of which consumers who were not previously covered by the financial services sector turn to official financial services, and not to the available informal alternatives.

The opposite phenomenon of financial inclusion is financial isolation and exclusion of some people and their groups from financial relations, inequality, and unavailability of financial services and products consumption.

The basic set of financial services according to which financial inclusion is defined in world practice today, is established by the Global Partnership for Financial Accessibility of the Group of 20 (G-20), and includes insurance, lending, savings, and payment services for both people and small businesses [5].

The emphasis in the Global Partnership for Financial Accessibility of the G20 is real access based on convenient and responsible provision of them at a reasonable cost by official financial institutions within the framework of their conscientious behavior. In this aspect, it is important: the activities of official financial service providers are regulated by the relevant supervisory authority to protect the rights and interests of their consumers; the official suppliers themselves conduct their activities responsibly, that is, transparently, reduce the consumer risks of their customers, take into account the solvency of customers, are focused on the convenience of consuming their services; consumers of financial services should have a sufficient level of financial literacy for effective consumption of services.

In our opinion, the inclusive development of financial services and products presupposes their production, rendering and provision in the future, in which all strata and groups of the population have real access to those whose consumption is individually effective and socially useful.

18.2 Methodology

The theoretical basis of the study was the work of foreign and domestic researchers in the field of theory and practice of financial inclusion [1, 2, 4–8], as well as works on the problems of accessibility of financial products and the peculiarities of financial behavior of the population [9]. The contributors applied O. V. Inshakov's meta-production function arguments [10] to systematize the factors of inclusive development of financial services and products.

The information and empirical base of the study was made up of official statistical materials of the Federal State Statistics Service of the Russian Federation and the Central Bank of the Russian Federation, analytical, review developments of researchers published in periodicals and the Internet.

The present research is based on the study and generalization of theoretical and factual materials on the problems of the development of financial services and products in modern Russia using various methods: systematic, comparative, as well as tabular and graphical methods of data processing and presentation.

18.3 Results

18.3.1 Accessibility is not Necessary for All Financial Services and Products

The study of financial inclusion in relation to the population should take into account that not all financial products and services are equally useful for mass and individual consumption, which means that financial inclusion by default should not extend to all existing financial services and products. That is, when assessing the availability of a financial product or service, it is necessary to understand that in one case it is good that they are available to all groups of consumers, and in another case mass availability is individually and (or) socially dangerous. According to the authors, financial accessibility should be considered and ensured in alignment with their socially and individually useful types.

Financial products and services consumed by the population can be divided into groups according to the criterion of purpose:

(1) savings and investment (they are an investment of available funds in order to preserve and increase them, including by obtaining additional income and (or) increasing their value. In Russia, this group includes bank deposits, savings certificates, foreign currency and precious metals, individual investment accounts, insurance policies, securities, products of collective forms of investment—mutual funds, non-state pension funds, consumer credit cooperatives and citizens' unions, individual trust management, speculation in financial markets);

(2) credit and loan products and services (carried out to finance the current monetary expenses of the borrower at the expense of borrowed funds. In fact, we are talking about financing current consumption at the expense of future income. In Russia, these include loans received from banks and microfinance organizations, interpersonal loans).

The usefulness and danger of mass availability of a financial service/product, in our opinion, is determined by the potentiality of their effective consumption. Financial products and services can be consumed efficiently and inefficiently.

Effective consumption, in our opinion, does not significantly worsen the current and significantly increases the future level and quality of life of the consumer and results in an increase in the level of his material well-being. In conditions of uncertainty, the efficiency of consumption of a financial product before the start of its consumption is considered as potential. Only with the course of time, retrospectively, it is possible to assess with confidence whether the consumption of specific financial products or services was effective or ineffective.

The potential efficiency of consumption of savings and investment financial products and services in conditions of uncertainty and Russian specifics is determined by their type: some products are initially ineffective (in Russian realities, these are bank deposits, most insurance policies, investments in non-state pension funds, investments in bonds), while others are potentially effective (individual investment accounts, individual trust management, investments in stocks, ETF funds, businesses and startups).

The potential efficiency of consumption of credit-borrowed financial products and services is determined by the type of loan and its burden on the budget of an individual or household. Obviously, inefficient consumption is inherent in expensive loans issued by microfinance organizations. Loans for business creation, mortgages, and educational loans have the highest efficiency, that is, those that are essentially investments in the future and not financing current consumption.

It is logical to assume that among any financial products and services, over time, appear "useful" and "dangerous" products and services to increase or decrease the material well-being, level, and quality of life of their consumers. The availability of "dangerous" financial services and products should be limited, and "useful" should be absolute.

Thus, achieving full financial inclusion for all financial services and products consumed by the population is impractical. It is more reasonable to ensure full accessibility for socially and individually useful financial services and products and limited accessibility for socially and individually dangerous ones.

18.3.2 Inclusiveness of Financial Services and Products in the World and Modern Russia

Conducting a cross-country analysis of individual indicators of inclusiveness of financial products and services allows identifying their typical and specific stoppers and drivers.

According to the latest surveys conducted by the World Bank in 2017 and published in the report "The Global Findex Database 2017. Measuring Financial Inclusion and the Fintech Revolution" [11] (Table 18.1), it can be seen that certain indicators of financial inclusion in a number of countries have similar values.

In particular, in the EAEU space, the Kyrgyz Republic has the highest value of the indicator characterizing the physical unavailability of financial institutions for the population due to territorial remoteness (17%). In Russia, Armenia, the Republic of Belarus, and Kazakhstan, a part of the population (6–9%) is also unable to use financial services due to the territorial remoteness of the place of residence from financial infrastructure facilities.

The high cost of financial products and services also reduces their accessibility for the population of the EAEU countries. In Armenia and the Kyrgyz Republic, almost 1/5 of the adult population notes this problem of financial inclusion. In Russia, 14% of citizens also note that the high cost of financial products and services does not allow them to be used. It is noteworthy that in China (not a member of the EAEU), this indicator is quite low—4%, and it is equal to the value of the indicator for the Republic of Belarus.

Table 18.1 Selected indicators of inclusiveness of financial services and products in the world (2017), %

Country	World	Russia	Armenia	Republic of Belarus	Kazakhstan	Kyrgyz Republic	China	Germany	France	Japan	USA
The proportion of the adult population that does not have an account with a financial and credit institution due to											
Territorial remoteness of a financial and credit institution	–	6	9	6	6	17	6	–	–	–	–
High cost of financial services	–	14	17	4	9	18	4	–	–	–	–
Lack of trust in financial institutions	–	14	13	6	8	19	2	–	–	–	–
Underfunding	–	21	37	13	18	40	17	–	–	–	–
Proportion of the adult population forming savings as a source of passive income for emergencies	35	19	10	35	12	3	38	53	63	87	62

(continued)

Table 18.1 (continued)

Country	World	Russia	Armenia	Republic of Belarus	Kazakhstan	Kyrgyz Republic	China	Germany	France	Japan	USA
Proportion of the adult population that used borrowed funds in the previous year	47	41	55	49	46	32	45	64	52	57	77
Proportion of the adult population paying for online purchases on the Internet	73	58	–	58	58	–	85	–	–	–	–
Proportion of the adult population who made electronic payments	45	62	32	69	38	27	61	97	90	89	89

Source Compiled by the authors based on [6]

A special problem of financial inclusion in the EEC countries is poverty. Almost 2/5 of the population of Armenia and the Kyrgyz Republic indicates the lack of funds as a key aspect of the inability to use financial products and services. In Russia and Kazakhstan, 1/5 of the adult population faces a similar problem.

The formation of savings as a source of income in emergency situations in the EEC countries is not developed. In particular, in the EAEU countries, the share of the adult population forming savings as a source of passive income for emergencies does not exceed 35% (the Republic of Belarus), and its minimum value is 3% (the Kyrgyz Republic). In contrast to the EAEU, in Germany, France, Japan, and the USA, this figure is 53%, 63%, 87%, and 62%, respectively.

In this regard, the analysis of indicators on the use of modern digital technologies in the financial sector, providing public access to financial products and services, is of particular relevance. The global indicator of the use of electronic payments by the population is 45%. There is an ambiguous picture in the EEC countries. In particular, in Russia and the Republic of Belarus, more than 60% of the adult population use electronic payments, which is higher than the global value. The Kyrgyz Republic has the lowest value of this indicator in the EAEU space—27%.

It should be noted that according to the World Bank, 97% of the adult population of Germany makes electronic payments. In France, Japan, and the USA, this indicator is also quite high and amounts to 90%, 89%, and 89%, respectively.

Analyzing individual indicators of inclusiveness of financial services and products in the world, we will deepen the analysis with data on modern Russia (Table 18.2). The data presented in the table demonstrate the ambiguous dynamics of financial inclusion indicators.

Against the background of a reduction in the number of financial and credit institutions, their concentration occurs in large cities and regional centers, which reduces the physical accessibility of financial services and products to the population, which was previously confirmed by World Bank data. However, this is smoothed over by the development of digital platforms and ecosystems of key players in the market of remote provision of financial services to the population.

The annual cost of financial services and products produced is also quite high for the population and can reach an average of 44% of the average per capita income of Russians. And the profitability of the most popular service, deposits, is rapidly declining. Therefore, about 85% of Russians annually refuse financial and credit services due to their high cost/low profitability.

However, against the background of the presented negative facts, there is an increase in the level of financial knowledge and activation of financial behavior. In particular, the low yield on deposits has reoriented the population to search for and choose new financial services, (for example, Individual Investment Account) resulting in a rapid increase in the volume of funds placed on these accounts (379% in 2020 compared to 2018).

Table 18.2 Selected indicators of inclusiveness of financial services and products in Russia in 2018–2020

Name of indicator	2018	2019	2020
Infrastructure of providers of financial services and products			
Number of active CO/MFOs/Insurance Business Entities/Professional securities market participants, units	484/2,002/275/297	442/1,774/255/263	406/1,385/232/260
Production of financial products and services			
Annual cost of a voluntary home insurance policy against standard risks, rub	n/a	6592.3	6748.8
Annual cost of the CASCO policy, rub	n/a	52,717.3	51,326.0
Annual cost of the CMTPL policy, rub	n/a	n/a	9,437.7
Payment for the use of consumer credit (interest rate in value terms), rub	n/a	15,356.5	14,835.9
Average weighted rate on deposits/loans (as of December of the corresponding year), %	5.62/17.87	4.66/14.83	3.38/13.41
Standard index of affordability of financial services (from 0 to 7, where 7–100% of the population refused to use all 4 financial services due to their high cost/low profitability)	n/a	6	6
Consumption of financial products and services			
Proportion of the adult population using at least one open account of an individual in the CO, taking into account deposit accounts, %	87.46	87.46	93.69
Number of accounts opened to individuals for which, since the beginning of the reporting year, operations have been carried out to write off funds, access to which is provided remotely, million units	236.2	254.5	282.9

(continued)

Table 18.2 (continued)

Name of indicator	2018	2019	2020
Proportion of the adult population with deposits in the CO and (or) funds placed in the NFO (MFO, CPC or SKPC) in the form of a loan agreement and (or) investment life insurance and/or in the IIA and (or) brokerage accounts and (or) in mutual funds/including CO/NFO, %	35.41/34.65/2.78	36.41/34.65/2.78	47.8/41.82/6.55
Percentage of the population using voluntary insurance, %	n/a	22.76	33.53
Per capita income of the population, thousand rubles	33.2	35.3	35.7
Share of expenses for savings in the structure of all private consumption, %	1.7	3.4	4.2
Volume of deposits of individuals in credit institutions, trn. rub	28.5	30.6	32.8
Volume of savings certificates issued by CO, thousand units	151.3	32.9	12.1
Volume of loans granted to individuals by CO, trn. rub	14.8	17.6	20.0
Amount of funds placed in the IIA, billion rubles	99	197,3	375,6
Financial Literacy Index	53	n/a	54
– Financial knowledge – Financial behavior – Financial setups	53 52 53	n/a n/a	54 54 52
Index of penetration of Fintech services	n/a	82%	n/a

Where CO—credit organizations, that is, commercial banks; NFO—non-credit financial organizations; MFOs—microfinance organizations; CPC—consumer credit cooperatives; SKPC—agricultural credit consumer cooperatives; CASCO—voluntary insurance of motor vehicles; CMTPL—compulsory motor third party liability; mutual fund—mutual investment fund; IIA—individual investment account.
Source Compiled by the authors based on [12–14]

18.3.3 Stoppers, Drivers, and Factors of Inclusive Development of Financial Services and Products in Modern Russia

Against the background of declining real incomes of Russians, rising unemployment, rapid development of technologies for remote provision of financial services, still low level of financial literacy, a surge in fraud in the financial sector, especially in digital format, gaps in the statutory regulation of consumer protection of financial services, the provision and consumption of financial services has a number of problems, including:

(a) high competition between official and alternative (unofficial) financial service providers in the microloan markets;
(b) the emergence of a number of unofficial specific market participants (collectors, anti-collectors, financial consultants (advisers), credit brokers, crowdfunding internet sites) and a number of financial services and products (electronic wallets and private electronic money, including cryptocurrencies) insufficiently regulated by the state;
(c) a high entry threshold for the consumption of truly profitable financial assets; etc.

Inclusive development of financial services and products in modern Russia, taking into account the announced problems and the current level of inclusiveness, should mainly involve:

(1) formalization of the status of domestic informal financial and credit institutions and their services and products, including financial and tax consultants, collectors, anti-collectors, credit brokers, electronic wallets, private moneylenders;
(2) overcoming the extractive nature of the provision of financial services;
(3) combating financial fraud and improving the security of financial transactions in the digital environment;
(4) the fusion of financial and digital literacy of Russians (now these are two different directions, little overlapping with each other);
(5) formation of an effective and efficient system of protection of the rights of financial services consumers;
(6) formation of a financial market infrastructure adequate to the structure of society, including the development of transparent technologies of over-the-counter financial markets and investment platforms for crowdfunding financing of startups whose activities will be supervised by the state directly or through supervision of self-regulatory organizations of such platforms and their participants;
(7) development of social partnership between business and the state as a basis for attracting Russians to voluntary forms of medical and pension insurance; etc.

Based on the current state of the financial market and its inherent problems, the stoppers of its inclusive development in modern Russia include: undeveloped forms of collective investment; undeveloped financial market; non-preferential taxation of investment income [7]; unsettled protection of the rights of financial services consumers, especially in terms of remotely consumed; lack of informative platforms about the state of the economy, the Russian joint-stock companies available to the general public; developed financial and digital fraud.

It should also be recognized that in Russia, the services and products of leading financial and credit organizations (banks, insurance companies, non-state pension funds, management companies, cooperatives, brokers, exchanges, etc.) are extractive, that is, they are focused on extracting resources from a few from many, their provision is biased, the playing field is uneven (different and selective) for people with different characteristics (income levels, place of residence, type of employment, age, health status), as well as for people and organizations (there are obvious distortions in the range of services, their cost, quality, maintainability for organizations and individuals).

The drivers, in our opinion, are high digital literacy of Russians, and its possible convergence with financial literacy; the need of all groups of economic entities (state, population, business, financial, and credit intermediaries) for investments for effective social protection of the population [3]; high level of Fintech development in Russia and remote financial services; and activation of financial self-awareness of Russians.

Based on the approach proposed by [10], we adapted [8] the meta-production function arguments proposed by him to highlight the factors of inclusive development of financial services and products in modern Russia to solve the problems announced above in the areas we identified earlier (Table 18.3). The authors have identified three main problems of production, provision, and consumption of financial services and products (a, b, c) and seven directions of inclusive development of financial services and products in modern Russia.

The presented factor matrix clearly illustrates the development pathways of financial inclusion in the context of existing stoppers and drivers of financial inclusion in Russia.

18.4 Conclusion

In the Expert Report "Global Standard-setting Bodies and Accessibility of Financial Services for the Low-income Population: Development of Proportionate Standards and Recommendations", issued in 2011 by CGAP on behalf of the G20 Global Partnership for expanding access to financial services (GPFI), the main direction of ensuring financial inclusion in the country is the joint development of proportionate standards and recommendations by five global regulatory authorities [11]. Today, five global regulatory authorities are represented by the Basel Committee on Banking Supervision (BCBS), the Committee on Payment and Settlement Systems (CPSS), the Group for the Development of Financial Measures to Combat Money Laundering (FATF), the International Association of Deposit Insurers (ISDA), and the International Association of Insurance Supervisors (IAIS). It is assumed that these organizations should jointly address issues of standardization of the activities of national financial and credit institutions supervised by them in order to formalize the statuses, including alternative providers of financial services; protect the rights of financial services consumers; remote provision of financial services with an emphasis

Table 18.3 Factor matrix of inclusive development of financial services and products in modern Russia (author's approach)

Factors	Factor description	Problem being resolved	Focus area
Humanitarian	Expanding the target audience: the low-income population, the elderly, people with limited capabilities or disabilities, certain religious communities, leveling geographical factors (coverage of remote, sparsely populated or hard-to-reach areas)	(c)	(2) (3) (4) (5) (6) (7)
Resource	Improving the quality and range of financial products and services, including through the creation of mechanisms for involving long-term cheap money held in the PFR and NPF in the investment process	(a) (c)	(1) (2) (3) (4) (5)
Technical	Development of digital business models, infrastructure platforms, improvement of elements of the Marketplace system, use of Internet and mobile access capabilities	(a) (b) (c)	(3) (4) (5)
Institutional	Changing norms and rules that activate and restrain the production, rendering and provision of individually and socially effective and useful financial services and products	(a) (b)	(1) (6)
Organizational	Expansion of institutions providing financial products and services, access points to financial services, adaptation of financial services distribution channels to serve customers with special needs, improving financial literacy in the context of using technologies to ensure financial accessibility, overcoming financial behavioral distortions	(a) (b) (c)	(1) (2) (3) (4) (5) (6)
Informational	Formation of a knowledge base about the main financial products and services, risks, rights of consumers of financial services and ways to protect them	(a) (b) (c)	(3) (4)

Where the PFR is the Pension Fund of Russia; NPFs are non-state pension funds
Source Compiled by the authors

on electronic money and agency models and similar innovations. The national mega-regulators of the financial and credit sphere, in turn, should gradually introduce these standards into banking, insurance, financial supervision, and regulation.

Russia needs to overcome the stoppers of inclusive development of finance in the following areas: formalization of the status of unofficial financial service providers, namely: collectors, anti-collectors, credit brokers, private moneylenders, crowdfunding Internet platforms, electronic wallets, independent financial consultants; protection of the rights and interests of financial services consumers, especially consumers of credit products of microfinance organizations, pawnshops; combating financial fraud; improving financial literacy of elderly consumers of financial services.

These areas correlate with the already existing factors of financial inclusion development, including the rapid development of remote forms of provision of financial services; the high level of digital literacy of Russians; the increasing financial literacy of the adult population; the increasing financial activity of Russians in the field of socially and individually useful types of financial services and products.

References

1. Dar, A.B., Ahmed, F.: Financial inclusion determinants and impediments in India: insights from the global financial inclusion index. J. Financ. Econ. Policy **13**(3), 391–408 (2021). https://doi.org/10.1108/JFEP-11-2019-0227
2. Figart, D.M.: Institutionalist policies for financial inclusion. J. Econ. Issues **47**(4), 873–894 (2013). https://doi.org/10.2753/JEI0021-3624470404
3. Lu, W., Niu, G., Zhou, Y.: Individualism and financial inclusion. J. Econ. Behav. Organ. **183**, 268–288 (2021). https://doi.org/10.1016/j.jebo.2021.01.008
4. Sarma, M., Pais, J.: Financial inclusion and development. J. Int. Dev. **23**, 613–628 (2011). https://doi.org/10.1002/jid.16981
5. Yarasheva, A.V.: Financial behavior of the population. In: Monitoring. ISEPN FNSC RAS, Moscow (2020)
6. Demirguc-Kunt, A., Klapper, L., Singer, D., Ansar, S., Hess, J.: Global Findex Database 2017: Measuring Financial Inclusion and the Fintech Revolution. World Bank, Washington, DC (2018). https://openknowledge.worldbank.org/handle/10986/29510. License: CC BY 3.0 IGO. Last accessed 26 June 2021
7. Gamze, O.-Y.: Financial inclusion and tax revenue. Cent. Bank Rev. **19**(3), 107–113 (2019). https://doi.org/10.1016/j.cbrev.2019.08.004
8. Shkarupa, E.A. Development of financial inclusion of agricultural producers. Reg. Economy. South Russ. **9**(3), 162–170 (2021). https://doi.org/10.15688/re.volsu.2021.3.15
9. Grigoryeva, L.V., Gorshkova, N.V., Mytareva, L.A.: Economic mechanisms of population protection against pension risks as a tool for socio-economic development of the region. J. Volgogr. State Univer. Ser. 3 Econ. Ecol. **1**(38), 31–44 (2017)
10. Inshakov, O.: Development nucleus in the light of the new factors of production theory. Econ. Contemp. Russ. **1**, 11–25 (2003)

11. Global standard-setting bodies and the availability of financial services for the low-income population. Development of proportionate standards and recommendations. In: Global Partnership for Expanding Access to Financial Services: Expert report prepared by CGAP on behalf of the G20 Global Partnership for Expanding Access to Financial Services (2011). https://www.cgap.org/sites/default/files/CGAP-White-Paper-Global-Standard-Setting-Bodies-Oct-2011-Russian.pdf. Last accessed: 17 Aug 2021
12. Financial Accessibility, the Central Bank of the Russian Federation (2021). https://www.cbr.ru/develop/development_affor/dic/. Last accessed:17 Aug 2021
13. Index of Penetration of Fintech Services 2019 Ernst & Young Global Limited (2019). https://assets.ey.com/content/dam/ey-sites/ey-com/ru_ru/news/2019/11/ey-fai-2019-rus.pdf. Last accessed: 17 Aug 2021
14. Standard of Living, Federal State Statistics Service of the Russian Federation (2021). https://rosstat.gov.ru/folder/13397. Last accessed 17 Aug 2021

Chapter 19
The Impact of Digital Technologies of Tax Administration on Increasing the Inclusiveness and Sustainability of Economic Development

Natalia V. Gorshkova, Viktoria M. Ksenda, and Irina V. Grigorenko

Abstract The modern world is at the next, innovative stage of its development, connected with mainstreaming of IT technologies and digitization of processes that have affected all spheres of human activity. Financial and economic relations are no exception. The digital economy paradigm is embedded in the financial system, providing convenience, time savings, and technology security, which increases not only the availability of financial services but also is a stimulating tool for ensuring and increasing the profitability of public finance. Contactless technologies have become an integral part of the tax administration system, which is designed to neutralize the negative perception of the tax process by the population and business entities, to make it open, accessible, which makes it possible to improve tax collection, increase inclusiveness and sustainability of economic development. The chapter formulates a modern concept of tax administration, critically analyzes existing electronic services and tax administration programs, presents their classification, and identifies shortcomings and prospects for optimizing this process.

19.1 Introduction

Taxes are one of the main financial levers of state regulation of economic development. The amount of financial support for the implementation of the tasks of the state depends on the efficiency of the tax administration system, which is complicated by the fact that the payers of fiscal payments tend to underestimate the tax base or conceal the object of taxation, and, accordingly, the amount of taxes, since for them

N. V. Gorshkova · V. M. Ksenda (✉) · I. V. Grigorenko
Volgograd State University, Volgograd, Russia
e-mail: KsendaVM@volsu.ru

N. V. Gorshkova
e-mail: GorshkovaNV@volsu.ru

I. V. Grigorenko
e-mail: GrigorenkoIV@volsu.ru

E. I. Inshakova and A. O. Inshakova (eds.), *New Technology for Inclusive and Sustainable Growth*, Smart Innovation, Systems and Technologies 287,
https://doi.org/10.1007/978-981-16-9804-0_19

it is a mandatory payment that is alienated from their own income. Therefore, one of the problems in tax administration is the evasion of mandatory payments and the use of illegal taxation schemes, which is a negative factor for the tax and budget system, and, accordingly, economic development. Thus, the main task of the state is to reach consensus in relations with taxpayers, while the digitization of the tax administration system will make the taxation process total, transparent, and inevitable.

19.2 Methodology

The world experience of digitization of certain areas of the tax sphere is considered in the works of foreign and Russian authors [1, 2, 3, 4]. The issues of disclosing the content and essence of tax administration, determining the problems and prospects of Russian tax administration are reflected in the results of scientific research by [5, 6, 7, 8, 9]. The technologies of tax administration are considered in the works of [10, 11, 12].

The data of the Organization for Economic Cooperation and Development [13], foreign tax administrations, as well as the Accounts Chamber of the Russian Federation [14], the Ministry of Finance of the Russian Federation [15], and the Federal Tax Service [16] were used in the paper.

When performing scientific research, general scientific methods of inquiry were used observation, analysis, induction, deduction, comparison, analogy, generalization; and private scientific methods the method of expert assessments, statistical analysis.

19.3 Results

The dominant technological trend of the XXI century—the Fourth Industrial Revolution, based on robotization (automation) and digitization of processes, has embraced the advanced world community, including Russia. The Russian economy is undergoing a digital transformation. The development of the digital economy assumes that "digitized data is a key factor of production in all spheres of socio-economic activity. This makes it possible to create conditions for economic growth, improve the quality of life of citizens, increase the competitiveness of the country and strengthen national sovereignty" [17].

The development of the digital economy is connected not only with the formation of key institutions and infrastructure, but also practically implemented, in particular, in the financial sector. The tax system was no exception.

Taxes are the main source of revenue for the budget system—the foundation for ensuring financial security and the viability of the state. However, the burden of paying them lies with taxpayers, some of whom consider taxes and fees to be excessive additional expenses and avoid paying them. These circumstances lead to the emergence of a hidden ("shadow") economy with the absence of accounting for taxpayers, objects of taxation, the tax base, and, accordingly, the emergence of tax arrears and shortfall in budget revenues (Fig. 19.1).

To eliminate and control such facts, a tax administration system is provided, understood by the authors of this study as a tax process management system aimed at implementing the tasks and achieving the goals of state tax policy [7].

The Federal Tax Service (FTS of Russia) is a federal executive authority that performs the functions of monitoring and supervising compliance with the legislation on taxes and fees, the correctness of calculation, completeness, and timeliness of taxes, fees, and insurance contributions, etc. to the relevant budget [18]. In addition, the service is the chief administrator (collector) of federal budget revenues. So, in 2020, it accounted for about 60% of the collection of total federal budget revenues (Table 19.1).

However, the functional responsibilities of the Federal Tax Service of Russia have become much broader and more voluminous. Now it is not only a controlling authority but also a service that provides a high level of public services for business organization and tax payment, which contributes to improving the efficiency of tax administration and inclusive development of Russia. This became possible due to the introduction and development of contactless information resources and programs allowing to systematize huge data streams.

Fig. 19.1 Indicator (Debt-to-income)—debt-to-income ratio, %. *Source* Compiled by the authors based on [16]

Table 19.1 Federal budget revenue receipts by federal budget revenue administrators, %

Name of the administrator of federal budget revenues	2016	2017	2018	2019	2020
Total	100.0	100.0	100.0	100.0	100.0
Federal Customs Service of Russia	32.7	30.3	31.2	32.0	25.4
Federal Tax Service of Russia	**51.5**	**60.7**	**61.3**	**62.5**	**58.6**
Ministry of Finance of Russia	–	1.8	1.7	1.9	8.1
Rosimushchestvo	7.1	1.8	1.7	2.3	2.4
Other federal agencies	8.7	5.3	4.1	1.3	5.5

Source Compiled by the authors based on the data of the Audit Chamber of the Russian Federation [14]

It is worth noting that the process of automating the activities of the Federal Tax Service of Russia began long before the digital economy became a priority for the development of the Russian economy. The stage of deep digitization started about 10 years ago, when the modern automated information system of the Federal Tax Service "Tax" (AIS "Tax") and Personal accounts of taxpayers started working. In our opinion, the beginning of automation, the so-called "zero stage", should be associated with the appearance in the 2000s of a pilot project on submitting tax reports via the Internet through specialized operators via telecommunication channels, which was successfully implemented and received its legal regulation with the appearance of the Order of the Ministry of Internal Affairs of Russia (at that time the Ministry of Taxes and Duties) on the procedure for submitting tax returns in electronic form. This expanded the possibilities of taxpayers to submit reports, in addition to postal items and personal presence in the tax authority.

In recent years, the following focus areas have been developed:

- technological (improvement of the VAT-2 Automated Control System, transition to Online cash registers, labeling of fur products and medicines, modernization of the Tax-3 AIS, introduction of the "My Tax" mobile application in connection with the introduction of the Self-employment tax (SET), introduction of the national goods traceability system, development of electronic document management, expansion of interaction between credit organizations and tax authorities);
- non-technological:

(1) managerial and organizational (expansion of interdepartmental interaction, international cooperation, application of a risk-based approach in planning, criteria for tax monitoring);
(2) legislative (submission of VAT tax reports only in electronic form, consolidation of the concept and status of "Taxpayer's Personal Account" in the Tax Code, automatic provision of tax benefits (without a taxpayer's application), reduction of the terms of desk tax audit, cancellation of tax returns on land and transport taxes).

These measures allowed the tax authorities to achieve better results in 2020 compared to 2018 [16]:

- additional charges for one on-site audit increased (over 30%);
- the number of on-site tax audits decreased (by more than 2 times);
- the number of court decisions on disputes that have passed pre-trial settlement of tax disputes has decreased (by more than 1.5 times);
- the number of documents submitted for state registration in electronic form has increased (more than 3 times).

Currently, the official website of the Federal Tax Service of Russia (www.nalog.ru) is a powerful web resource that hosts 66 electronic services: personal accounts; COVID-19; business registration; information about the individual taxpayer number

Table 19.2 Indicators of the functioning of the services of the Federal Tax Service of Russia in 2020

Service	Value
My taxpayer account as an individual entrepreneur (IE)	2.6 mln IE
My taxpayer account as legal entity	1.1 mln companies
My taxpayer account for individuals	34.4 mln persons
Information about the TIN of an individual	1.164 billion requests
Information about legal entities and individual entrepreneurs in respect of which documents for state registration are submitted	513 mln requests
Check yourself and the counterparty	2.828 billion requests

Source the Federal Tax Service of Russia [16]

(TIN); payment of taxes and duties; business risks; information from state registers; tax calculators; tax accounting; international taxation; background information; feedback/assistance (Table 19.2). In addition, 86 open data sets, 12 state registries, 5 mobile applications are presented there [16].

Referring to the research addressed to information support of tax administration processes, it should be noted their descriptive nature or analysis of advantages and disadvantages [5, 7, 8, 9].

We propose to classify software products in this way:

(1) for the work of tax authorities (AIS "Tax-3", ASK "VAT-2", ASK "KKT", PC "Region", "Svod-2000", "Local level EDI system");
(2) for accounting management and interaction with taxpayers:

- by companies ("My taxpayer account as legal entity", "My taxpayer account as a foreign company", "Business registration", "Online cash registers");
- by individual entrepreneurs ("My taxpayer account as an individual entrepreneur", "Business registration", "Online cash registers");
- by individuals ("My taxpayer account for individuals", "My taxpayer account for self-employment taxpayer", "Business registration").

The functioning of such resources is based on technologies: BigData (AIS "Tax-3"), Blockchain (ASK "VAT-2"), and Api ("My tax"). Their use makes it possible to efficiently extract, process, store voluminous data, build chains of relationships, and program [10, 11].

To assess the effectiveness of automation and the introduction of digital resources, let us analyze the quantitative indicators of the use of software products and the level of tax revenues.

Until 2020, federal budget revenues demonstrate a steady growth trend (Table 19.3).

According to the estimates of the Ministry of Finance of the Russian Federation [15], the main reasons for the decline in revenues in 2020 were, first of all, restrictive measures caused by the spread of coronavirus infection, as well as lower prices

Table 19.3 Actual execution of federal budget revenues in 2016–2020

Indicator	2016	2017	2018	2019	2020
Income, % of GDP	15.7	16.4	18.7	18.5	17.6
Oil and gas revenues, % of GDP	5.7	6.5	8.7	7.3	4.9
Non-oil and gas revenues, % of GDP	10.1	9.9	10.0	11.2	12.7

Source Ministry of Finance of the Russian Federation [15]

for Urals crude oil, which affected oil and gas revenues. However, non-oil and gas revenues demonstrate positive dynamics, which is associated with the indexation of excise tax rates, and the rise in business activity of businesses, which affected the increase in the amount of value added tax.

Starting in 2019, the structure of federal budget revenues began to change, which is also associated with changes in oil prices and, as noted in the Ministry of Finance of the Russian Federation [15], with an increase in the quality of tax administration aimed at improving the efficiency of administration of non-oil and gas revenues. Thus, compared to 2018, in 2020, the share of non-oil and gas revenues in the total volume of federal budget revenues increased by 18% and amounted to 72%. Accordingly, the share of oil and gas revenues decreased from 36% in 2016 to 28% in 2020. The factors that influenced such changes include prices for hydrocarbon raw materials; the exchange rate of the US dollar against the ruble; the completion of the "tax maneuver" in the oil industry; the introduction of a tax regime on additional income from the extraction of hydrocarbons; the expansion of benefits for hydrocarbon production and a decrease in oil production and exports as a result of the OPEC + agreement.

There have also been changes in the structure of oil and gas revenues of the federal budget. By 2020, the share of the mineral extraction tax has increased, and the share of export customs duties has decreased.

In the structure of non-oil and gas revenues of the federal budget, more than half is accounted for by value added tax, the share of which shows positive dynamics from year to year (about 58% in 2020). And, despite the force majeure circumstances that arose in 2020, the restoration of business activity by the end of 2020 allowed VAT receipts not to be reduced. Corporate income tax receipts also showed positive dynamics until 2020 [15].

It is hard if not impossible to evaluate the functionality and effectiveness of the use of such software products as the Automated Information System "Tax-3" (AIS "Tax-3") and the Automated control System "VAT-2" ASC "VAT-2": their users are employees of tax authorities, and the data posted and stored there belong to tax secrecy and are intended for official use. At the same time, analyzing regulatory documents and scientific publications [6, 11], we should note that AIS "Tax-3" is a unified information repository of taxpayer data, as well as a resource for collecting, processing information, and compiling reports. Despite the fact that the current version is more advanced than all the previous ones, there are still some shortcomings of functioning in the form of errors in the operation of the software and duplication of information in other resources. The main scientific and Innovative Implementation Center (GNIVC) is the developer of AIS "Tax-3".

ASK "VAT-2" is also an internal software product used by the Federal Tax Service of Russia. Its essence lies in the control of illegal VAT deductions along the chain: at the stage of product creation and to the final consumer by identifying multi-risk ("green", "yellow", "red") taxpayers. If any have been found, the tax authority denies the taxpayer a tax deduction for VAT. The main drawback of this program is that the result of the data analysis that is carried out cannot be used when considering tax disputes, only exclusively for the purpose of conducting tax audits. In addition, the collection and analysis of documents are labor-intensive and are already carried out directly by employees of tax authorities, which reduces the effects of this software product. However, the VAT-2 Automated Control System has proven itself quite effective. So, at present, the share of doubtful VAT deductions does not exceed 0.4% [16], and as we have already noted, the share of tax revenues has increased.

Taxpayers' personal accounts are services that have proven themselves extremely effective. The most popular and functional service is one for individuals, in which about 30 million people are registered. In addition, it is also possible to use this service through the public services portal (a reference and information Internet portal that provides individuals and legal entities with access to information about state and municipal services in the Russian Federation). Personal accounts for legal entities and individuals are less in demand and functional. Such taxpayers, as a rule, use other software products integrated with accounting. In addition, the main disadvantage of these services is the impossibility of electronic filing of tax reports, which is very important for individual entrepreneurs working without employees or with a small number of them, since accounting is minimal and does not require the use of special accounting and tax accounting programs.

One of the advanced technologies and innovative solutions in the field of taxation is the introduction of a self-employment tax and the application "My Tax". This application is intended for "self-employed" citizens, for those who have logged in through their Personal Account on the official website of the Federal Tax Service of Russia or downloaded the application. The usual tax registration is not required; the very fact of authorization confirms that an individual or an individual entrepreneur who does not attract employees has expressed a desire to carry out a certain type of activity and therefore pay taxes. The application, like the taxation system itself, has no analogues and is in demand by taxpayers. The results of its implementation since 2019 are presented in Table 19.4.

These data indicate the demand for this tax regime, which attracts, first of all, by the simplicity of application: the absence of a separate registration as a taxpayer, free access to the mobile application "My Tax". It is only necessary to register and deposit the received income, or in case of non-cash receipt, the system will recognize them itself and the tax will be calculated automatically. In addition, various categories of citizens are involved in the taxation process, which solves several main tasks: ensuring tax revenues, recruitment and employment problems, and social adaptation of the population. Thus, it can be said that the development of tax inclusivity is

Table 19.4 Results of the application of the self-employment tax

Factor	Value
Number of self-employed	5 million people (including 25,000 minors)
The amount of income received	327 billion rubles
Amount of taxes paid	8 billion rubles
Age of the self-employed	30–40 years
Employment indicators (gender indicator)	60%—men, 40%—women
Main activities	Consulting, construction and marketing services, renting of apartments, taxi
Types of activities carried out (age indicator): up to 30 years from 40 years	Marketing and advertising, delivery services; consulting and tutoring, renting of dwellings
Types of activities carried out (gender indicator): women men	Hairdresser, cosmetologist, manicure master, babysitter, speech therapist; car repair and maintenance master, taxi driver, loader, electrician, plumber

Source Compiled by the authors based the Federal Tax Service of Russia [16]

in evidence, which assumes the simplicity of participation in tax relations and the availability of fulfillment of tax obligations, which expands the possibilities of each person to do what he knows without unnecessary difficulties and at the same time pay taxes without hindrance.

According to OECD experts, "the Federal Tax Service of Russia has a high degree of digital development of administration" [1]. The Russian Tax Service is a member of the OECD group on digital transformation, where the best world practices are studied, and the global paradigm of digital transformation of tax administration is being formed. In addition to Russia, this group includes Australia, Great Britain, Denmark, Spain, Canada, Kenya, Finland, New Zealand, Norway, and Singapore.

Thus, in the OECD document "Tax Administration 3.0: Digital Transformation of Tax Administration" [13], the tax monitoring system introduced in Russia has become an example of world practices. This is a reliable, secure authentication, and authorization of taxpayers to provide the tax authority with remote access to the taxpayer's accounting and tax reporting systems. This system can be called "soft" remote tax control. In 2019, the number of participants in tax monitoring increased 6 times to 44, in 2020, 52 more major taxpayers were added. Together with the current participants, about 30% of federal budget revenues are provided.

Along with Russia, there were the following examples of world practices:

- Singapore, where the National Digital Identity System (NDI) has been created, which provides all residents with a digital identity card for individuals, a corporate pass, a corporate digital identity card for businesses and other organizations. SingPass provides authenticated access to government electronic services, ViaCorp pass can be used by companies to interact with customers, securely make transactions with government agencies and other organizations;

- Norway has developed a new loan application scheme: instead of receiving information from applicants, banks can now receive the necessary information digitally from the tax service;
- Kenya has expanded the range of payment channels to include mobile money payments via M-Pesa (Digital Tax Payments). Kenya is a world leader in mobile money. As of December 2019, there were 58.4 million mobile money subscribers in the country (with a population of 47.5 million inhabitants). Digital tax payments allow taxpayers to pay taxes quickly and conveniently using their mobile phones. After entering the uniquely generated payment registration number and phone number into the KRA payment application, the taxpayer receives a message on his mobile phone asking him to approve the tax payment;
- Australia has launched the Single Touch payroll, which calculates wages, taxes, and generates reports to the Australian Taxation Office in real time;
- Spain has developed a Virtual VAT Assistant. This is an artificial intelligence system (chatbot) that provides information about the registration and correction of invoices, obligations, tax benefits, and tax rates. The system provides 24/7 support, which reduces the administrative burden and increases the mobility of information;
- Finland has replaced more than 70 old outdated systems with new COTS software, an information system that generates taxpayer data and dossiers [13].

The OECD Report "Tax Administration 2021: Comparative Information on the OECD and other countries with developed and Developing economies" [13] provides comparative data on aspects of tax administration in 59 countries of the world, including the effectiveness of tax administrations in collecting taxes; the processes of registration and identification of taxpayers, processing and desk tax control of declarations; digital services that facilitate the smooth execution of tax obligations, etc.

The world's leading examples in the development of tax administration technologies are Australia (Digital Partnership Office), Canada (implementation of a platform combining digital IT services, API management, interaction capabilities, automation, and cloud technologies), and the Netherlands (Development of reliable online ecosystems, paperless exchange). The tax administrations of these countries are developing new technological conditions for servicing taxpayers with the introduction of third-party services, which will speed up the process of calculating and paying taxes. In addition, one of the transformative innovations of the global tax administration is the use of chat-bots (electronic assistants) and mobile applications. The countries that have implemented these innovations include Australia (Improvements to the online digital assistant—Alex), Canada (Live agent chatbot), Costa Rica (TRAVI chatbot and online chat review), Peru (Virtual assistant—SOFIA), Russia (Virtual assistant (chatbot)—TAXIK), Singapore (Filing chatbot), Spain (Virtual assistance—Personal income tax and personal information), and Great Britain (Webchat and other digital services). Mobile applications are relevant for use by taxpayers in Brazil (Mobile app tax and customs "Normas") and Russia (online service solution "My tax"). The number of states whose tax administrations use application programming interfaces (API—application programming interfaces) is growing: Israel (Zero VAT on hotel accommodation services), Norway (modernization of the Norwegian VAT system), Russia (Tax Monitoring), Singapore (Collaboration with software developers), and the United Kingdom (Making tax digital). However, there are countries that have long integrated API technologies into their services and are engaged in their in-depth implementation, creating entire centers: Australia (The ATO Strategic direction for APIs) and Canada (API Centre of Excellence) [13].

The system of tax administration of foreign countries is not identical because it depends on the development of the state, the construction of the budget and tax system, and public administration. Digital transformation has comprised many developing countries. In countries such as Singapore, Chile, and South Africa, there are already informational web pages that provide taxpayers with information about their obligations, answer many questions from taxpayers, and ensure the submission of documents. Brazil has managed to expand the coverage of electronic registration of all its taxpayers [1, 2, 3, 4]. As the authors note [4], the Russian system of tax administration, in comparison with foreign ones, has a repressive character and is aimed at subsequent control. However, one cannot agree with this, since preventive mechanisms and methods have been introduced in the tax administration system, and tax sanctions are the result of non-compliance with the requirements of laws.

The analysis allows us to conclude that the modern national system of tax administration, functioning in a digital environment, forms a solid financial basis for achieving national goals and priorities of state development (Fig. 19.2).

Digital technologies in tax administration expand the functionality of tax authorities in the control of dishonest taxpayers and provide an affordable way for the executors of tax obligations to implement them.

The development of the digital environment of interaction between the subjects of tax relations—tax authorities and taxpayers—encourages not only to improve the digital tax infrastructure but also to increase the technical capabilities and increase the tax literacy of the executors of tax liabilities. Scaling of these conditions will contribute to the active tax behavior of taxpayers and effective tax administration.

It is supposed that it will be possible to ensure tax collection and promote the competitiveness of Russian business through the development of a national system of traceability of goods, customs monitoring, and the launch of a new tax regime "USN.online", the introduction of the institute of "single tax payment", improving the quality of tax administration of excise taxes on tobacco, modernization of resource rent taxation [15].

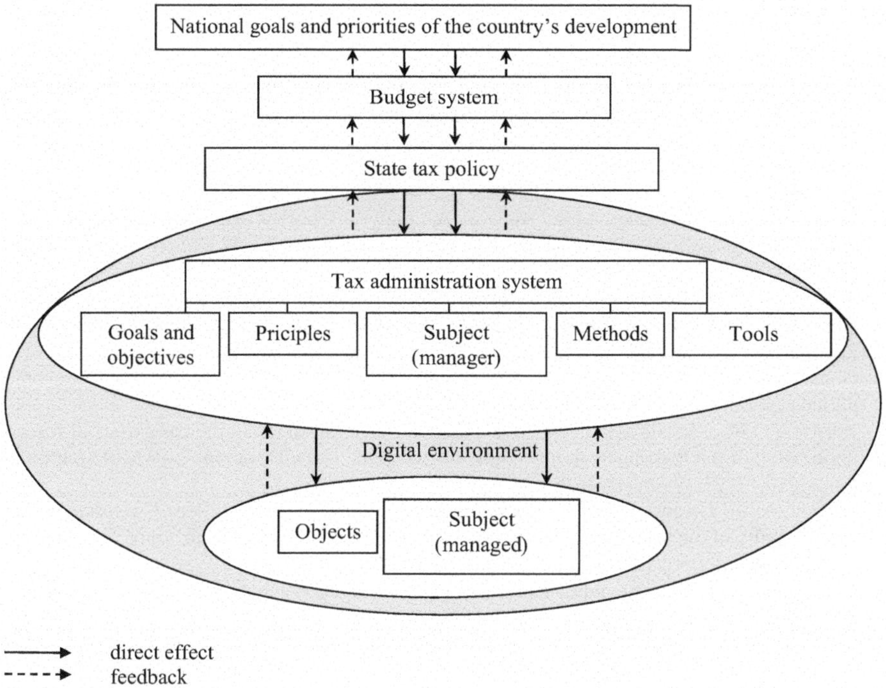

Fig. 19.2 Model of the tax administration system. *Source* Developed and compiled by the authors

19.4 Conclusion

The new era of the information society and advanced technologies has affected not only the sphere of production, financial services and products, but also tax relations. The Russian tax administration system has become invisible and convenient for taxpayers, functioning in real time, which increases the performance indicators of tax authorities, reduces the risks of questionable transactions, and expands the circle of taxpayers. Now the functions of tax authorities are centered not only on control but also on the provision of information and consulting services; a favorable atmosphere is created for the fulfillment of tax obligations, which increases tax collection and reduces the level of tax offenses. The analysis made it possible to classify the software and service products of the Federal Tax Service of Russia, evaluate the results of their use and propose a modern model of the tax administration system. The advantage of this model is the effective implementation of budget, tax, and customs tariff policy, based on the principles of transparency of the interaction of subjects of tax relations, the functional trend chasing of the fulfillment of tax obligations, the quality of services provided in the conditions of digitization and aimed at increasing inclusiveness and sustainability of socio-economic development of the country.

References

1. Slemrod, J. Sexing up tax administration. J. Tax Adm. **1**(1), (2015). http://jota.website/index.php/JoTA/article/view/7/0. Last accessed 20 Sept 2021
2. Bird, R.M.: Fiscal Decentralization and Decentralizing Tax Administration: Different Questions, Different Answers (2015). Rotman School of Management Working Paper No. 2694651. https://doi.org/10.2139/ssrn.2694651. Last accessed 23 Nov 2015
3. Ekimova, K.V., Gordienko, M.S., Aliev, A.A.: Ogly Tax policy of foreign countries in the context of digitization of the economy. Bus. Educ. Law **4**(19), 47–52 (2019). https://doi.org/10.25683/VOLBI.2019.49.434
4. Totikova, T.E. Modernization of tax administration systems in foreign countries. Proc. Saratov Univ. New Ser. Ser. Econ. Manag. Law **16**(4), 445–448 (2016). https://doi.org/10.18500/1994-2540-2016-16-4-445-448
5. Goncharenko, L.I., Malkova, Y.V., Advokatova, A.S.: Actual problems of the tax system in the digital economy. Econ. Taxes Law **2**, 166–172 (2018)
6. Karpova, O.M., Maiburov I.A. Transformation of value added tax in the conditions of forced digitization of the Russian economy. Bull. Tomsk. State Univ. Econ. **46**, 7–19 (2019). https://doi.org/10.17223/19988648/46/1
7. Ksenda, V.M., Grigorenko, I.V., Shkarupa, E.A., Turgaeva, A.A.: (2018) Compliance and serviceability of the tax administration system. Bull. Astrakhan State Tech. Univ. Ser. Econ. **4**, 86–93. https://doi.org/10.24143/2073-5537-2018-4-86-93
8. Molchanov, E.G.: Innovative tax administration tools in the context of the digital economy in the Russian Federation program. Account. Stat. **3**(55), 55–65 (2019)
9. Fedotova, G.V., Gontar, A.A., Ilyasov R.H., Ksenda, V.M. The strategy of provision of tax security of the state in the conditions of information economy. In: Optimization of the Taxation System: Preconditions, Tendencies, and Perspectives. Switzerland, pp. 217–228 (2019). https://doi.org/10.1007/978-3-030-01514-5_25
10. Lyutova, O.I., Fialkovskaya, I.D.: Application of blockchain technology in tax administration. RUDN J. Law **25**(3), 693–710 (2021). https://doi.org/10.22363/2313-2337-2021-25-3-693-710
11. Mandroshchenko, O.V., Bogachev, S.V.: Tax administration in the digital economy: development prospects. Econ. Syst. **14**(52), 152–158 (2021). https://doi.org/10.29030/2309-2076-2021-14-1-152-158
12. Rodygina, V.E.: Digitalization of VAT tax control: legal consequences and prospects. Financ. Law **9**, 31–34 (2018)
13. The Organization for Economic Cooperation and Development (OECD). https://www.oecd.org (2021) Accessed: 04 Oct 2021
14. Audit Chamber of the Russian Federation. https://ach.gov.ru/. Last accessed 04 Oct 2021
15. Ministry of Finance of the Russian Federation (2021). https://minfin.gov.ru/ru/. Last accessed 04 Oct 2021
16. Federal Tax Service (FTS of Russia) (2021). https://www.nalog.ru/rn34/. Last accessed 04 Oct 2021
17. The Digital Economy of the Russian Federation Program (2021). Decree of the Government of the Russian Federation. Last accessed 28 July 2017. (No. 1632-r)
18. On Approval of the Regulations on the Federal Tax Service: Decree of the Government of the Russian Federation. Last accessed 30 Sept 2004. (No. 506)

Chapter 20
The Internet as a Special Information Space for Attracting and Implementing Investments

Denis E. Matytsin

Abstract The study considers the modern information space as a set of electronic websites for the turnover of investment objects. This set of platforms is available to an unlimited number of investors and recipients of investments, for these persons to make remote investment transactions and make a profit. Within the framework of a systematic approach, the concepts of the investment Internet space and the turnover of investment objects in the Internet space are formulated. The classification of investment websites by the level of qualification of investors by recipients of investments has been carried out. The investment Internet space is continuously and intensively used by the individuals and legal entities. The parties interact on websites to make remote investment transactions. There are no restrictions on the free will of a person, who intends to enter the specified space as an investor or a recipient of investments. This freedom has no legal restrictions that can be implemented effectively and with complete certainty. The activity in the remote transactions by millions of individuals—the unqualified investors in modern Russia—is growing at an accelerated pace. This requires immediate preventive measures at the federal legislative level.

20.1 Introduction

Encyclopedic volumes reveal the term Internet as an international (worldwide) computer network of electronic communication, uniting regional, national, local, and other networks [1]. This concept is widely discussed in both Russian and foreign scientific literature. For example, S.V. Petrovsky proposed the following definition: "An international public telecommunication network intended for the exchange of machine-readable messages (data), that is, information about the surrounding world, its objects, processes and phenomena, objectified in a form that allows for their direct

D. E. Matytsin (✉)
Volgograd State University, Volgograd, Russia
e-mail: Dmatytsin@ya.ru; gimchp@volsu.ru

Volga Branch of the International Law Institute, Volzhsky, Russia

E. I. Inshakova and A. O. Inshakova (eds.), *New Technology for Inclusive and Sustainable Growth*, Smart Innovation, Systems and Technologies 287,
https://doi.org/10.1007/978-981-16-9804-0_20

machine processing" [2]. Researcher Maksurov A.A. sees the Internet as a kind of human communicative activity, because of which all its main features fully belong to him [3].

The legal definition of the term Internet is not contained in the regulations. In particular, we do not find it in Federal Law No. 149-FZ of July 27, 2006 "On Information, Information Technologies and Information Protection" [4]. However, some features of the Internet as a legal phenomenon can be identified from regulatory legal acts. This is a special system based on communication technology; the main purpose of the Internet is to receive and transmit information in various forms and in large volumes; you can use the Internet only through computer devices [5]. Here we disagree with N.V. Rachmanina, who apparently means computers, since at present it is possible to use the Internet quite effectively through mobile telephone communication devices (smartphones). At the same time, mobile telephone communication allows a person to connect with one or another addressee on the Internet via radio waves, and computers provide the same via wires (cables) [6]. Over the past 5–8 years, technologies that allow computers to connect to the Internet without wires have been developing quite intensively. Such options provide for the "transformation" of a computer into a mobile phone when a so-called modem is used, to which a standard telephone SIM card is connected [7].

For the purpose of our research, the investment Internet space should be understood as a system of interconnected electronic machines—servers and other devices that store, process, receive, and transmit information through wires and radio waves in the form of special computer codes. Data and commands are transmitted at the will and in the interests of investors in contact with each other, investment intermediaries, and recipients of investments, whose circle is not limited. This option of remote interaction is very rational and in demand: there are no restrictions in distance, in time of day; the data exchange rate is very high. During the last 5–8 years, a huge leap has occurred due to the development of computer technology and software, which currently allows huge amounts of information to be transmitted at great speed. As a result, it is now possible to receive images and sound of several persons on a computer monitor at once and conduct a live dialogue with them regardless of the geographical location of the persons in contact with each other. It should be noted that the equipment itself, electronic devices, servers, tangible, and intangible assets that make up the Internet belong to specific individuals, who for the most part are not public (state) entities. The main tools of this development process are digital platforms, with the help of which the transition from the order of interaction "one–to-one", "one-to-many" to the system "many -to-many" is built. We agree that in modern Russia, the change in the technologies of interaction of economic entities in combination with the created consulting and technical business centers provides real progress toward an innovative economy [8].

In this chapter of the monograph, as part of the regulatory framework, Federal Law No.149-FZ of 27.07.2006 "On Information, Information Technologies and Information Protection"; Federal Law No.325-FZ of 21.11.2011 "On Organized Auctions" were studied. The study of doctrinal sources covers the scientific works of a number of Russian and foreign scientists, among these authors: Inshakova A.O., Goncharov

A.I., Ershova I.V., Petrovsky, S.V., Maksurov, A.A., Rachmanina N.V., Budarina N.A., Akkaev A.A., Kurasova S.E., Afanasyeva E.V., Amelin R.V., Khabrieva T.Ya., Chernogor N.N., Fadeeva A.Yu., Tamm V., and Tapscott D. At the same time, at the scientific and methodological level, the issues of using the Internet as a special information space for attracting and making investments have not received much attention from researchers until recently.

20.2 Methodology

The development of the content of this chapter is based on the materialistic worldview and the universal scientific method of historical materialism. General scientific methods of cognition are applied: dialectical, hypothetical-deductive method, generalization, induction and deduction, analysis and synthesis, and empirical description. The research also uses private scientific methods: dogmatic, comparative-legal, hermeneutic, structural–functional, etc.

20.3 Results

As noted above, it can be assumed that the legislator leaves us the opportunity to independently reveal the concept of the Internet. Moreover, to do this based on the definition that a special system of technical devices integrated into an information and telecommunications network functions to transmit information through communication channels, which is accessed using computers [1]. This network functions as a global conglomerate of websites, each of which has its own unique domain name and address. Domain names and (or) network addresses allow you to identify sites on the Internet. At the same time, the legislator interprets the website as a complex of computer programs and information accessed via the Internet [4]. Users receive and transmit information to each other, precisely focusing on these sites. In our opinion, the first brick—the key link of the entire web of the Internet is precisely the website belonging to a specific person (persons). A website with its unique domain name and its highly specialized subject combines a certain number of necessary files and related archival resources. These resources contain relevant information that is accessible via the Internet. The World Wide Web erases the boundaries of national jurisdictions and time constraints, providing the widest possible space for finding both recipients of investments and investors anywhere in the world. The flexibility and initiative of this method significantly reduces the costs of contacting persons and increases the effectiveness of communications.

Many researchers also claim that the website is the key tool of Internet communications. At the same time, some researchers insist that face-to-face meetings between partners are much more effective than communication via the Internet, since the perception of partners is much better during face-to-face meetings [9]. However, as noted above, modern video technologies used when making transactions on the Internet almost completely neutralize the disadvantages of the absence of meetings of individuals at an agreed time in the same room. In our opinion, the adaptation of Internet users to the remote method of video communication is happening very rapidly, the widespread transition to the preparation, execution, and implementation of transactions on the Internet will definitely happen before the end of the 2020s. The necessity and expediency of meeting individuals in the same room for transactions as well as for interaction during the execution of such deals will decrease to zero over the next 5–8 years.

Modern society quite consciously increases the role and expands the functionality of the information space in public relations from year to year [10]. It is natural that the investment sphere is also moving leaps and bounds into the information space. It should be clarified that direct investments involving investments in tangible assets of specific property complexes of business entities within the framework of the implementation of relevant development projects still require the personal presence and active participation of investors, and this form of investment is still being successfully implemented. At the same time, transactions with certain objects are massively reproduced in the investment Internet space and from year to year the varieties of such transactions are constantly expanding; in our opinion, it is reasonable to say that there is a turnover of investment objects. We believe that the turnover of investment objects in the Internet space should be understood as systematically executed transactions with certain assets for their purchase and sale, as a rule, at changing prices, through remote interaction of persons through websites in order to make a profit.

At the public level in the Russian Federation, the Agency for Strategic Initiatives has introduced an Investment Standard that combines 15 successfully used practices at the regional level (http://investstandart.ru/o-standarte/). These practices were applied by the most economically developed regions, and later they were extended to other regions. This standard has become the main project that should help economic entities in the regions of Russia to attract investment. With the help of the standard, the business community can evaluate the actions of authorities in the field of investment development and even influence some of their decisions. The specified standard, one of the mandatory requirements, provides for the availability of a website on the Internet in two languages (Russian and English, maybe in other languages used in the region or by potential investors), with a description of investment opportunities. Taking into account the fact that digitalization and information technologies are increasingly penetrating both business and the public sectors, the website seems to be an excellent presentation of the potential that this subject of the Federation has. The website must be up to date so that potential investors can fully assess investment opportunities and build interaction.

As you can see, the investment website is a hub from which potential investors begin to get acquainted with a wide range of offers from investment recipients. Next, the investor can focus his attention on the website of a single recipient of the investment [11]. Facebook, Instagram, etc., which are used to attract partners, will also be mentioned here about the existence of social networks, such as Vkontakte, Facebook, etc., which are used to attract partners. In addition, there are specialized social networks for finding business partners, such as LinkedIn, My Circle, Professionals.ru, Viadeo, etc.

We believe that the websites that exist on the Internet to attract investment by initiators of business projects should be differentiated into two groups according to the level of training of participating investors. The first group is investment platforms designed for the execution of transactions by qualified investors. The second group is investment platforms where unqualified participants can invest their capital. The first group of investment sites is characterized, first of all, by the large amounts of money for which transactions are carried out; the specificity of the objects themselves for investment; and the complexity of remote investment technologies. The second group of platforms is intended for unqualified participants and conducting simple, standardized transactions by the widest range of investors (people). It is significant that the investment issues concerning the second group of websites relatively widely are covered in the scientific literature. However, we practically do not find developments on the functioning of the first group of websites in the scientific literature.

In addition, in our opinion, remote Internet sites related to the investment sphere should be differentiated into three groups by initiators of interaction: banking websites, websites of investment intermediaries, and websites of direct recipients of investments. Let us consider these sites further.

The first group is the Internet sites of commercial banks. Undoubtedly, banks are the largest and main subjects of the capital market, acting as investment intermediaries and direct recipients of investments. The classic investment option—a bank deposit—has been used for several hundred years and it is quite in demand and satisfies the interests of many owners of capital. In addition, banks are quite actively attracting funds from a wide range of investors for their turnover by placing various securities, for example, bonds. Along with this, banks as intermediaries provide services to clients, for example, for the acquisition and (or) sale of certain investment instruments, for example, investment fund units.

The second group is the Internet sites of investment intermediaries. In this group, we include only legal entities with special legal capacity—these are licensed investment companies (brokers, dealers, asset managers); operators of investment platforms included in the relevant Register of the Bank of Russia; and exchanges.

Traditionally, the largest and most well-known investment platforms around the world are exchanges, they carry out the bulk of the turnover of investment instruments. Federal Law No.325-FZ of 21.11.2011 “On Organized Auctions” provides for the functioning of legal entities of special legal capacity in the Russian Federation—trade organizers [12]. Exchanges are possible only in the organizational and legal form of a joint-stock company with its own funds the amount of 100 million rubles. Six exchanges are listed in the Register of licenses issued by the Bank of Russia (https://www.cbr.ru/registries/rcb/): PJSC “Moscow Exchange MICEX-RTS”; PJSC “St. Petersburg Exchange”; JSC “St. Petersburg International Commodity Exchange”;

JSC "St. Petersburg Currency Exchange"; JSC "Stock Exchange "St. Petersburg"; and JSC "National Commodity Exchange" (there are no trading systems in Russia).

In Russia, PJSC "Moscow Exchange MICEX-RTS" (hereinafter referred to as MICEX) dominates, of course. The website presents hundreds of different services that contribute to both the implementation and attraction of investments: for bidders, private investors, asset managers, international investors, and issuers.

In 2013, the Bank of Russia issued Exchange License No. 045–002 to PJSC Saint Petersburg Exchange (hereinafter referred to as PJSC SPB). The document states that the license was issued to carry out activities for conducting organized auctions (https://spbexchange.ru/ru/futures/files/licenzia_birji.pdf.). The website of PJSC SPB states that PJSC SPB is an organizer of trading in foreign securities within the Russian jurisdiction, which provides opportunities for private investors to make transactions with foreign financial instruments. It is also indicated that individuals can carry out independent transactions only through professional participants of the securities market admitted to trading on the Stock Exchange, who have licenses from the Bank of Russia for brokerage activities, of which 46 companies are represented on the website of PJSC SPB (https://spbexchange.ru/ru/stocks/participants.aspx.).

One of the services on the website of PJSC SPB provides for the procedure for remote connection of a potential investor to remote exchange services. First, it is necessary to choose a broker from among the above accredited bidders and contact him via the Internet. Next, it is necessary to conclude a contract with the broker—you can also make a remote transaction. Secondly, the investor is invited to choose an online trading system from the list of certified systems and get access to transactions on the exchange. It should be clarified that it is possible to connect to PJSC SPB remotely only through a licensed brokerage company, from among the legal entities admitted to trading (https://investcab.ru/ru/inmarket/connect/).

For remote interaction, technical and network services are provided by PJSC SPB Association of Financial Market Participants "Non-Profit Partnership for the Development of the RTS Financial Market" (hereinafter NP RTS). There are three possible options: (1) dedicated communication channel—own fiber-optic line (cable) built from the client to the NP RTS data processing center, there may be a dedicated cable provided by NP RTS. Two (or more) parallel channels increase the reliability of access. To service each channel, the client needs to purchase special equipment and transfer it to the management of NP RTS; (2) encrypted channel via the Internet "network-to-network" (LAN-to-LAN) from the NP RTS network to the client's network segment. To service the channel, the client needs to purchase special equipment and transfer it to NP RTS; (3) Internet connection using a VPN client via a secure channel to connect one client computer does not require the purchase of additional equipment (PPTP client software package) (https://nprts.ru/ru/services/it-service/net/.).

The third group is the websites of legal entities—issuers offering their securities to a wide range of investors directly. Russian legislation does not prohibit a legal entity that has issued securities, for example, bonds, from distributing them to investors without contacting investment intermediaries or the stock exchange. It is also possible to subscribe directly to additionally issued shares, issued depositary receipts of the company, etc. To clarify, we are not aware of issuers that carry out direct remote transactions with investors. First of all, this concerns the regime of commercial secrecy, and, in addition, most issuers consider the organizational and financial costs of direct conduct and registration of transactions impractical. For example, on the website of LLC "Leasing Company Simple Solutions" it is indicated that at the beginning of 2021 there are four issues of PR-Leasing bonds in circulation on the MICEX, which can be purchased through a broker. On the PR-leasing website, there is a service for an investor to open an account with a broker and conduct a transaction for the purchase of bonds of the specified issuer.

For example, on the website of the Joint-Stock Company "DOM.RF" it is indicated that this JSC attracts investments, improves the quality and affordability of housing through the development and implementation of federal standards for the profitable purchase of housing in a mortgage. The website also states the "DOM.RF". The Russian Federation has issued 25 bond issues worth 195.6 billion rubles. At the same time, all AHML 2018 Eurobonds (issue for RUB 15 billion) were repaid in full and on time earlier, in 2018. From the "DOM.RF" website can easily switch to the website of its subsidiary Mortgage Agent. The exclusive subject of its activity is, firstly, the acquisition of claims on loans (loans) secured by mortgages and (or) mortgages, and secondly, the issuance of mortgage-backed bonds.

According to our estimates in 2022, individuals and legal entities on a scale of at least 5 billion users use the Internet. The modern citizen, as well as employees of legal entities, implements the absolute majority of communications on the Internet. The investment sphere, moreover, is simply repleted with opportunities for making investments, for example, you can specify several websites that facilitate remote investment transactions [13]:

- towave.ru—an online portal for startups, investors, where news and recommendations for entrepreneurs are published;
- napartner.ru.—the website for initiators of business projects and investors provides transaction support services. There is information about how successful companies were financed, how they received investments;
- investgo24.com.—the website with investment projects, proposals for the purchase and sale of businesses, analytical articles about investments;
- townmoney.ru.—the difference of this website is the opportunity to get a loan for the development of your business project;
- other websites, such as investors.partners, start2up.ru, business-platform.ru, innovationportal.ru, investclub.ru, rusinvestproject.ru, avito.ru, fbip.ru, beboss.ru an angel.cooperation, simex.global.ru, etc.

20.4 Conclusion

The pandemic of 2020–2021 caused a natural increase in remote investment activity. According to the National Association of Stock Market Participants, deposits of individuals in banks decreased from 22.9 to 21.2 trillion rubles in 2020 [14]. The number of Russian citizens who opened brokerage accounts on the stock exchange doubled in 2020 and reached 9.8 million people. Investments of individuals in securities and other investment instruments increased to 6 trillion rubles.

We believe that by 2025–2027 at least 30 million private investors will participate in investment turnover from the territory of the Russian jurisdiction, which sharply exacerbates the urgency of preventive measures by the relevant state organs. Russian President Vladimir Putin is rightly concerned about the rapid development of events: "Other forms of financial investments are being developed, including the purchase of securities. Moreover, this is by definition more risky than bank deposits. It is necessary to protect the interests of people who invest their money in the securities market, but are not of course professional investors" (https://www.rbc.ru/finances/01/02/2021/6017e70f9a7947cf2022efa4).

Thus, the modern information space (billions of individual computers connected via cables to supercomputers, radio waves of billions of smartphones on the air) is continuously used by individuals and legal entities. In particular, it is used as a set of electronic websites for the turnover of investment objects. The free will of a person who intends to enter the specified investment Internet space as an investor is not limited. As well as the freedom of entry for the recipient of investments is not limited. These persons can perform legally significant actions without having legal restrictions, while restrictions cannot be implemented effectively and reliably [15]. Activity in remote transactions by millions of individuals—unqualified investors—is growing at an accelerated pace in modern Russia, which requires urgent preventive measures from the Federal Center. We propose to make a decision at the legislative level during 2023–2025 to develop and implement the State Program "Investment Literacy of the population".

Acknowledgements The work was supported by Russian Science Foundation (project No. 20-18-00314).

References

1. Large encyclopedic dictionary: https://dic.academic.ru/dic.nsf/enc3p/139894(2021)
2. Petrovsky, S.V.: Internet services in Russian law. Publishing Service, Moscow, p. 272. (2003)
3. Maksurov, A.A.: On methodological foundations of legal regulation of Internet relations. Legisl. Econ. **2**, 59–61 (2012)
4. Federal Law No.149-FZ: On Information, Information Technologies and Information Protection. Rossiyskaya Gazeta No.165 (2006)
5. Rachmanina, N.V.: The concept of the Internet and the problems associated with the provision of financial Internet services. Legal Concept, 4 (2015)
6. Tapscott, D.: The Digital Economy: Promise and Peril in the Age of Networked Intelligence, pp. 45–48. McGraw-Hill, New York (1995)
7. Budarina, N.A., Akaev, A.A.: The impact of the Internet on changing the structure of investments. Int. J. Human. Nat. Sci. **9**(2), 106–109 (2019)
8. Kurasova, S.E., Afanasyeva, E.V.: Digital economy: features and trends. Probl. Dev. Modern Soc. **4**, 328–331 (2020)

9. Kamelin, R.V.: Legal regulation of public relations in the field of information systems: civilistic and information-legal approaches. Actual Probl. Russian law **12**, 68–77 (2017)
10. Khabrieva, T.Y., Chernogor, N.N.: Law in the conditions of digital reality. J. Russian Law **1**, 85–102 (2018)
11. Fadeeva, A.Y.: The role of the regional investment portal in increasing the investment attractiveness of the region. Manager, **2**(60) (2016)
12. Federal Law No.325-FZ: On Organized Auctions. Rossiyskaya Gazeta, No.266s. (2011)
13. Tamm, V.: TOP 15 sites for investor search and investment projects. https://promdevelop.ru/business/top-15-sajtov-dlya-poiska-investora-i-razmeshheniya-investitsionnyh-proektov/ (2021)
14. Inshakova, A.O., Goncharov, A.I., Ershova, I.V.: Digital blockchain registration of legally significant stages of complex good's export-import supplies by business entities of the EAEU and BRICS jurisdictions. In: Inshakova A., Inshakova E. (eds.) Competitive Russia: Foresight Model of Economic and Legal Development in the Digital Age. CRFMELD 2019. Lecture Notes in Networks and Systems, vol 110, pp. 328–336. Springer, Cham (2020). https://doi.org/10.1007/978-3-030-45913-0_37
15. Inshakova, A.O., Goncharov, A.I., Salikov, D.A.: Electronic-digital smart contracts: modernization of legal tools for foreign economic activity. Lect. Notes Netw. Syst. **91**, 3–13 (2020)

Chapter 21
Investment Technologies for Ensuring the Inclusive Growth of the National Economy

Inna M. Shor, Dildarakhon A. Shelestova, and Lubov I. Galamyan

Abstract The study aims to devise methodical and practical recommendations on using investment technologies as a means to ensure the inclusive growth of the national economy. Based on the authors' approach to the notion and the role of investment technologies for ensuring the inclusive growth of the national economy, we provide recommendations for implementing economic methods, approaches, and instruments at relevant stages of the investment process. At the stage of determining the goals and areas of investment, it is advisable to invest in physical and human capital, knowledge economy, housing construction, and export development. At the stage of garnering resources to invest in investment projects, it is important to pay attention to the government support of investment projects and public–private partnership. At the stage of the realization of investment projects, it is recommended to secure a weighted growth of trade protectionism. Under Industry 4.0, there are broad opportunities to form new assessment tools for investment technologies. Thus, we introduce the author's methodology to assess investment technologies based on the matrix method using the digital technology of Big Data.

21.1 Introduction

The COVID-19 pandemic has had adverse effects on the key indicators of social and economic development in every country. In this way, the restrictive measures concerning the labor force, decrease in consumer demand, and change in consumption patterns have led to the deterioration in the financial situation of the business. In

I. M. Shor (✉) · D. A. Shelestova
Volgograd State University, Volgograd, Russia
e-mail: ShorIM@volsu.ru

D. A. Shelestova
e-mail: ShelestovaDA@volsu.ru

L. I. Galamyan
The Russian Presidential Academy of National Economy and Public Administration, Moscow, Russia

E. I. Inshakova and A. O. Inshakova (eds.), *New Technology for Inclusive and Sustainable Growth*, Smart Innovation, Systems and Technologies 287,
https://doi.org/10.1007/978-981-16-9804-0_21

some cases, businesses closed and it evoked the growth of unemployment. With the increasing mortality rates, there is a growing discrepancy between the economies of developed and underdeveloped countries as well as between rich and poor. Moreover, there is an increase in gender inequality what makes women more vulnerable given the peculiarities of their employment and income as compared to men's.

Overall, it is possible to refer to the current situation as a syndemic which is a combination of epidemiologic, social, economic, and other processes during the coronavirus pandemic. So, it requires a complex approach to rethink the priorities in countries' social and economic development. The authors suppose that one of those elements is investment technologies. They can help overcome the negative influence of the pandemic through having beneficial effects on improving the population's potential, rate of productivity growth, stability of the economy, and through providing the population and state with great benefits. Therefore, it is necessary to follow the rule in implementing investment technologies which is enhancing the quality of life is the main criterion of the national economy's development, so the assessment of economic reforms should be based on the improvement of welfare and standards of living.

The aim of this study is to devise methodical and practical recommendations on using investment technologies as a means to ensure the inclusive growth of the national economy.

21.2 Methodology

Several works [1–10] focus on the notion, nature, and factors of inclusive economic growth. Meanwhile, without any intention to compromise the aforementioned works, it is crucial to highlight that there seems to be no research on investment technologies as a factor of inclusive economic growth. What is more, such research would be highly relevant under the given coronavirus crisis and the development of new technologies in Industry 4.0.

In this research, we used general scientific methods that ensure the objectivity, comprehensiveness, and exhaustiveness of the study. We also applied the methods of observation and expert estimation that helped identify relevant issues in implementing investment technologies. The SWOT analysis helps to identify strengths, weaknesses, opportunities, and threats to investment technologies implemented in Russia. The statistical method supplements our research with the analysis of the Russian investment practices describing the obtained results in the implementation of investment technologies. The matrix method for evaluation of investment technologies combined with Big Data technology helps operationalize the influence of factors in investment technologies at any level of their implementation as well as to conduct factor analysis of investment technologies in use and the assessment of the risks of the reduction of the role of investment technologies in the inclusive growth of the national economy.

21.3 Results

Investment technologies are a set of economic methods, tools, and techniques that are used in the investment process with regard to its stages (the stage of determining the goals and areas of investment, stage of garnering resources, stage of the realization of investment projects).

At the stage of determining the goals and areas of investment, it is important to invest actively in physical and human capital, knowledge economy, housing construction, and export development since they are the major drivers of inclusive growth of the national economy. However, we should mention that, according to the World Bank estimations, Russia ranks 55 out of 146 countries in the knowledge economy. The result of this gap can be seen in the export of high-tech products: exports in Russia equals USD 10 billion, while it is much higher in China (USD 740 billion), Germany (USD 207 billion), the USA (USD 156.4 billion), and Singapore (USD 155.4 billion) [11]. Moreover, the development of the knowledge economy, which is based on the education system and innovation ecosystem, is connected with new technologies (i.e., Industry 4.0) and evolves under global social issues. Thus, in order to unleash the full potential of the new economy, it is necessary to create a digital basis for the economy with the use of advanced information and telecommunication technologies as well as through investing into high-quality human capital [12].

At the stage of garnering resources and their further investment, it is crucial to ensure that there is government support for investment projects that is aimed at, firstly, improving the level and quality of life (since nowadays they are a social challenge and an obstacle for economic growth), and secondly, changing the quality of production capital and employment (since majority of employment opportunities are low tech and low paid which leads to low wages in those sectors of the economy). Meanwhile, we would like to highlight the importance of government support in the investor's cost recovery under the current budget law of the Russian Federation and/or in providing tax deduction under the tax legislation of the Russian Federation.

We can address the federal targeted investment program for 2021 and for the plan period of 2022 and 2023 as an example of the government's participation in projects' financial security. According to this program, the government plans to invest RUB 846.3 billion for 1,100 investment projects in 2021, RUB 885.5 billion in 2022, and RUB 925.8 billion in 2023. As for the structure of expenditure, the construction of road facilities ranks first with RUB 579.2 billion, maritime transport ranks second with RUB 371.8 billion, and housing and public works rank third with RUB 361.1 billion. It is also planned to invest in the construction of more than 40 healthcare, sport, and social welfare facilities. Moreover, the additional focus is on disease prevention under the sanitary and epidemiological conditions and its forecasted change [13].

Under the COVID-19 pandemic, it is highly important to preserve the government systematic support of investment projects under a predictable and stable investment environment. It will facilitate the achievement of strategic indicators of the country's economic development, particularly regarding the growth of annual rate of capital investment in Russia no less than 70% in real terms by 2030 as compared to 2020 [14].

Overall, a shift toward a situation when the government actively stimulates inclusive growth requires changing the current budget law in Russia, particularly concerning the right to exploit no less than half of international reserves. Nowadays, Russia's international reserves exceed the aggregate reserves of Germany, the UK, France, and Italy (while the aggregate population in these countries is about 50 million citizens more than in Russia). Thus, in order to ensure the financial security of the inclusive growth of the national economy, it is necessary to earmark USD 300 billion [15]. The areas to use these reserves that imply long-term low-interest investment loans are as follows: technological reequipment, new transportation, logistics infrastructure, and new capacities for high-tech industries based on the concept and technologies of Industry 4.0.

At the stage of garnering resources to invest in investment projects, it is also important to enhance public–private partnership. The crisis, which was caused by the pandemic, has demonstrated that public–private partnership projects are the most resilient in difficult social and economic situations. Despite the fact that some project implementation rates of public–private partnership worsened, there was no default in projects in this sphere, according to InfraOne [16]. It is crucial to highlight that public–private partnership schemes help to spread the dispersion of risks between the private investor and the government allowing a more flexible response to a changing market situation, caused by the pandemic in this case.

At the stage of the realization of investment projects, it is important to secure a weighted growth of trade protectionism that promotes manufacture relocation to Russia. One of the critical areas of economic development, which has become highly relevant under the pandemic situation, is the minimization of trade barriers for Russian products and goods. In this regard, the government measures taken to support export are highly relevant (e.g., the simplification of foreign exchange regulation and monitoring, customs regulations, export control; saving time and reducing costs for exporters; the digitalization of public procedures, etc.). Also, we would like to emphasize the introduction of a complex digital ecosystem "single export window" that can help reduce time costs and cost implications when submitting the same documents to different state bodies.

As it regards the three stages of the investment process, SWOT analysis is applicable to estimate the strengths, weaknesses, opportunities, and threats to the current investment technologies in Russia. Some of the elements of the SWOT analysis are included in Table 21.1.

This SWOT analysis helps to develop a complex of appropriate measures to improve investment technologies in order to ensure the inclusive growth of the national economy.

As it regards the peculiarities of modern development under Industry 4.0, it is possible to state that digital technologies allow introducing a new set of tools to assess investment technologies. In particular, Big Data technology can be used in the methodology for investment project assessment based on the matrix approach. A

Table 21.1 SWOT analysis of the current investment technologies

Strengths	Weaknesses
The country's beneficial economic and geographic location	The competitive disadvantage among Russia's regions for attracting investments as compared with the leading regions
The vast supply of resources (natural, material, information resources, workforce, etc.) that is crucial for the investment process	The infrastructural, normative, bureaucratic, and other obstacles in developing economic methods, approaches, and tools for the investment process
The extensive experience in the development and implementation of economic methods, approaches, and instruments in the investment process	The lack of effective use of the advantages of the country's economic and geographic location
The government support for investment activity	The low liquidity of physical funds of economic entities and, consequently, it is impossible to employ advanced economic methods, approaches, and tools for the investment process
The extensive ties to intraregional, interregional, and international relationships under the investment process	The low level of investment activity among economic entities
Opportunities	Threats
The proactive development and implementation of economic methods, approaches, and tools for the investment process as a result of the Industry 4.0 technologies and the improvement of the investment climate	A discriminatory mechanism of the coordination of investment flows that favors the federal economic zones, the largest regions, particular economic entities
The increase in sustainable development practices in private–public partnership as a means to promote investment activity	An increase in the competitive disadvantage among Russia's regions and economic entities in the struggle to attract investment resources with the leading regions and particular economic entities
	The decrease in the investment opportunities in the budgets of Russia's regions due to the increased burden of the social policy commitment
	A high level of dependence in intraregional, interregional, and international relationships upon the internal and external social, economic, and political situation

Source Compiled by the authors

matrix can be regarded as a logical system that represents a synthesis of the factors determining the content of investment technologies regarding their stages [17].

The matrix in Fig. 21.1 reflects consolidated groups of factors at consolidated stages of the implementation of investment technologies. On the one hand, they encourage the effective use of economic methods, approaches, and tools in the

		Factors that determine the essence of investment technologies			
		Administrative and legal (AL)	**Financial and economic (FE)**	**Social (So)**	**Environmen tal (E)**
The level of the implementation of investment technologies	**National economy (NE)**	NE/AL	NE/FE	NE/So	NE/E
	Sector (S)	S/AL	S/FE	S/So	S/E
	Economic entity (EE)	EE/AL	EE/FE	EE/So	EE/E

Fig. 21.1 A matrix based on the principle "factors–levels" of investment technologies. *Source* Compiled by the authors

investment process. On the other hand, they are obstacles and barriers to the development of investment technologies. Some of them have already been mentioned in the SWOT analysis presented above. For instance, government support, as a financial and economic factor, plays an important role in resourcing investment activity at any stage of the implementation of investment technology. Meanwhile, a discriminatory mechanism of the coordination of investment flows that favors the federal economic zones, the largest regions, and particular economic entities increases the competitive disadvantage in the investment development of other economic entities.

It is important to highlight that the matrix combines a factor and a risk approaches as it regards the research on investment technologies. It is highly relevant under the current economic conditions that are characterized by a high level of instability and active implementation of new Industry 4.0 technologies [18].

In scientific and practical terms, the quality of the aforementioned methodology for the assessment of investment technology based on the matrix method relies heavily on the objective formation of factors, which determine the essence of investment technologies, and speed of response to the risks of the effect of these technologies on the inclusive growth of the national economy. Thus, given the modern tendency toward digitalization, it is important to pay attention to the use of Big Data technology for analytics based on structured and unstructured data. As a result, it is possible to identify what factors and at which level of the implementation of investment technologies decrease the positive impact of these technologies on the inclusive growth of the national economy. This information can help devise and implement a complex of necessary measures to eliminate or mitigate the aforementioned risks.

21.4 Conclusion

The use of investment technologies to ensure the inclusive growth of the national economy requires the implementation of the most advanced economic methods, approaches, and tools at relevant stages of the investment process. At the stage of determining the goals and areas of investment, it is advisable to invest in physical and human capital, knowledge economy, housing construction, and export development. At the stage of garnering resources to invest in investment projects, it is important to pay attention to the government support of investment projects and public–private partnership. At the stage of the realization of investment projects, it is recommended to secure a weighted growth of trade protectionism that promotes manufacture relocation to Russia.

Under Industry 4.0, there are broad opportunities to form new assessment tools for investment technologies. Thus, we introduce the author's methodology to assess investment technologies based on the matrix method using the digital technology of Big Data. It helps reflect the effects of the factors, which determine the essence of investment technologies and which are formed at any stage of their implementation, and perform a factor analysis of data on investment technologies and assess the risks of the reduction of the role of investment technologies in the inclusive growth of the national economy.

References

1. Andrievskaya, V.B.: Public administration efficiency as necessary prerequisite of inclusive growth of economy. Ideas and Ideals **1**(23) (2015). https://doi.org/10.17212/2075-0862-2015-1.2-90-101 https://ideaidealy.nsuem.ru/storage/uploads/2015/04/Andrievskaya_23_%d0%a2.2.pdf Last accessed 1 Sept 2021
2. Bodrunov, S.D.: Rossijskaya tragediya–deindustrializatsiya otechestvennoj ehkonomiki. Institut novogo industrial'nogo razvitiya, Sankt-Peterburg (2013). https://inir.ru/wp-content/uploads/2014/02/01_Российская-трагедия–деиндустриализация-отечественной-экономики.pdf. Last accessed 1 Sept 2021
3. Chorn, S., Siek, D.: The impact of foreign capital inflow on economic growth in developing countries. J. Finance Econ. **5**(3), 128–135 (2017). https://doi.org/10.12691/jfe-5-3-5.
4. Cichowicz, E., Rollnik-Sadowska, E.: Inclusive growth in CEE countries as a determinant of sustainable development. Sustainability **10**, 3973 (2018). https://doi.org/10.3390/su10113973
5. Glaziev, S.Y.: On the urgent measures to enhance the economic security of russia and set the russian economy on course to priority development. Report. Institute of Economic Strategy. Russian Biography Institute, Moscow (2015)
6. Inshakov, O.V.: Strategy and tactics of nanoindustry development in the Russian Federation. Econ. Reg. **3**, 246–250 (2011). https://doi.org/10.17059/2011-3-29
7. Li, M., Zhang, Y., Fan, Z., Chen, H.: Evaluation and research on the level of inclusive green growth in Asia-Pacific region. Sustainability **13**, 7482 (2021). https://doi.org/10.3390/su13137482
8. Mau, V.A., Yasin, E.G.: Dvadtsat' let rynochnykh reform i novaya model' ehkonomicheskogo rosta: doklad k XIII Aprel'skoj mezhdunarodnoj nauchnoj konferentsii po problemam razvitiya ehkonomiki i obshhestva, Moscow, April 2012; National research University "Higher School of Economis". HSE Publ. House, Moscow (2012). http://www.hse.ru/data/2012/04/01/1265143363/mau.pdf Last accessed 1 Sept 2021
9. Shelest, S.A.: Problemy izmereniya inklyuzivnogo ehkonomicheskogo rosta dlya Rossii i stran SNG. Rossiya: tendentsii i perspektivy razvitiya. Moscow, Institute of Scientific Information for Social Sciences of the Russian Academy of Sciences (INION RAN), pp. 517–520 (2017)
10. Yiew, T.H., Lau E.: Does foreign aid contributes to or impeded economic growth. J. Int. Stud. **11**(3), 21–30 (2018). https://doi.org/10.14254/2071-8330.2018/11-3/2
11. Kalashnikov, V. :Itogi SPEHK-2020. Na puti k ekonomike znanij. (2021). https://spbvedomosti.ru/news/financy/na-puti-k-ekonomike-znaniy-itogi-spek-2020/ Last accessed 1 Sept 2021

12. Drobyshevskaya, L.N., Popova, E.D.: Development of Russian knowledge economy in the era of digital transformations. Kreativnaya ekonomika [Creative Economy] **12**(4), 429–446 (2018). https://doi.org/10.18334/ce.12.4.39019
13. Morozova, T. Gosudarstvo vlozhit v investitsionnye proekty 2.7 trln rublej za tri goda. [The state will invest 2.7 trillion rubles in investment projets over 3 years]. (2021). https://www.vedomosti.ru/economics/articles/2020/12/11/850573-gosudarstvo-vlozhit-v-investitsionnie-proekti Last accessed 1 Sept 2021
14. Decree on the national development goals of Russia until 2030 (2021). http://www.kremlin.ru/events/president/news/63728 Last accessed 1 Sept 2021
15. Strategiya dlya Rossii: (2021). http://www.veorus.ru/события/хроника-мероприятий/strategiya-dlya-rossii/. Last accessed 1 Sept 2021
16. Investments into infrastructure: market of projects to crisis: (2021). https://infraone.ru/sites/default/files/analitika/2020/rynok_proektov_v_krizis_infraone_research.pdf Last accessed 1 Sept 2021
17. Aleksandrov, G.A., Skvortsova, G.G.: Transition to methods for assessing the investment attractiveness of enterprises based on digitalization. Creat. Econ. **14**(11), 2811–2822 (2020). https://doi.org/10.18334/se.14.11.111099.
18. Skvortsova, G.G., Kondratyeva, O.A.: Modern methods to assessing the country's investment climate: opportunities and disadvantages. Bull. Tver State Univ. Ser. Econ. Manag. **3**(47), 24–31 (2019)

Chapter 22
Efficiency of Human Capital Investments as a Factor of Innovative Technologies Growth and Sustainable Development

Veronika V. Antonenko, Olga S. Oleinik, and Ekaterina V. Stepanova

Abstract Ineffective funds and resources investment in basic elements of human capital is increasingly observed across Russia currently, which is unexceptionally relevant for all the subjects of such investments from individuals and households to the state, in general. Therefore, in the course of 2013–2020, the authors of the chapter conducted a series of studies aimed at assessing the current investments efficiency in national human capital, as well as at detecting and developing means, methods and institutional mechanisms to increase such efficiency. Statistical analysis of large data sets was used as methodology for assessing investments effectiveness; data sources included, firstly, publications of official state, regional, and departmental statistics, and secondly, massive representative surveys of various economically active population groups, heads of government bodies, and business organization, in-depth and expert interviews with individual representatives of all the respondents groups. This assessment method of investment effectiveness is completely authentic and has not previously been used in such studies. Furthermore, tackling the urgent for Russian Federation problem of investment efficiency increase in human capital will enable the state to develop and introduce transformational new technologies, scale successful investment solutions more quickly, in order to implement the sustainable development goals.

V. V. Antonenko (✉)
Volgograd State University, Volgograd, Russia
e-mail: avv1@mail.ru

O. S. Oleinik · E. V. Stepanova
Branch of Russian Presidential Academy of National Economy and Public Administration, Volgograd Institute of Management, Volgograd, Russia
e-mail: oleyniko@vlgr.ranepa.ru

E. V. Stepanova
e-mail: stepanova.volg@mail.ru

E. I. Inshakova and A. O. Inshakova (eds.), *New Technology for Inclusive and Sustainable Growth*, Smart Innovation, Systems and Technologies 287,
https://doi.org/10.1007/978-981-16-9804-0_22

22.1 Introduction

The main reason for human capital investment is its constant expanded reproduction which is indicated by the positive dynamics of numerical values of the key indicators characterizing this process in relation to the funds spent. In particular, the essence of education is transfer of various knowledge and skills from previous generations to subsequent ones. However, the primary goals of economy, determining current business practices, embrace profits increment by efficiency improvement and volume increase of production, enhancing the quality of life of population, etc. [15]. Nonetheless, transition of knowledge itself from one generation to another does not necessarily lead to its increment and, accordingly, expanded reproduction of human capital does not occur. In order to achieve the expanded accumulation goals, not only the transfer of existing knowledge but also acquisition of new expertise is required in each generation with further subsequent transfer to the following ones. In modern society, this responsibility is fulfilled by science, being a constant companion of education.

Science is an essential tool in constant and massive knowledge acquisition process, which results are eventually introduced in production, management, social, and everyday life, and are subsequently transmitted to further generations through education. This is an essential condition for human capital expanded reproduction, but only in case of effective functions performance by key institutions involved. However, at the current stage of global economy and its systems development, we face the problem of exponential growth of information and knowledge available and their unprecedentedly fast increment which is ever accelerating. As a result, it is now progressively difficult to pass on expertise to new generations in equality structured and quality manner as at earlier stages of economic system. The era of conveyor production arguably now concerns not only goods and services, but also human capital, including transfer of experience and knowledge. And just as the consumers of goods and services are no longer able to absorb all the volumes produced, hence, the crisis of overproduction; so much so, modern recipients of education and knowledge cannot assimilate all of the expertise even within the framework of one major or training specialization.

Therefore, the key conditions for human capital reproducibility at the present stage of global economic system evolution are, firstly, further differentiation and specialization of knowledge transmitted to new generations, and secondly, precise selection of those individual representatives who by their predisposition are more suitable for assimilation of this data type, and subsequently more successful in corresponding professional activity. This is especially true for critically important industries, spheres of economic practice and science, because a certain level of qualified, competent and efficient individuals in these areas ensures their further development; secures competitiveness in the national and international arena [2, 3]. It should be pointed out, that this trend is typical for Russian society where individuals since early childhood are encouraged to develop numerous skills and are increasingly involved in diverse educational and extracurricular activities, sometimes to the extant where their health deteriorates due to constant overloads, both physical and emotional.

22.2 Methods

Despite the long evolution and a significant number of concepts and theories of Human Capital, researchers and followers do not share a unified perspective over it nor has a «canonical» or commonly universal definition of the concept been formulated. Therefore, every new economic research of the subject in the past or present, analyzing various aspects of the issue of formation, reproduction, or efficiency of human capital, brings into its essence and contents its own distinctive insights, concepts, and ideas which correspond to goals and objectives of the research. Only the core of content, which was formed by the key elements of human capital, remained unchanged. In the course of the present research, we were primarily interested in those components of human capital, the volume and structure of investments in which it is fundamentally possible to calculate and analyze, as well as assess the return on them, which ultimately allows us to determine how these investments are economically or socially effective, and to formulate a complex measures and institutional mechanisms to improve the efficiency of investment in human capital at the level of economic systems of comparable and statistically dominant regions of modern Russia. These key components are education and healthcare.

Nonetheless, currently existing approaches to assessing investment effectiveness and investment appraisal techniques, entitled "discounting methods", are exceptionally applicable to measure corporate performance level and omit other structures of the entire multi-level national economic system. For the most part, at all other tiers hierarchically located both above and below [12, 13], there are critically significant restrictions that do not allow to accurately determine the costs of human capital formation; specifically, in the present study, the investment volume in education and healthcare nor does it enable to calculate even approximately the income change of an economic entity under consideration or prove that the income increase is an outcome of investments made in human capital.

At the level of individuals or households, the factors of such restrictions are the long periods of return on investment in human capital, at which any accurate nano- and micro-economic forecasts are no longer possible, as well as the prevailing non-economic motivation of most investment subjects. For instance, the economic effect of higher professional education, as compared to secondary vocational education, is either not obvious at all or is perceived decades after graduation from university. At the regional level of Russian economy, the direct economic effect from total budgetary and other investments in education and healthcare is almost unfeasible to assess due to the huge number of factors that, in addition to these investments, affect changes in the incomes of economic entities of the country at any level. Moreover, the onset of the effects under consideration may refer to ultra-long-term processes, for example, the return on investments in infrastructure projects for protection of maternal and infant health occurs decades later, when such effects can no longer be distinguished from accomplishments of numerous other factors.

Therefore, application of specific previously unused methods for assessing investments effectiveness in human capital justifies development and implementation of fundamentally new measures, institutions, and practices for the reproduction of human capital in an average constituent entity of the Russian Federation, as well as carry out a comprehensive monitoring of the results of such measures for a number of key parameters specially selected for this purpose. The authors of the chapter have proposed methodology for statistical assessment of the investment return under consideration which implies correlation of costs with the dynamics of the indicators characterizing it within a specific territory on the basis of intergenerational analysis for sufficiently long periods in order to effectively assess the received outcomes as statistically significant. The suggested approach enables to evaluate efficiency of human capital reproduction at the level of individuals, households, enterprises, industries, and settlements typical for a particular territory, as well as in general for Russian municipal districts, regions, and the country as a whole.

In general, in the course of 2013–2020, within the framework of the research carried out by the authors, the following methods of collecting and processing information were applied:

1. Data analysis from official state, regional, sectoral, and departmental statistics, both for the Russian Federation as a whole, specifically for the Volgograd region, as well as selectively for other constituent entities of the Russian Federation.
2. Sociological surveys on representative samples of young people in the Volgograd region (N = 374),[1] their parents and other older relatives (N = 345), various categories of economically active population of the Volgograd region (N = 604), heads of enterprises, organizations, institutions of regional municipalities (N = 240). The sociological surveys were related to collection real data on the structure and volume of investments in human capital at all levels of the national economy as well as on the return received.
3. In-depth semi-formalized interviews with young generation of the Volgograd region representatives (N = 30).
4. Expert interviews with public and municipal government officials, regarding their official duties, factual expenditures of the federal, regional and municipal budgets for formation, development and maintenance of various elements of human capital particularly in younger age groups (N = 20), as well as with the heads of enterprises and organizations (N = 20) to find out their opinions as employers, in particular, about the real level of expertise and skills of young graduates in dynamics.
5. Processing of primary data was carried out using the computer program "Statistica"; classical correlation-regression and factor analysis, as well as descriptive statistics and graphical method were applied.

[1] N – Number of Respondents.

22.3 Research Outcomes

Throughout the evolution period of both capitalist economic systems and economics as a science, attempts were made to assess investments effectiveness of human capital, bypassing the above-mentioned discounting methods, initially due to lack of those, and later as a result of limited application options; first and foremost, such evaluation practices were applied at national or regional levels. For this, it was also necessary to calculate the costs of reproducing human capital (in the sense of it that prevailed at the corresponding stage of economic evolution and the development of the political economy science, including those within which the term "human capital" itself did not yet exist), the traditions of which were laid even W. Petty, A. Smith, A. Marshall and other representatives of the classical and neoclassical approaches in economic theory [8, 20, 21].

However, the calculated costs of human capital reproduction should correlated with some results, and preferably, with those expressed in monetary terms, which poses the main difficulty in this type of research. In the most detailed and elaborated form, the solutions to this problem are presented in the works of world economists [4, 14, 17, 20] and in Russian economic science by [3, 6, 10, 14, 17, 18, 20, 22]. Nevertheless, none of the listed above and other authors has completely resolved the issue of correlating the calculated costs of reproducing human capital with useful effects in the form of subsequent returns, including non-monetary and generally noneconomic ones.

As noted above, any forms and types of investment in human capital provide mainly its simple reproduction, including in terms of the formation of education capital. For expanded reproduction, it is not enough to simply transfer existing knowledge (no matter how efficiently it is conducted), but it is also necessary to form new expertise. Then and only then will its constant increment be realized. Science is responsible for such an increase, and it is unambiguous that this institution has no opportunity to exist and function effectively without significant investments. Thus, it is exceptionally important that in addition to the transfer of knowledge to new generations, a clear differentiation of the carriers of human capital is also required according to the ability to perceive this knowledge (due to their huge volume and diversity at this stage of the global economy evolution), both in quantitative and in a qualitative sense, meaning ability to learn in general and to acquire specific skills and expertise.

Generally in modern Russia, the function of learners' differentiation according to academic capacities is performed using school grades system, the Uniform state exams, as well as entrance, intermediate and final tests in universities, community colleges and specialized educational institutions. However, identification of carriers of human capital inclined to a particular type of professional, scientific, or creative activity, including early vocational guidance, still functions ineffectively if at all in this country today, which negatively affects human capital reproducibility. In addition, government is to ensure not only the transfer and augmentation of knowledge, including through its regional and municipal authorities and institutions, but also enhancement of conditions for economically and socially effective use of knowledge gained in economic, scientific, or other activities [5, 7]. Only in this case is it possible to expand the reproduction of human capital at all levels of national economy.

This is directly related to another significant element of ensuring the investment efficiency in human capital, that is, achieving its expanded reproduction and, as a result, a strategic, long-term return and increase of invested funds. A correct, accurate identification and selection of carriers of human capital according to their psychophysiological, social, and other congenital or acquired characteristics, ability to perceive and assimilate information of this particular type, as well as the corresponding general knowledge and professional skills and expertise should be conducted. The system is not just vocational guidance, although it is exceptionally important for ensuring investment efficiency in human capital at any level, but first of all, it is the identification of the propensity of a particular individual to any type of activity in comparison with the approved list of critical types, on which the industrial, scientific, economic performance, and safety of the country directly depend. Therefore, national defensive capacity and strategic sustainable development are as well involved, that is why, the earlier the selection system of individual approach to human capital is introduced, the more effective and accurate such choosing is.

This does not mean that from early childhood it is necessary to orient an individual to engage in a strictly defined type of professional activity in future. The point is only that investing in higher professional education, primarily, at the state level, as well as other corporate subjects of investing in human capital, should be conducted in carriers ensuring a higher degree of probability to guarantee the return and increase of the invested funds which, without such investments, would not be able, through their own efforts or through the efforts of their families, to reveal their physical, intellectual, or other potential to the fullest. In other words, it is necessary to pay for expensive sports or scientific equipment (its purchase, maintenance, etc.) as a priority in relation to those carriers of human capital who are clearly capable of achieving significant results for the investor in this area, which does not at all exclude the financing of general developmental sports or creativity for all categories of teenagers and young people. But the primary task here seems to be "not to miss" a leader who is potentially able to carry out in the future an increase in human capital (national, regional, etc.) in the form of obtaining new knowledge and skills or successful practical application of existing ones, potentially in new ways.

Another topical issue in Russia today is the lack of a single centralized system for the effective and accurate identification, maintenance, and development of innate talents and abilities in gifted individuals from early age through adolescent and youth periods. Currently existing institutions, as well as regularly held activities in this area do not represent a single system, or a common database or databank, and are not sufficiently effective. On the contrary, they often contribute to the dissipation and irrational use of material, financial and other means and resources of all actors involved in this process, from family to state, and also lead to excessive expenditure of time and effort both for children and adolescents themselves, and for their parents [1, 9]. Therefore, a thorough development of government strategy in relation to gifted individuals and talented youth with differentiation by levels and subjects of their implementation is necessary, which should be implemented for the most complete and accurate disclosure of the potential of gifted children, adolescents, and youth of this country.

This system introduction will significantly increase the competitiveness of domestic human capital including in the international arena, especially in those areas of activity where the relevant talents and abilities are of key importance. It should be emphasized that events, for example, for talented and active children and adolescents at the level of all municipalities and regions are held regularly, including academic Olympiads, sports and creative competitions, etc. Institutionally, organizational infrastructure for such activities in Russia is sufficiently developed, since considerable efforts are made by local and federal governments to largest extent in terms of stimulation of involvement of children and adolescents in academic, sports, and creative activities. It is officially believed that this contributes to the development of talents, disclosure of abilities, and finally, those individuals at adult age will be able to work more productively, bringing a higher performance or simply additional income for themselves, their employers and the state in general. In fact, those are indicators of a competitive human capital, as well as the economic efficiency of investment in it [16].

Nevertheless, in practice this is evidently far from being the case: colossal investments in human capital in the form of effort, time, and money at all levels, from family to state, do not reveal high efficiency as many children and adolescents when they grow up, do not always demonstrate the declared high professional expertise or intellectual qualities. Moreover, at universities, the level of knowledge and skills of students has been constantly decreasing for at least fifteen years. Moreover, when fresh graduates enter the labor market, employers indicate the decline in the intellectual, creative, and professional potential of graduates of higher and secondary specialized educational institutions. One might assume that institutions aimed at formal development of various skills and talents in children do not actually contribute to progress, but arguably lead to excessive waste of time, energy, and health, both for children and their families. For the latter, as well as for the state, this also involves significant material and financial costs which will not be compensated in foreseeable future, that is, investment in human capital is characterized by zero or negative efficiency. The conclusions drawn are confirmed by the results of mass and expert polls, formalized and in-depth interviews conducted by the authors. The decline in the state of the nation children and young people's wellbeing and health, especially toward the senior grades at high school is convincingly demonstrated by the data of medical and demographic statistics.

Figure 22.1 clearly reveals that after 2004, even in the years between the crises of 2008 and 2014, budgetary spending on public health increased, many times faster than inflation and consumer prices. Consequently, a smooth growth in such costs from 1992 to 2004 and an exponential increase after 2004 can be witnessed equivocally. In this study, initial assumption is made that the share of expenditures on children's health care in the overall structure of budgetary investments in health care is constant, or deviations in the values of this share in some years are mutually offset, which is most likely true [19]). Therefore, we can accept as a reliable fact a similar increase

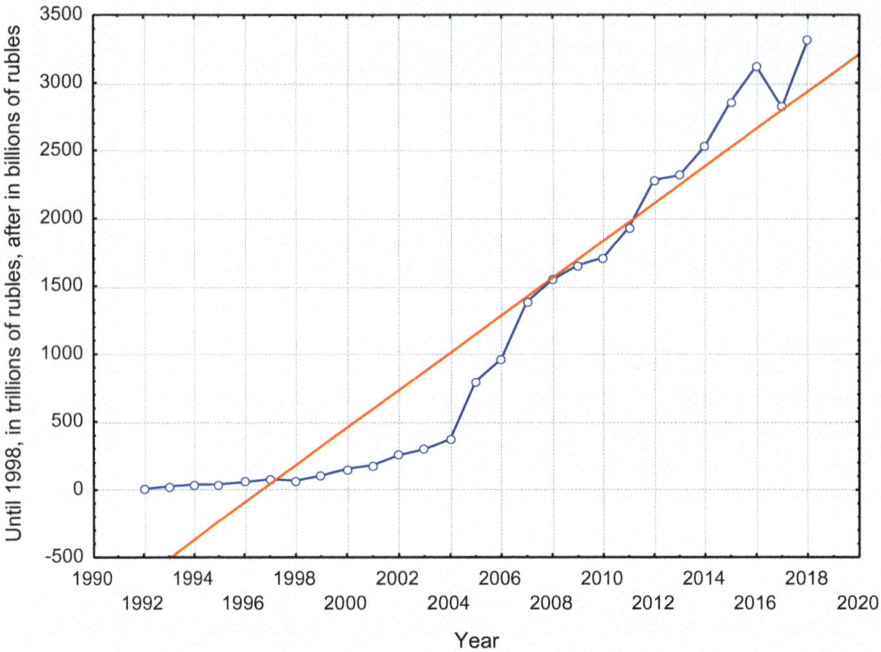

Fig. 22.1 Budgetary investments in Russian healthcare system over the period of 1998–2018. *Source* Compiled by the authors based on [11]

in expenditures specifically for maintaining the health of children whose share in the total structure of the considered expenditures most likely remained approximately the same.

However, at the same time over the specified period in Russia, the qualitative characteristics of reproduction of human capital have been steadily decreasing. In particular, disease incidence in children, which consists of diseases in the main classes and groups of illnesses, is growing. It is obvious that the dynamics of the indicator presented in Fig. 22.2 has a tendency to increase, although it demonstrates a different degree of its intensity and uniformity in certain years. The trend is still unsteady until 2007–2008, yet after the indicator values unwaveringly go up. In general, the trend is unambiguously positive, which refutes any claims about the effectiveness of investments at the state level in children's health. This example is typical, since a similar situation is observed in most cases with other indicators characterizing both health and education of carriers of human capital, namely: with a multiple increase in investments over the period under review, the statistical characteristics of the reproduction of human capital, on the contrary, decrease. At the same time, a comparative analysis of expenditures on education and healthcare in the Russian Federation and most global countries of shows that, in this country, spending do not exceed those of the most developed economies. The corresponding data are presented in Table 22.1.

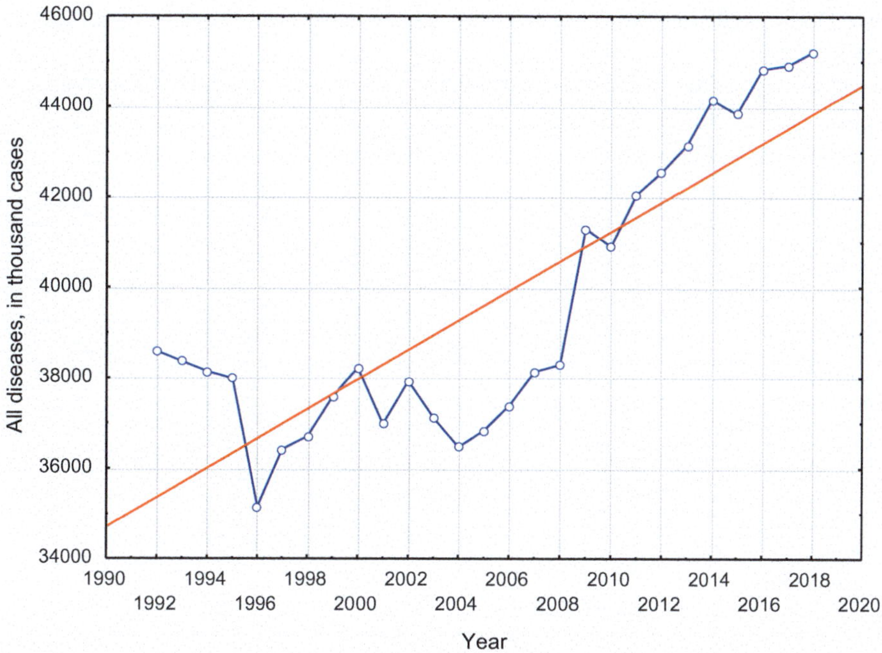

Fig. 22.2 Dynamics of total morbidity in children (all diseases) in Russia over the period of 1992–2018. *Source* Compiled by the authors based on [11]

The data in Table 22.1 demonstrates that government spending on education and healthcare in Russia is quite comparable to those in most global economies. Despite the fact that the represented nations vary considerably in their economic development, population size, and territorial location of the state, in none of them the corresponding costs do not show numerical values that are many times more or less than in Russia. It should be noted that although the Russian Federation according to the presented indicators is characterized by such values rather closer to the minimum; nevertheless, it is important here that they are not absolute, but represent a share of the GDP, according to which Russia is now the 6th in the world, that is, it is a perfectly acceptable and at the same time typical indicator. We might assume that it no longer makes economic sense to increase government spending on education and health care to any significant extent, but it is their structure that needs to be optimized, that is, the investment resources reserved for the purpose of achieving expanded reproduction of human capital cannot be spent chaotically, scattering them throughout the social economic and demographic space, without considering all the possible benefits and risks. On the contrary, it is necessary to develop precisely those skills and abilities of carriers of human capital that will guarantee in future the development of new technologies, introduction best economic practice as a contribution to sustainable economic growth of corporations, enterprises, regions, and country as a whole.

Table 22.1 Education and healthcare public expenditures as the main elements of human capital, percentage of GDP

Country	Year	Education expenditures	Year	Healthcare expenditures
Russia	2019	3.7	2019	3.4
Belarus	2019	4.9	2019	4.2
Germany	2017	4.9	2017	8.7
Italy	2017	4.0	2017	6.5
Poland	2017	4.6	2017	4.5
France	2017	5.5	2017	8.7
Switzerland	2017	5.1	2017	3.8
Azerbaijan	2019	2.7	2019	1.1
Israel	2017	6.1	2017	4.7
India	2013	3.8	2017	1.0
South Korea	2016	4.3	2017	4.4
Turkey	2015	4.3	2017	3.3
Japan	2017	3.2	2017	9.2
South Africa	2019	6.5	2017	4.4
Argentina	2017	5.5	2017	6.6
Brazil	2017	6.3	2017	4.0
Mexico	2017	4.5	2017	2.8
Chile	2017	5.4	2017	4.5
USA	2014	5.0	2017	8.6
Australia	2017	5.1	2017	6.3

Source Compiled by the authors

Thus, modern technologies development and implementation in economic practice requires availability of new knowledge, skills, and expertise, as well as expressed abilities for this, which is unattainable without effective investments in the human capital of their carriers. Furthermore, professional knowledge and abilities subsequently develop competently and accurately, diverting significant amounts of funds from current consumption over long periods of time. Today in this country, huge funds are allocated for the development of a wide variety of qualities of preschoolers, schoolchildren, and students, but "at the exit" we only get a decrease in the level of health of the younger generations (due to physical and emotional overload in preschool and school age), as well as formally higher level of education, skills, and abilities, which in practice do not always correspond to real expectation of employers in the labor market from the human capital of both young people and specialists of older ages. Consequently, deep institutional transformations are required for the entire sphere of investment in human capital, at all its levels and for all the subjects involved.

22.4 Conclusion

As a result of the conducted study, the following conclusions were obtained:

1. Currently in Russia, there is no national strategy for gifted children and talented youth, affecting all subjects of its implementation at the federal, regional, and municipal levels. In order to establish it, it is required to build a unified system for identifying, maintaining, and developing the human capital of gifted individuals from early to adolescent age, starting from the youngest children and ending with university graduates. Furthermore, direct and active involvement in the process is required of preschool, general education, secondary special and higher, additional education institution institutions, as well as of authorities, potential employers, and business representatives. In addition, in order to master new technologies and achieve sustainable growth, the national economic system of Russia requires a complete revision regarding volume and structure of investments in human capital.
2. Over the past 30 years in the Russian Federation, there has been a steady growth in real volumes of investments in the human capital of children from the main subjects of such investments, but at the same time, statistical indicators characterizing the human capital of consecutive (by year of birth) generations of children and youth are falling. In particular, during this period, wellbeing and health of children, adolescents, and youth is constantly deteriorating, as well as do their level of knowledge, education, erudition, and general culture. The above-mentioned facts indicate ineffectiveness of investments in human capital particularly in children in Russia, and their benefits continue to decline. Consequently, a significant, and for specific components of human capital and subjects of investment, a complete revision is required in terms of volume and structure of corresponding investments, as well as their immediate and sustainable goals and expectations.

References

1. Abankina, I.V., Kravchenko, I.A., Filatova, L.M.: Early development and preschool education: financing in Russia and abroad. Educ. Policy **4**(84), 36–47 (2020). https://doi.org/10.1787/19991487
2. Antonenko, V.V.: Competitive human capital as a factor and a result of the forming and development of smart cities. Lect. Notes Netw. Syst. **155**, 522–529 (2021). https://doi.org/10.1007/978-3-030-59126-7_59
3. Bagieńska, A.: Measurement and analysis of the efficiency of human capital in a small enterprises in Poland. Finan. Int. Quart. "e-Finance" **11**(2), 1–9 (2015). https://doi.org/10.14636/1734-039X_11_2_001
4. Becker, G.: Human Capital: A Theoretical and Empirical Analisis with Special Reference to Education, 3d edn. The University of Chicago Press, Chicago (1993)
5. Benabou, R.: Equity and efficiency in human capital investment: the local connection. Rev. Econ. Stud. **63**(2), 237–264 (1996). https://doi.org/10.2307/2297851

6. Dedov, S.V., Kovalevskij, I.J.: Theory and methodology for assessing the effectiveness of innovative resources. Finan. Credi **34**, 69–75 (2007)
7. Dedov, S.V., Shishkin, A.F., Gordienko, N.S.: Theoretical and methodological approaches to human capital assessment. Finan. Credit **22**(310), 64–69 (2008)
8. Denison, E., Poullier, J.: Why Growth Rates Differ. Postwar Experience in Nine Western Countries. The Brookings Institution, Washington, D.C (1967)
9. D'jachkova, K.Ju., Proskura, M.A.: Parental investments in children as a way to increase the child's chances of success in modern society. Econ. Bus. Banks **S6**, 39–46 (2018)
10. Elizarov, V.V., Kvasha, A.J., Tatevosov, R.V.: The Economic Foundations and Principles of the Demographic Policy of the USSR. Publishing house Moscow State University, Moscow (1981)
11. Healthcare in Russia: Statistical Compendium. Federal State Statistics Service, Moscow (2019)
12. Inshakov, O.V.: Systems-level analysis of the object, subject and method of economics. Bull. St. Petersburg Univ. Econ. Finan. **4**(40), 5–18 (2004)
13. Inshakov, O.V.: Development nucleus' in the light of the new factors of production theory. Econ. Contemp. Russia **1**, 11–25 (2003)
14. Kendrick, J.: Expanding imputed values in the national income and product accounts. Rev. Income Wealth **25**(4), 349–363 (1979)
15. Masters, A.: Efficiency of investment in human and physical capital in a model of bilateral search and bargaining. Int. Econ. Rev. **39**(2), 477–494 (1998). https://doi.org/10.2307/2527303
16. Mičiak, M.: The efficiency of investment in human capital in IT enterprises. Transp. Res. Procedia **40**, 1134–1140 (2019). https://doi.org/10.1016/j.trpro.2019.07.158
17. Mincer, J.: Schooling, experience, and earnings. Nat. Bur. Econ. Res., N.Y (1974)
18. Nesterov, L.I., Ashilova, G.T.: National wealth and human capital. Econ. Issues **2**, 103–110 (2003)
19. Oleinik, O.S., Oleynik, A.B., Stepanova, E.V.: Features of measuring poverty indicators: international and Russian experience. Regional Econ. South Russia **8**(4), 53–63 (2020). https://doi.org/10.15688/re.volsu.2020.4.5
20. Schultz, T.: Investment in human capital. Am. Econ. Rev. **51**(1), 1–17 (1961)
21. Shhetinin, V.D.: Human capital and the ambiguity of its interpretation. World Econ. Int. Relat. **12**, 44–48 (2001)
22. Strumilin, S.G.: The time factor in capital investment projects. news of the Ussr academy of sciences. Dep. Econ. Law **3**, 195–215 (1946)

Part V
Enhancing Cognition Technology to Achieve Inclusive Human Development in the Digitalizing World

Chapter 23
Multilinguality of Digital Platforms as a Factor of Inclusive Growth in the Global Economic Space

Vera A. Mityagina, Irina D. Volkova, and Stephan Walter

Abstract The concept of inclusive sustainable growth is an important element of the emerging philosophy of inclusiveness. Its primary aim is to provide equal opportunities for all members of society. These opportunities are associated with a personal identity—cultural, social, gender, national, and linguistic. Multilinguality of information resources as a manifestation of linguistic pluralism is one of the ways to ensure sustainable economic development. Leading multinational companies have successfully used the strategy of glocalization for many years by means of self-presentation on localized versions (in the languages of partner countries) promoting values and ideas around the globe. The issues of translation and pragmatic adaptation of digital content have become a relevant subject of up-to-date linguistic research. Multilingual websites of various governmental and corporate organizations are the best manifestation of the concept of linguistic inclusiveness especially in the time of active migration processes. Numerous digital platforms provide content available in a number of languages which ensures understanding of information in general and in detail. This chapter presents an analysis of effective communication-based solutions for creating multilingual websites, pragmatics of translation, strategies for making up content for smooth online communication in the modern age of forced isolation of heterogeneous ethnic groups and multicultural societies.

V. A. Mityagina (✉) · I. D. Volkova
Volgograd State University, Volgograd, Russia
e-mail: mityagina@volsu.ru

I. D. Volkova
e-mail: idvolkova@volsu.ru

S. Walter
Johannes Gutenberg University, Mainz, Germany
e-mail: walters@uni-mainz.de

E. I. Inshakova and A. O. Inshakova (eds.), *New Technology for Inclusive and Sustainable Growth*, Smart Innovation, Systems and Technologies 287,
https://doi.org/10.1007/978-981-16-9804-0_23

23.1 Introduction

Informatization is one of the key tools for development of the global economic space. It has affected all elements of social infrastructure, including education, healthcare, trade, leisure, etc. The concept of the information society, which has been actively used since the 1970s, most accurately reflects the essence of modern life influenced markedly by advances in information and digital products and services, which are communicated through a wide range of media, the Internet in particular.

23.2 Methodology

According to W.J. Martin, the information society is a society in which the quality of life, as well as socioeconomic development in general, are directly dependent on information resources and their use [18, p. 16]. Digital transformations in the field of informatization, such as the creation of web platforms and technologies of Industry 4.0, are one of the main factors in the development of companies, industries, national economies, and the global economic space in general [25, p. 58]. Based on expert estimates, the expected results of the Russian economy's digitalization include an increase in the country's GDP by 19–34% by 2025 [26, p. 63]. International institutions, the European Union in particular, highlight the importance of promoting multilingualism of the information space [12]. The EU language policy considers multilingualism as an important competitive advantage.

Global and national practices of using online services by governmental and profit-making organizations show that advances in information and communications technologies promote the development of new socioeconomic and sociocultural patterns and provide equal access of citizens to various e-services, such as electronic education, electronic government, teleconferencing, networking, teleworking, and e-commerce transaction services. The European sector of information and communications technologies of Industry 4.0 has the highest financing rates in the period up to 2020 as compared to such important sectors as manufacturing, engineering, and electronics [11, p. 14].

23.3 Results

The problem of development and adoption of these technologies is intercultural in nature, since the realities of the electronic global village assume the creation of globally oriented information resources placed on various online platforms. Availability of online content for various linguocultural communities is a prerequisite for inclusive growth in the global economic space. To comply with this condition, it is necessary to solve a number of current problems associated with information and communications technologies and linguistics.

First of all, it should be noted that for many regions of the world the problem of digital inequality remains relevant. This problem primarily consists in the lack of access to the Internet as the main source of information. According to the World Economic Forum data, about 3.9 billion people on Earth—more than half of the world's population—are still not online due to economic, political, legal, and other factors [14, p. 1]. The ways of dealing with this problem are being actively searched for at the global and national levels within the framework of the countries' economic development strategies. The project of the World Economic Forum *Internet for All* has become widely known today. It aims to provide global access to the broadband, as well as create local content and information infrastructure [14, p. 4]. The problem of digital inequality is also covered by national initiatives. For example, the task of its elimination between the urban and rural population of Russia is outlined in the Federal Law "On Communications" (amend. of 2021) [10]. In 2013, the leaders of 30 African countries adopted the SMART Africa Manifesto, which sets out the principles of promoting digital inclusivity and usage of information and communications technologies [22]. Microsoft's Airband Initiative advances digital equity in many countries across the globe, including the USA, Ghana, Kenya, India, and others [2].

Secondly, the problem of online content availability for linguistic minorities requires special attention, since, as a rule, websites of international organizations have only a few language versions, English one being a global option. According to the World Bank, 80% of online content is available in only 1 in 10 languages, despite of the fact that these languages are native to 3 billion people [20]. At the same time, the UN report emphasizes that restrictions on the provision of information and services in minority languages disable citizens to exercise their language rights [15]. There are many cases of effective implementation of inclusive language policy in the Internet by certain countries. For example, in the UK, healthcare providers are required by law to provide translation of web materials, including documents and videos, for non-English speaking patients. In Iceland, the authorities inform the public about social services in seven languages thus providing equal access to information for all citizens.

Unlike domestic initiatives of inclusive language policy, global tasks require the use of the latest technologies and translation programs to ensure availability of Internet platforms for speakers of as many languages and dialects as possible around the world. As Irina Alekseeva rightly notes, globalization is possible under the condition of a high level of translation activities [3, p. 8]. Translation has a great potential for radical reform of the global economy, and "by the end of 2021, the world market of linguistic services will reach $ 56 billion" [24]. Website localization is the most frequently used procedure for translating verbal content and adapting extralinguistic parameters of digital platforms. It is an effective strategy for international companies development. According to the definition of one of the largest language service providers Language Line Solutions, localization is the process of adapting digital content to a specific region's language and cultural sensitivities so that it seems natural to customers [17].

Today, a huge number of Internet sites of profit-making and governmental organizations have a multilingual user interface. Links to localized versions are usually indicated by the names of languages or countries. Let's consider the cases of website multilingual support using the examples of various online platforms. The French State Agency for the Promotion of French Higher Education Abroad and International Student Mobility 'Campus France' is an online resource which aims to inform foreign students about French higher education, administrative procedures when traveling to study in France, as well as to provide assistance in choosing educational programs and organizing a stay in the country. The website contains 95 localized versions. The wide geographic coverage indicates the global scope of this organization and its aim to provide complete information about educational services to citizens of different countries. Table 23.1 shows fragments of localized texts for citizens of Russia, Australia, and Germany.

The use of the names of corresponding cities and countries and country-specific information in the localized versions indicates high quality of the website's multilingual support. It should be noted that the website also contains a global English-language version called 'International Website'. Its content is neutral in terms of culture and geography and is intended for users not covered by localized options. As a matter of fact, it is impossible to pursue a website's inclusive language policy and take into account the world's linguistic diversity due to the high cost of localization services. As a rule, the possibilities of culture-sensitive multilingual support are severely limited, and therefore it is advisable to focus on several leading markets or target groups. Besides, user-generated content (blogs, forums, podcasts to audio, and video content) cannot be cost-effectively rendered in multiple languages using human translation.

Table 23.1 Fragments of localized texts from the Campus France website for citizens of Russia, Australia, and Germany

For citizens of Russia	• *В России существует большое количество учебных центров, которые финансируются посольством Франции в Москве* • *Свём жилья во Франции «заочно» из России на долгий период или практически невозможен, или упирается в заведомо невыполнимые требования его владельца* • *Водительские права, полученные в России ("международные права"), действительны во Франции только в случае вашего пребывания во Франции не более года*
For citizens of Australia	• *Many Australian universities offer courses in French language, for example by enrolling in a Bachelor of Languages or a part-time Diploma of Languages* • *New Caledonia is the closest French-speaking country to Australia*
For citizens of Germany	• *WGs sind in Frankreich weniger üblich als in Deutschland. Es ist auf jeden Fall sinnvoll, vor Ort zu suchen, um sich ein Bild von der Wohnung machen zu können* • *Eine Steuererklärung muss in Frankreich jedoch ab dem 18. Lebensjahr erfolgen (entweder alleine oder gemeinsam mit den Eltern), unabhängig davon, ob Steuern gezahlt werden*

Source [6]

Today, many large corporations, such as Twitter, WeChat, Foursquare, TensorFlow, P&G, and others, use machine translation software to provide multilingual support for user-generated content. The latest developments in the field of machine translation are associated with improving the quality of translated texts through the use of neural networks (Neural Machine Translation—NMT). The Google company offers innovative NMT-based technologies Google Cloud (Translation API, AutoML Translation, and Media Translation API) for instant translation of website content into more than 100 languages from Afrikaans to Zulu. In addition to the standard function of embedding machine translation software into online platforms, new technologies allow for building custom models for domain-specific needs, i.e., applications learn to work with texts of specific genres and topics.

Figure 23.1 shows an example of a bilingual post from 'The UK in Russia' community of the Twitter social network. The first part of the post is a translation of the tweet performed by the built-in Google Cloud program in accordance with the user's default language settings. It is followed by the text in the original English language.

Online platform of the largest electronic payment system PayPal is another example of Google Cloud-based website which has localized versions for 202 countries. New translation technology enables PayPal to help people around the world to make financial transactions using simple, convenient, and secure online services [13]. As in the previous example of the Twitter post's machine translation, the translated content of PayPal multilingual versions is generated according to the principle of semantic equivalence with the preservation of translation invariants and the absence of human-made pragmatic adaptation. Table 23.2 shows the main heading of the section 'How PayPal Works' in the Russian, English, German and French languages to prove the aforesaid.

Let us also give a few examples of multilingual websites of German cities with a special emphasis on social and marketing aspects. Multilinguality pursues two goals:

Fig. 23.1 Bilingual post from 'The UK in Russia' community of the Twitter social network. *Source* [27]

Table 23.2 Text fragments of the section 'How PayPal Works' in the Russian, English, German and French languages

Russian	English	German	French
С PayPal вы можете отправлять переводы, совершать покупки и управлять средствами с помощью одного интернет-кошелька	*Millions of people trust PayPal to buy, sell, and send money – without sharing their financial information*	*Mit PayPal können Sie einkaufen, Geld senden und anfordern und vieles mehr. UnddasmitnureinemKonto*	*Des millions de personnes font confiance à PayPal pour acheter, vendre et envoyer de l'argent, sans avoir à partagerleurscoordonnéesbancaires*

Source [21]

Fig. 23.2 Posters of Emden and Wuppertal promoting vaccination. *Source* [8, 19]

internal and external. The internal, inclusive goal is associated with a focus on urban migrants who do not have sufficient language skills to participate in city life. In this case, multilinguality is becoming a "language mode" out of digital communication—in various institutions, on large construction sites, in school classrooms or entire city districts [5]. The external goal is to attract the attention of a global audience of interested users who are not proficient in German and provide them with the information about a city's economy, tourism, and other sectors. Cities are showing a growing interest in place branding and self-promotion in the global Internet competition.

Multilinguality can be observed in specific campaigns. For example, the cities of Emden and Wuppertal (Fig. 23.2) conduct campaigns for vaccination against coronavirus infection through the use of posters depicting local landmarks (for example, the cable railway in Wuppertal).

Slogans with the word "vaccinated", such as "*Wuppertal residents are vaccinated*" or "*Emden is getting vaccinated*" are given in many languages—Russian, Turkish, Polish, Italian, Portuguese, Greek, Romanian, Farsi, and Arabic. These are the native languages of the majority of migrants. Ensuring the availability of relevant information by overcoming language barriers has become a generally recognized state and public task. This is convincingly proven by the educational activities of various institutions. For example, informational materials on vaccination are presented on the web portal of the Robert Koch Institute in the 23 most common languages of migrants [4]. The website of the German Federal Government for Migration, Refugees, and Integration (Fig. 23.3) also provides information about anti-coronavirus measures in 23 languages [28]. Many higher education institutions of Germany are socially active and offer guides for creation of multilingual websites. Such activities are another promising opportunity for inclusive growth.

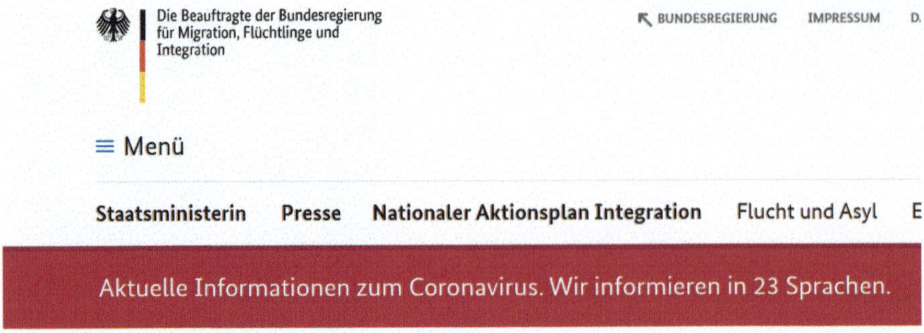

Fig. 23.3 Homepage of the website of the German Federal Government for Migration, Refugees, and Integration. *Source* [28]

Multilinguality is an important component of semiotic and linguistic landscapes which provides a variety of communication opportunities and increases attractiveness of places [16]. Regional factors often play an important role along with the social function of inclusion. In Germany, 4 minority languages (Frisian, Danish, Sorbian, and Romanic) are legally protected by the Council of Europe. In this connection, the official website of the city of Flensburg [23] has a Danish-language version along with the German and the English ones. In this way, the federal state of Schleswig–Holstein supports its Danish-speaking residents and focuses on the citizens of the neighboring state.

The website of the border town of Görlitz is presented in German, English, Czech, and Polish [9], since Czechs and Poles being the neighbors are among the frequent website users. To date, the website does not have a Sorbian-language version, despite of the fact that Görlitz belongs to the Sorbian-speaking region. Only the website of the city of Cottbus has a version in Sorbian (or rather, in Lower Sorbian) and is also localized into border Polish [7]. It should be noted that the city of Zittau, located at the "junction of three countries" (Dreiländereck), also uses the multilingual official website with a localized version in Czech to draw attention to the Dreiländereck economy [29]. This website does not have an English-language version. The website of the city of Aachen has four language options besides German—English, Chinese, Dutch, and French [1].

Thus, cities' multilingual websites do not only ensure the global availability of information necessary in economic and tourism communication, but also strengthen local / regional identities and promote friendly relations with neighboring countries. Integration of the Google Translate system into websites to transfer content in a large number of languages is also a friendly one-click step on the way of multilinguality becoming the standard of digital platforms focused on social communication accessibility and efficiency.

23.4 Conclusion

On the one hand, multilinguality of digital platforms is an important component of the inclusive language policy, and on the other hand, it is a competitive advantage of websites from the marketing point of view. The importance of translation grows with the increase in the volumes of information, its digitalization, and the speed of distribution, which leads to the emergence of innovative translation technologies. Website localization aims to create local culture-sensitive versions for the key target audiences, while the capabilities of neural machine translation can significantly expand the number of potential users due to numerous language options. At the same time, localization of texts, performed by human translators with a focus on pragmatic characteristics of an audience, assumes a high degree of adaptation of source texts and requires a significant investment of time and money. In contrast, the latest machine translation applications are able to instantly process large amounts of data at a lower price. In general, it is worth noting that the ability of translation programs to self-study and accumulate information is an important element in building a global Industry 4.0.

References

1. Aachen Tourist Service. https://www.aachen-tourismus.de/en/. Accessed 14 Aug 2021
2. Airband Initiative. https://www.microsoft.com/en-us/corporate-responsibility/airband. Accessed 22 July 2021
3. Alekseeva, I.S.: Introduction to Translation Studies. St. Petersburg, Moscow (2004)
4. Aufklärungsmerkblatt zur COVID-19-Impfung mit mRNA-Impfstoff. https://www.rki.de/DE/Content/Infekt/Impfen/Materialien/COVID-19-Aufklaerungsbogen-Tab.html. Accessed 14 Aug 2021
5. Busch, B.: Local actors in promoting multilingualism. In: Hogan-Brun, G., Mar-Molinero, C., Stevenson, P. (eds.), Discourses on Language and Integration. Critical Perspectives on Language Testing Regimes in Europe, pp. 129–152. John Benjamin, Amsterdam (2009)
6. Campus France. https://www.campusfrance.org/ Accessed 15 August 2021
7. Cottbus. https://www.cottbus.de/. Accessed 14 Aug 2021
8. Die offizielle Website der Stadt Emden: Das Meer an Leben. https://www.emden.de. Accessed 14 Aug 2021
9. Die Stadt Görlitz – Erlebnis Görlitz. https://www.goerlitz.de. Accessed 14 Aug 2021
10. Federal Law "On Communications" (as amended on July 2, 2021): Electronic fund of legal and regulatory technical documents. https://docs.cntd.ru/document/901867280. Accessed 20 July 2021
11. Geissbauer, R., Schrauf, S., Coch, V., Kuge, S.: Industry 4.0 – Opportunities and Challenges of the Industrial Internet (2014). https://www.pwc.nl/en/assets/documents/pwc-industrie-4-0.pdf. Accessed 22 July 2021
12. Hériard, P.: Sprachenpolitik. Kurzdarstellungen zur Europäischen Union. https://www.europarl.europa.eu/factsheets/de/sheet/142/sprachenpolitik. Accessed 15 Aug 2021
13. How PayPal empowers people and businesses in a global marketplace. https://cloud.google.com/customers/featured/paypal. Accessed 15 Aug 2021
14. Internet for All: A project of The World Economic Forum's Digital Economy and Society System. https://www3.weforum.org/docs/WEF_Internet_for_All_4_pager.pdf. Accessed 20 July 2021

15. Izsak-Ndiayne, R.: Linguistic Rights of Language Minorities: A Practical Guide by the United Nations Special Rapporteur on Minority Issues (2013). https://www.ohchr.org/Documents/Issues/Minorities/SR/LanguageRightsLinguisticMinorities_RU.pdf. Accessed 22 July 2021
16. Jaworski, A., Thurlow, C.: Introducing Semiotic Landscapes. Semiotic Landscapes. Language, Image, Space. London (2010)
17. Localization Services. Language Line Solutions (2017). https://www.languageline.com/localization-services. Accessed: 15 Aug 2021
18. Martin, W.J.: The Global Information Society. New York
19. Mobile Impfaktion im Haus der Integration. https://www.wuppertal.de/microsite/integrationsportal/aktuell/mobile-impfaktion-im-haus-der-integration.php#&gid=1&pid=1. Accessed: 14 Aug 2021
20. Orcutt, M.: The Online Language Barrier (2015). https://www.technologyreview.com/2015/03/06/168966/the-online-language-barrier/. Accessed 16 Aug 2021
21. PayPal. https://www.paypal.com. Accessed 15 Aug 2021
22. Smart Africa. https://smartafrica.org. Accessed 22 July 2021
23. Startseite. Flensburg. https://www.flensburg.de. Accessed 14 Aug 2021
24. Statistics and facts on the language services industry in the U.S. https://www.statista.com/topics/2152/language-services-industry-in-the-us/. Accessed: 13 Aug 2021
25. Tarasov, I.V.: Industry 4.0: Concept, Theories, Development Trends. Business Strat. **5**, 57–63 (2018). https://doi.org/10.17747/2311-7184-2018-5-43-49
26. Trachuk, A.V., Linder, N.V., Ubeiko, N.V.: Formation of dynamic business models by e-commerce companies. Manager **4**(68), 61–74 (2017)
27. UK in Russia. Twitter. https://twitter.com/ukinrussia. Accessed 15 Aug 2021
28. Was Sie über das Corona-Virus wissen müssen. https://www.integrationsbeauftragte.de/ib-de. Accessed 14 Aug 2021
29. Wirtschaftsstandort Zittau. https://zittau.de/de/wirtschaft. Accessed: 14 Aug 2021

Chapter 24
Linguistic Diversity: Institutional Mechanisms, Language Policy, and Inclusive Economic Development

Larisa A. Kochetova

Abstract Inclusive economy is closely related to the digital transformation of society and has a linguistic dimension that requires the implementation of specific language policy ensured by state institutions and public organizations. The paper considers digitalization processes as having potential for preserving and maintaining small-numbered languages and defines the set of targets to provide their maintenance and preservation such as machine translation that can support the urgent task of documenting the world's minority languages; funding corpus-building projects that could be used for language development and documentation purposes, developing Internet resources that would help to create a learning environment and unite speakers of minority language from different locations. The results obtained would be useful when estimating long-term inclusive economic development as well as planning language policies, education policies aimed at preserving cultural and language diversity as an indispensable tool for the implementation of sustainable economic growth.

24.1 Introduction

Human capital as applied to knowledge of languages constitutes an important theme of the relatively new discipline of Language Economics that emerged at the cross of linguistics, economics, and sociolinguistics. Recently, analysis of language policy has started to consider the effect of language skills on income and trade, and the costs and benefits of language planning options, preservation of minority languages, etc. The human capital dimension represents a large part of the new discipline, as is explained in an edited volume on the subject by Gazzola and Wickström [7].

The idea of interaction between economy and language was first proposed by German sociolinguist Florian Coulmas who discussed 'how economic developments

L. A. Kochetova (✉)
Volgograd State University, Volgograd, Russia
e-mail: kochetova@volsu.ru

E. I. Inshakova and A. O. Inshakova (eds.), *New Technology for Inclusive and Sustainable Growth*, Smart Innovation, Systems and Technologies 287,
https://doi.org/10.1007/978-981-16-9804-0_24

influence the emergence, expansion, or decline of languages; how linguistic conditions facilitate or obstruct the economic process; how multilingualism and social affluence are interrelated; how and why language and money fulfill similar functions in modern societies; why the availability of a standard language is an economic advantage; how the unequal distribution of languages in multilingual societies makes for economic inequality; how the economic value of languages can be assessed; why languages have an internal economy and how they adapt to the demands of the external economy [5]. According to Coulmas, languages can behave like economic systems and their spread depends on economic conditions. This means that assigning an economic value to a certain language in the linguistic marketplace means vesting it with some of the privileges and power related to that language.

It has long been recognized that economic development puts pressure on local languages. The mechanism producing this pressure has to do with the fact that economic development is based on specialization and trade, which in turn requires a common language as a means of communication. Thus, as economic development proceeds more and more people take on a common language thereby reducing the importance of local languages and contributing to their extinction.

'Language planning refers to deliberate efforts to influence the behavior of others with respect to the acquisition, structure, or functional allocation of their language codes' [4, p. 45]. Language Policy can be implemented either officially through legislation, court decisions, or policy that determines how languages are used, cultivates language skills needed to meet national priorities, or establishes the rights of individuals or groups to use and maintain languages.

The concept of language policy (or language planning) offers a possible common framework among disciplines and a shared research area for linguists, economists, and social scientists in general. Language policies are increasingly acknowledged as being a necessary component of many decisions taken in several policy areas. For instance, language policy in education affects the supply of linguistic skills available in a given labor market, and it can influence workers' transitional mobility decisions; language policies adopted by universities, condition the access to knowledge for students; language planning at the state level can modify the rights of ethnic minorities, etc. Hence the type and the range of questions raised by some of the trends discussed above call upon innovative, efficient, and fair language policies to manage linguistic diversity at several levels, and therefore upon interdisciplinary research to support decision-making.

As Gazzola notes the potential contribution of economics and policy analysis to the study, the design, and the evaluation of language policies were already clear to sociolinguists in the 1970s, but the role of economics in language planning remained marginal for roughly 20 years. All this began to change in the 1990s, when a growing number of social scientists started to be more often involved in research on language in society and in the economy, and where different applied linguists felt the need for new inputs from the social sciences [7].

Methods used to measure the correlation between linguistic diversity and economy have been suggested in the literature. François Grin presents a broad-based overview of language economics and he proposes a theory-based typology of research and highlights the connections between economics and language policy, by focusing on the selection, design, and evaluation of language policies [9].

The approach developed by Victor Ginsburgh and Shlomo Weber expands the existing measurements of diversity that traditionally focus on the size or the number of different groups of speakers by the notion of distance or dissimilarity. A number of different indicators have been proposed to measure dissimilarity between languages based on the notions of 'fractionalization' and 'polarization.' It gives some examples of the relationship between, on the one hand, the indexes of linguistic fractionalization or polarization and, on the other hand, different economic and sociological outcomes such as growth, the quality of government, public good provision, and redistribution. Hence indexes of linguistic diversity can contribute to our understanding of the connections between languages and welfare [8].

There is a lack of consensus on the relation between linguistic diversity and GDP per capita. Some researchers take a negative view stating that 'a country that is linguistically highly heterogeneous is always undeveloped or semi-developed, and a country that is developed always has considerable language uniformity' [14, 15]. On the other hand, other researchers claim that, when using enough other explanatory variables, besides GDP, linguistic heterogeneity ceases to affect the level of economic development [10]. Laitin and Ramachandran come to a similar conclusion: once they account for linguistic distance from the official language, diversity no longer influences GDP per capita [13].

Linguistic Diversity Index (ILD) [20] is a new quantitative measure of trends in linguistic diversity. At the global level, the ILD measures how far, on average, the world's languages deviate from a hypothetical situation of stability in which each language is neither increasing nor decreasing its share of the total population of the grouping. The ILD can also be used to assess trends at various sub-global groupings. Key findings are:

- Globally, linguistic diversity declined 20% over the period 1970–2005.
- The diversity of the world's indigenous languages declined by 21%.
- Regionally, indigenous linguistic diversity declined over 60% in the Americas, 30% in the Pacific (including Australia), and almost 20% in Africa. According to some linguists, this process may eliminate up to 90% of the world's living languages before the end of this century.

This hypothesis was tested in the study [6] that using a sample of more than 160 countries, and controlling for a number of other factors, found that economic development (measured by per capita income) has a significant and powerful effect on linguistic diversity. By simulating a counter-factual scenario in which the question was asked what the number of languages would be if all the countries in the world had the same per capita income as the average European country, it was found that in this counter-factual scenario the number of languages in the world would be approximately 20% lower. The effect was found to be most pronounced for Africa where the number of languages would decline by 35%. Using forecasts of per capita income for 2100 assuming a yearly growth of per capita income of 2.5%, it was defined that under those conditions the number of languages would decline by approximately

32%, the figure much beyond the predicted 90%. This study paints a much brighter future for linguistic diversity, which should be indispensable to an inclusive economy, in which there is expanded opportunity for more broadly shared prosperity.

Another argument that supports linguistic diversity comes from the study [3] according to which there appears to be a close link between linguistic diversity and national identity. The results of a study on the impact of language skills on national identification (as opposed to ethnic identification) in different Sub-Saharan countries show that speaking more than two languages increases the chances of identifying in national terms rather than in ethnic (or tribal) terms. In other words, multilingual citizens are more committed to the nation. In former French African colonies, there is a positive relationship between feelings of national identification and the probability of having some language in common, a variable captured by the 'index of communication potential' (IPC), which is defined as the probability that an individual can communicate with another randomly selected person within the society [3].

It should be noted that in the studies of the relationship between language skills and the feeling of national identification, the focus is not on a well-defined economic variable such as income or growth, but on an essentially cultural variable as identity. Developing a feeling of collective identity can be viewed as a form of immaterial capital (or 'cultural capital') that can contribute to the well-being or welfare of a community.

As is shown above, besides global scale, linguistic diversity has one more dimension related to the variety of languages within a certain state, of which Russia as a home to many nationalities is a good example. Russia is a federal state consisting of units with equal status, which are called the *subjects of the federation.* They have different designations (okrugs, krais, etc.). Language policy in Russia is regulated by the Federal Law 'On the Languages of the Peoples of the Russian Federation' of 1991 [21], and the Federal Law 'On the State Language of the Russian Federation' of 2005 with amendments adopted in 2014 [23]. According to Article 68 of the Constitution of the RF, Russian is defined as the state language of Russia throughout the territory of the Russian Federation. However, Article 69 of the Constitution reads that the state protects the cultural identity of all peoples and ethnic communities of the Russian Federation, guarantees the preservation of ethnocultural and linguistic diversity. Thus, although there is the absence of direct enshrinement in the Constitution of the right of autonomous okrugs and regions to establish their own state languages, these subjects of the Russian Federation establish the official status of these languages by their own norms and laws.

Taking all into consideration, the paper seeks to propose an outline for language policy that would maintain linguistic diversity as an indispensable attribute of national identity that as a non-economic variable contributes to building of an inclusive economy giving priority to human capital.

24.2 Materials and Method

The data for this study were obtained from the Information Materials on the Final Results of the National Population Census 2010, Federal State Statistics Service [20], a previous sociolinguistic research [1], data on income throughout the regions of the Russian Federation provided by the Russian information Agency.

24.3 Results

According to the Linguistic Diversity Index (LDI, 2017), with the value 0.283, the Russian Federation is ranked 143 out of 209 countries. On a list of the number of languages by country, Russia is ranked 18 with as many as 115 languages and 44 immigrant languages, which makes 2.24% of the total number of the world's languages [21].

An analysis of the Russian regions on income and language diversity showed that according to the 2010 census, the regions of the Russian Federation with high linguistic diversity are the Komi Republic, The Tumen Oblast, Khanty-Mansi Okrug, the Yamalo-Nenets Okrug, The Republic of Adygea, The Republic of Sakha, the Republic of Dagestan, the Kabardino-Balkarian Republic, the Karachay-Cherkess Republic, the Republic of North Ossetia, The Republic of Tatarstan, etc. It should be noted that this list includes regions with the economic indicators that are among the highest in the country (the Khanty-Mansi and the Yamalo-Nenets Autonomous Okrugs, The Tumen Oblast), as well as regions with low economic performance [25].

Sociolinguistic research shows that minority languages are spread in many regions of the Russian Federation, some of which do not have the status of republic, autonomous okrug, or autonomous republic. According to the 2010 census [20], in Tymen Oblast, small-numbered people demonstrate high loyalty to their ethnic language as compared to other regions of the Russian Federation. For instance, in the Tymen Oblast, 84% of the Nenets claimed the Nenets language as the native whereas in the Arkhangelsk Oblast, of which the Nenets Autonomous Okrug is a part, this figure comprised only 56%; 70% of Tatars in Tymen Oblast indicated Tatar as their native language, whereas in other regions this proportion is much smaller. According to the 2010 census, out of 460,971 of Yakuts who indicated their language proficiency, 416,780 people claimed their command of Russian, and 401,240 claimed mastery in Yakut. At the same time, representatives of the Yakut nationality called their native language, basically, the Yakut language (438,664 people), a much smaller part noted Russian (27,027 people).

Table 1 shows the results of the four censuses for Kalmyk people who claim Kalmyk as their native language (the results of the three censuses are adapted from [1]).

We see that native language loyalty among Kalmyks that was in a small decline between 1970 and 1980s shows signs of recovery as more people claimed the status of Kalmyk as the mother tongue.

Considering the facts, the reasons behind low economic development must be sought elsewhere, and it would be misleading to conclude that higher linguistic diversity is the cause of poverty. T. Romaine argues that economic growth in the hotspots should go hand in hand with the protection and promotion of biodiversity and linguistic diversity. An increase in the socioeconomic well-being of indigenous populations could be achieved if such populations were allowed to preserve their traditional cultural and economic activities, which usually are intertwined with traditional knowledge and therefore transmitted with local languages. On the contrary, assimilation to the dominant culture and language creates cultural alienation, undermines the social cohesion of indigenous populations, and it promotes the abandonment of traditional economic activities that are usually linked to a deep knowledge and respect of the natural environment in which such activities are embedded [17].

The major role in preserving and maintaining small-numbered languages belongs to institutions, which Prof. Inshakov defined as functional organizations that ensure a realization of the specific system of homogeneous institutions. 'The variety organization forms for socially fixed functions are embodied in the system of institutions of society. In the consciousness of each person, this system forms "imaginary communities" to which a person assigns himself, and behaves accordingly ….' [11].

Ethnic identity, which is an indispensable part of social inclusion, is ensured by institutional mechanisms that act at the federal and local levels and complement each other to support and preserve minority and indigenous languages and promote inclusive economic growth. In a study of the Nenets language conducted by Zmyvalova [19], the institutional mechanisms that ensure the right of indigenous children to learn their mother tongue at school are discussed, and the core elements of this right were identified. The author points out that despite the fact that most of the international treaties in the sphere of regulation of this right have been implemented in Russia, it has become clear that formal implementation does not mean that implementation is carried out in fact. One of the problems the researcher names is the absence of a learning environment.

It is necessary to say that digital transformation of societies, which has a great impact on economic growth, can contribute to a great extent to the maintenance and preservation of minority languages, which is essential for inclusive economic development. Let us look at a few examples of how digitalization can help preserve linguistic diversity and provide social inclusion for people who belong to ethnic minorities. It is known that statistical machine translation has been remarkably successful for the world's well-resourced languages, and much effort is focused on creating and exploiting rich resources such as treebanks and wordnets. Machine translation can also support the urgent task of documenting the world's endangered languages. The primary object of statistical translation models, bilingual aligned text, closely coincides with interlinear text, the primary artifact collected in documentary linguistics. Even though there are many technical and logistical problems to be addressed, starting with the problem that—for most of the languages in question—no texts or lexicons exists, it ought to be possible to exploit this similarity in order to improve the quantity and quality of documentation for a language [2].

Another opportunity to document endangered languages and create resources that can be successfully used in education, lexicography, translation, etc. is corpus linguistics that provides tools for language documentation, processing, and development. The Corpus of Kalmyk language comprises 800 thousand tokens of various written genres such as novels, novellas, short stories, essays, newspaper articles (second half of the twentieth—beginning of the twenty-first centuries). The texts are provided with morphological markup that allows searching by lexeme, grammatical characteristics, translation, as well as by combinations of word forms at a given distance [24]. To date, the launch of the Yakut language Corpus [25] has been pronounced that would include samples of fiction, journalism, spoken texts, and folklore. The Corpus of the Yakut language would rely on the design of the National Corpus of the Russian language, and it would include samples of teenage speech and text samples of modern writers. Also, the Institute for Humanitarian Research is planning to develop morphological analyzers, which would be a large step toward the creation of artificial intelligence in Yakut, which is essential for inclusive digital economy that provides equal access to goods and services for every individual regardless of the language they speak. Earlier, in honor of the 100th anniversary of the Udmurt Republic, the national corpus of the Udmurt language was launched. The corpus of the Bashkir language is being compiled as well.

Digitalization and rapid Internet penetration in all spheres of life can help secure languages and create a new inclusive linguistic environment. As Kuzmin points out 'Use of ICT has both positive and negative consequences. On the one hand, it may decrease linguistic and cultural diversity, on the other hand, it opens new prospects for preserving and even developing languages and cultures in cyberspace' [12].

To promote multilingualism in cyberspace, goal-oriented measures need to be taken to create a multilingualism-friendly environment. These goals can include 'developing public use of the Internet for minority languages; elaborating information resource development programs in minority languages; promoting training in ICTs and information, especially in local languages; promoting the creation of content in local languages; encouraging translation of the world's classics into minority languages, and giving access to these works by posting them online; developing integrated multilingual information resource networks; introducing electronic documentation and record management in at least two languages; and promoting development of operating systems, search engines and Internet browsers in minority languages, developing online dictionaries, reference books, thesauruses, semantic nets in minority languages' [12].

So, we can conclude that the digitalization process can be advantageous for minority languages as it can provide a platform for communication, create a learning environment and unite people who share a minority language from different geographical locations thus maintaining a sense of ethnic identity and contributing to social inclusion and economic welfare. As one of the main challenges in preserving linguistic and cultural diversity is to encourage small indigenous nations to use their own languages at all possible circumstances, web may provide a communication model which can support solidarity within local community as well as mutual exchange with global community [16].

24.4 Conclusion

The 'economics of language', the field of research, which is grounded in the discipline of economics, and studies the ways in which linguistic and economic processes influence one another, provides analytical tools for displays a strong interdisciplinary orientation, which places it on the fringes of mainstream economics. It is also well placed to contribute to the evaluation of public policies regarding language because it offers analytical tools for the systematic identification and measurement of the advantages and drawbacks of policy alternatives and their contributions to the development of an inclusive economy via enhancing cultural identity.

It has been found that economic growth is not correlated with linguistic diversity, but the last need to be maintained to support inclusive economic development. The economic approach to language policy, and focus on an application to education policy, detailing the economics of second or foreign language education emphasizes the need to combine disciplines to develop an inclusive methodology for the selection, design, and implementation of language policy.

We think that the application of economic theories and research methods to the study of languages in society and multilingualism can be stimulating and bring about several insights. As languages are inextricably embedded in social interactions, an adequate knowledge of socioeconomic phenomena and of the relationship to language and communication may enhance our understanding of language behavior and language change.

Several aspects of such relationships need to be analyzed, notably: the impact of language diversity on economic outcomes, the link between language knowledge and national or ethnic identification.

The data show that ethnic populations in the regions with economic stability and ethnic variety demonstrate a strong sense of ethnic identity that is reflected in the number of people who claim ethnic language other than Russian their mother tongue. It seems that an inclusive economy would encourage language and cultural diversity.

Digitalization processes can be viewed as beneficial for maintaining language diversity and ethnic identity as they provide several opportunities for social and economic inclusion for numerous ethnic groups, whose representatives can be separated geographically. Availability of accessible ethnic languages corpora, machine translation technologies, Internet websites using languages other than Russian would enhance opportunities for social inclusion and contribute to sustainable and inclusive economic development.

Even though this study lacks quantification of the data we think that linguists and language policy scholars may benefit from becoming more aware of the potential relevance and import of formal models and quantitative analysis to the study of

Table 24.2 The proportion of the Kalmyk claiming Kalmyk as a native language

The year of census			
1970	1979	1989	2010
98.67	95.11	93.57	94.6

languages in society. The interdisciplinary dialogue between economists and other social scientists requires intense conceptual work and a sound understanding of complex variables such as language. Language policy can be a suitable common ground for interdisciplinary research. In order to understand existing language policies (either explicit or implicit) and to design and evaluate new policies and their effects on inclusive economic development, we need to take into account both patterns of language use, and people's linguistic attitudes, ideologies, on the one hand, and notions such as costs, benefits, incentives, and welfare, economic growth, on the other hand.

References

1. Belikov, V.I., Krysin, L.P.: Sociolingvistika (Sociolinguistics). Moscow (2001)
2. Bird, S. Chiang, D.: Machine translation for language preservation. In: Proceedings of COLING 2012: Posters, Mumbai, pp. 125–134. COLING (2012)
3. Buzasi, K.: Languages and national identity in Sub-Saharan Africa: a multilevel approach. In: Gazzola, M., Wickström, B.-A. (eds.) The Economics of Language Policy, pp. 225–264. The MIT Press (2016)
4. Cooper, R.L.: Language Planning and Social Change. Cambridge University Press, Cambridge (1989)
5. Coulmas, F.: Language and Economy. Blackwell Publishers, Oxford (1992)
6. De Grauwe, P.: Language Diversity and Economic Development Manuscript, Katholieke Universiteit Leuven (2006)
7. Gazzola, M., Wickström, B.-A.: The Economics of Language Policy. The MIT Press (2016)
8. Ginzburgh, V., Weber, S.: Linguistic diversity, standardization, and disenfranchisement: measurement and consequences. In: Gazzola, M., Wickström, B.-A. (eds.), The Economics of Language Policy, pp. 95–140. The MIT Press (2016)
9. Grin, F.: Language Planning and Economics. Curr. Iss. in Lang. Plan **4**(1), 1–66 (2003). https://doi.org/10.1080/14664200308668048
10. Fishman, J.: Some contrasts between linguistically homogeneous and linguistically heterogeneous polities. In: Fishman, J., Ferguson, C., Das Gupta, J., (eds.), Language Problems of Developing Nations, pp. 53–68. Wiley, New York (1968)
11. Inshakov, O.: Social nature and specifics, typology and classification of economic institutions. Sotsiologicheskie Issledovaniya [Sociology research] **9**, 42–51 (2003)
12. Kuzmin, E.: Social institutions supporting linguistic and cultural diversity in cyberspace: roles, functions, responsibilities. In: Linguistic and Cultural Diversity in Cyberspace. Proceedings of the 2nd International Conference, Yakutsk, Russian Federation, 12–14 July, 2011. pp. 32–45. Interregional Library Cooperation Centre, Moscow (2012)
13. Laitin, D., Ramachandran, R.: Language Policy and Human Development. Am. Polit. Sci. Rev. **110**(3), 457–480 (2016). https://doi.org/10.1017/S0003055416000265
14. Nettle, D.: Linguistic fragmentation and the wealth of nations: the fishman-pool hypothesis Reexamined. Econ. Dev. Cult. Change **48**, 335–348 (2000)

15. Pool, J.: National development and language diversity. In: Fishman, J., (ed.), Advances in the Sociology of Language, vol. II (The Hague: Mouton), pp. 213–230 (1972).
16. Rhee, Y.: A web-based participation model to preserve linguistic and cultural diversity. In: Linguistic and Cultural Diversity in Cyberspace. Proceedings of the 2nd International Conference (Yakutsk, Russian Federation, 12–14 July, 2011). pp. 131 – 135. Interregional Library Cooperation Centre, Moscow (2012)
17. Romaine, S.: Language, Education and Development. Urban and Rural Tok Pisin in Papua New Guinea. Oxford Univ Press, Oxford (1992)
18. Samsonova, V.: Preservation of cultural heritage of indigenous small-numbered peoples of the Russian North, Siberia and the Far East. In: Linguistic and Cultural Diversity in Cyberspace. Proceedings of the 2nd International Conference, Yakutsk, Russian Federation, 12–14 July, 2011. pp. 240–245. Interregional Library Cooperation Centre, Moscow (2012)
19. Zmyvalova, E.: Indigenous children's right to learn their mother tongue at school: implementation and realization in Russia. Arctic Rev. Law Polit. **6**(2), 155–158 (2015)
20. Final Results of the National Population Census 2010. Federal State Statistics Service. http://www.gks.ru/free_doc/new_site/perepis2010/perepis_itogi1612.htm. (2010). Accessed 29 Sept 2021
21. Linguistic Diversity Index (LDI). https://www.nationalgeographic.org/maps/language-diversity-index/. Accessed 12 Sept 2021
22. On the languages of the peoples of the Russian Federation: Federal Law №. 1807-I of October 25, 1991. Moscow, 1991. - 11 sheets. - Electronic version of the normative-legal act (11 files, 300 dpi, JPEG, 8.6 MB). President of Russia, the official website. http://www.kremlin.ru/
23. On Amendments to the Federal Law 'On the State Language of the Russian Federation' and certain legislative acts of the Russian Federation in connection with the improvement of legal regulation in the sphere of the use of the Russian language: Federal Law No. 101-FZ of May 5, 2014: adopted by the State Duma on April 23, 2014: approved by the Federation Council on April 29, 2014. Moscow, 2014. 11 sheets. RIA Rating. https://riarating.ru/infografika/20210706/630203876.html. Accessed 12 Sept. 2021
24. The Kalmyk Language Corpus. http://web-corpora.net/KalmykCorpus/search/index.php?interface_language=ru. Accessed 21 Sept 2021
25. The Yakut Language Corpus. http://adictsakha.nsu.ru/corpora/corp. Accessed 21 Sept 2021

Chapter 25
Host Country Language Teaching: Theoretical and Practical Aspects of Providing Inclusive Education for International Students

Nikolay L. Shamne, Marina V. Milovanova, and Elena V. Terentyeva

Abstract The contributors aim at substantiating the need to create and test universal methodologies ensuring inclusive educational environment for non-native speaker learners in universities of host countries. The study shows that inclusive educational environment is impossible without the use of digital technologies (mobile learning, learning using VR and AR, artificial intelligence). The chapter proposes a universal methodology for organizing teaching on the basis of modular training using mobile technologies of industry 4.0. The authors present the results of the implementation of modules in experimental groups of non-native speaker learners and trainees of pre-study courses of Volgograd State University in the 2020–2021 academic year. The academic results of the course participants taking part in the experiment were processed using the Fisher angular transformation method, which allows comparing samples with high accuracy of calculations. Analyzing the participants' satisfaction with the training results the researchers used a questionnaire based on Likert scale and followed by data processing. The study showed that the use of the proposed training module based on mobile technologies of industry 4.0 increases the level of satisfaction with learning, which plays a crucial role in the process of adaptation of non-native speaker learners to a foreign-language learning environment.

25.1 Introduction

Currently, the ideas of the fourth technological revolution are being actively discussed in various research fields, including education from the standpoint of both theory and practice. In the field of education these ideas are primarily related to the introduction and active use of experimental teaching methods based on digital technologies, which in many ways make it possible to facilitate the inclusion of foreign students in the educational environment and to reduce to a certain extent the time of social and cultural adaptation [1–3, 12]. The possibilities of using digital technologies in the

N. L. Shamne (✉) · M. V. Milovanova · E. V. Terentyeva
Volgograd State University, Volgograd, Russia

E. I. Inshakova and A. O. Inshakova (eds.), *New Technology for Inclusive and Sustainable Growth*, Smart Innovation, Systems and Technologies 287,
https://doi.org/10.1007/978-981-16-9804-0_25

process of teaching foreign students are becoming particularly relevant at the present time—the period of the COVID-19 pandemic has determined the demand for distance and mobile learning. The challenges of the modern global world require higher education to develop and test universal methods that ensure inclusive teaching of non-native speakers in host educational institutions. It is obvious that in the conditions of the fourth industrial revolution inclusive education should be organized using m-learning, Augmented Reality (AR) and Virtual Reality (VR) technologies, and artificial intelligence. The authors present an analysis of the experimental activities of universities to solve the problems of organizing inclusive teaching of the Russian language to listeners and students based on Industry 4.0 technologies on the example of Volgograd State University in the framework of "projects main task of which is to digitalize education and maintain a unified language educational space using the state language of the Russian Federation" [11].

25.2 Methodology

While discussing the digital educational environment, and in particular, the inclusive training of foreign students, the researchers emphasize the need to encourage motivation to study, while noting a number of positive aspects: saving educational content, matching it with the professional needs of students, focusing on the student himself, and the intensity of the educational process, etc. [5; 7]. Volgograd State University is seriously interested in increasing the motivation to study the basics of Russian as a state language. In this regard, in the 2020–2021 academic year the administration of VolSU initiated an experimental study aimed at finding forms of effective modular learning. The study consisted of three stages. At the first stage, in February 2020 an entrance test in the Russian language was conducted on the topics of the training module "Business Communication" for non-native students of the 2nd year of the 45.03.01 Philology course and foreign trainees of the VolSU preparatory courses. The second stage included training in the basics of business communication in Russian as part of the module in the period from March to May 2020. At this stage, it was assumed that experimental and reference groups would be formed from each of the tested groups of students and trainees to evaluate the results of the inclusive training on the topics of the training module "Business Communication", however, the transition to a distance learning format in March 2020 in connection with the COVID-19 pandemic required changes to the initial plan of the experimental study. Studying of the "Business Communication" module was organized remotely on the basis of the electronic information and educational environment (EIOS VolSU) using the mobile learning functionality. The use of m-learning technologies has become possible due to the 100% availability of smartphones for students. Students and trainees were offered several training sets created using the Quizlet platform [10]. This platform has proven itself well enough in learning foreign languages [8; 13]. The material of the training module "Business communication" was taken as a basic course, which "was organized according to the thematic principle and included several standard situations "Business conversation in the office", "Specifying information by phone", "Filling out a questionnaire", "Writing an application". Lexical and grammatical material and a system of exercises were offered for each situation [9].

The third stage included students and trainees' polling and their impression appraisal after training.

In the course of the work reference and experimental groups of students and trainees were formed to compare the results of the inclusive modular and traditional training. We formed the groups randomly without taking into account the level of students' knowledge. The students of the experimental groups completed modular distance learning course using m-learning technology; the reference groups completed distance learning course according to the traditional system within the framework of the courses "Culture of Communication" and "Culture of communication in various fields of communication" and the Russian language course with the section "Business Communication" for non-native speaker trainees of the VolSU preparatory courses.

After the training the participants of the experiment completed verification tests. In order to provide a statistically confirmed evaluation of the obtained learning results, we used the Fisher angular transformation method to take advantage of the possibility of comparing small samples with high accuracy calculations using the formula $\varphi = 2*\arcsin()$, where P is the percentage expressed in fractions of one. To evaluate the results obtained students and trainees were asked to fill out a questionnaire about their impressions of modular training at the end of the semester.

25.3 Results

The first stage of the study, within which the entrance testing of students of reference and experimental groups in the Russian language was carried out, showed that all the test subjects demonstrated approximately the same level of proficiency in Russian. It is obvious that the second-year non-native speakers studying within the program "Philology" had slightly better results than the trainees of the preparatory courses. Next, we started the second stage, which involved teaching the module "Business Communication" in a distance format applying mobile learning technologies. Mobile training was organized as follows: learning study sets—interactive work with lexical and grammatical material in a distance format—work with authentic business texts (forms, samples of applications, questionnaires; texts of short business conversations on the phone, etc.), feedback, evaluation. As criteria by which it was possible to evaluate the effectiveness of modular learning using mobile learning based on the Quizlet platform, the following indicators were determined: the time spent on studying the training set, the percentage of correct answers when performing lexical and grammatical exercises and tasks for texts. In order to evaluate the results obtained, students and trainees were asked to fill out a questionnaire about their impressions of the training. We organized a survey using the classic Likert scale, consisting of five ratings: disagree—1; partially disagree—2; don't know—3; partially agree—4; agree—5 [4, 6].

Mobile training in experimental groups was included in the module "Business communication". When studying each of the situations, students and listeners referred to the author's study sets created specifically for this module. The functionality of the Quizlet platform allows one to use various illustrations to facilitate the perception of the learning situation when studying lexical and grammatical material. For example, the training sets "Specifying information by phone", "Filling out a questionnaire", "Writing an application" were provided with accessible images that visually reinforce the given lexical minimum and the lexical and grammatical constructions necessary for memorization. The academic results of mobile learning had to be demonstrated to students and trainees when performing tasks and tests at electronic information educational environment of VolSU on the site of the modular course. Of course, we are aware that the inclusive modular training in business communication in Russian using distance learning and m-learning technologies was a kind of pilot project, and we had to change our initial plans in the situation of the COVID-19 pandemic, actively adapt the training material to new conditions, therefore the obtained training results need to be additionally verified and studied more detailed. However, we still believe that the conducted experiment is a valuable experience that can be used in the future when organizing and combining different training formats.

So, proceeding to discussing the effectiveness of the inclusive modular training we represent the table below showing the results of the final testing on all topics of the inclusive modular distance learning applying m-learning technology in the experimental and using traditional methods in the reference groups of non-native speakers studying in the bachelor's degree program "Philology", the data are calculated using the Fisher angular transformation method. The column "Effect obtained" indicates the number of subjects who successfully coped with the final test (83.3 %) and showed results above 60 points on a 100-point scale (Table 25.1).

The obtained result φ empirical 2.621 is higher than the accepted critical value of φ 2.11, which is statistically reasoned proof of the effectiveness of the training.

Proceeding to the consideration of the results of the training of students of the Russian language courses (see Table 25.2) it can be seen from the table, the value of the empirical 2.314 is higher than the accepted critical value of 2.21, which is also a statistically sound proof of the effectiveness of the training.

Thus, the data of φ*empirical confirmed the hypothesis concerning the effectiveness of the inclusive modular training in Business Russian course, since the empirical value of the Fisher criterion ($\varphi^{*}emp = 2.621$ and $\varphi^{*}emp = 2.314$) exceeds the critical value when comparing the results of training in experimental and reference groups. In order to get feedback from students and trainees from participating in the experiment, we asked the participants of the experimental groups to fill out a questionnaire where they needed to evaluate modular learning using the Likert scale. The survey was attended by students in the number of 15 people (trainees) and 12 people

Table 25.1 The results of the effectiveness of modular training in the 1st and 2nd groups of 2nd-year students of the bachelor's degree program "Philology"

Groups	Effect obtained	No effect	Sums
	Number of subjects	Number of subjects	
1 experimental group	10 (83.3%)	2 (16.7%)	12 (100%)
2 reference group	4 (33.3%)	8 (66.7%)	12 (100%)
Answer: $\varphi^*_{emp} = 2.621$			

Source Compiled by the authors

Table 25.2 Results of the effectiveness of modular training in the 1st and 2nd groups of students of Russian language courses

Groups	Effect obtained	No effect	Sums
	Number of subjects	Number of subjects	
1 experimental group	12 (80.0%)	3 (20.0%)	15 (100%)
2 reference group	6 (40.0%)	9 (60.0%)	15 (100%)
Answer: $\varphi^*_{emp} = 2.314$			

Source Compiled by the authors

(students) who expressed their assessment with " + " or "v" signs. An example of the questionnaire will be shown in Table 25.3.

Basing on the results of the survey the authors complied two summary tables for experimental groups indicating the total score for all judgments for each student. Then the obtained data were statistically processed with the formation of a conjugacy table, on the basis of which the correlation coefficient for each judgment was calculated, followed by the calculation of Spearman's rank correlation coefficient using the Excel package. The results are presented in Table 25.4.

To interpret the data obtained the authors used a standard scale, according to which coefficient values less than 0.3 are an indicator of weak tightness of the connection of signs; values greater than 0.3, but less than 0.7 are an indicator of average tightness of the connection of signs, and values 0.7 and more are an indicator of strong tightness of the connection of signs. Thus, the results of statistical data processing show the presence of a strong connection for the judgment 3 "I liked learning business Russian on the Quizlet platform." A fairly high indicator of the closeness of the connection of the signs demonstrates judgment 2 "I liked to study business Russian remotely, the tests allowed me to monitor my academic results," the closeness of the connection of the remaining judgments is in the average values. In general, the results obtained indicate that the students positively evaluated the modular training and its results. We also found out that more than 80% of students and trainees who have completed the included modular training highly appreciated the use of the Business Communication training module, they noted that training within this module allowed them to adapt

Table 25.3 Questionnaire form for evaluating judgments on learning and its results on the Likert scale

Judgment	Disagree	Partly disagree	Don't know	Partly agree	Agree
	points				
	1	2	3	4	5
1. Studying the module "Business communication" in Russian was useful for me					
2. I liked learning business Russian remotely, the tests allowed me to monitor my academic results					
3. I enjoyed learning Business Russian on the Quizlet platform					
4. I am satisfied with my academic results					
5. Studying "Business Communication" module has increased my motivation to learn Russian					

Source Compiled by the authors

Table 25.4 Results of the Spearman coefficient calculation

Judgement	Coefficient r_s
1. Studying the module "Business communication" in Russian was useful for me	0.42
2. I liked learning business Russian remotely, the tests allowed me to monitor my academic results	0.67
3. I enjoyed learning Business Russian on the Quizlet platform	0.71
4. I am satisfied with my academic results	0.32
5. Studying "Business Communication" module has increased my motivation to learn Russian	0.45

Source Compiled by the authors

to the conditions of study at a Russian university better, increased their motivation to study Russian and professional disciplines in Russian.

25.4 Conclusion

Thus, the effectiveness of modular training of non-native speaker learners in a distance format (m-learning technology) indicates the feasibility of the proposed approach. The results obtained can be used to develop other modules within inclusive training of non-native speaker learners with the expansion of the set of digital platforms and technologies.

Currently, the large-scale changes caused by the development of the fourth industrial revolution (Industry 4.0) cover all spheres of society. The introduction of digital technologies in education is a kind of revolution in the field of access to knowledge, as well as a rethinking of traditional learning concepts. Training becomes more interactive and individual, since it allows, on the one hand, to involve the student in the learning process more actively, on the other hand, it makes it possible to determine individual performance more clearly. Applying distance format of modular training of foreign students increases their motivation to a certain extent, since it sets clear goals for them, allows them to "immerse themselves" in a certain thematic area, and also gives them the opportunity to clearly see their mistakes and achievements.

Further development of the proposed methodology for teaching non-native speaker learners in the context of the development of new technologies in society (Industry 4.0) is a promising linguodidactic direction focused on the development of individual competencies of students in the context of digitalization of the educational process.

Digitalization of education in its various formats contributes to the inclusive growth of Russian society, since it makes the educational environment accessible to speakers of any languages and cultures, regardless of origin, gender, physical abilities, etc.

References

1. Anger, C., Plünnecke, A., Schüler, R.M.: INSM-Bildungsmonitor 2018. Participation, prosperity and digitization. Study on behalf of the Initiative New Social Market Economy (INSM) (2018). http://Bildungsmonitor:KaumnochFortschritte-iwd.de. Accessed 16 June 2021
2. Ayupova, G.T., et al.: Tsifrovizatsiya kak drayver rosta nauki i obrazovaniya: monografiya (Digitalization as a compiler of the growth of science and education). ICNP "New Science", Petrozavodsk (2020)
3. Borgwardt, A.: Digitization in Science: Series of Network of Excellence at German Universities (2018). http://library.fes.de/pdf-files/studienfoerderung/14620.pdf Accessed 16 June 2021
4. Croasmun, J.T., Ostrom, L.: Using Likert-type scales in the social sciences. J. Adult Educ. **40**, 19–22 (2011)
5. Hoffmeyer-Zlotnik, P., Grote, J.: Recruiting and retaining international students in Germany. Study by the German national contact point for the European Migration Network (EMN) (2019). http://11b_germany_international_students_de.pdf (europa.eu). Accessed 16 June 2021

6. Kvon, G.M., Vaks, V.B. Pozdeeva, O.G.: Ispolzovanie shkaly Laykerta pri issledovanii motivatsionnykh faktorov obuchayushchikhsya (The use of Likert scales in the study of motivational factors of students). Sci. Methodol. Electron. J. "Concept" **11**, 1039–1051 (2018). https://doi.org/10.24411/2304-120X-2018-11086
7. Martin, F., Bolliger, D.U.: Engagement matters: student perceptions on the importance of engagement strategies in the online learning environment. Online Learn. **22**(1), 205–222 (2018). https://doi.org/10.24059/olj.v22i1.1092
8. Martin, F., Ertzberger, J.: Here and now mobile learning: An experimental study on the use of mobile technology. Comput. Educ. **68**, 76–85 (2013)
9. Milovanova, M.V., Terentyeva, E.V.: Modulnyy printsip obucheniya migrantov russkomu yazyku (Modular printouts of teaching Russian to migrants). Nizhny Novgorod Educ. **3**, 182–186 (2013)
10. Quizlet: http://quizlet.com Accessed 20 March 2020
11. Shamne, N.L.: Metodologicheskie aspekty issledovaniya problemy adaptatsii migrantov (Methodological aspects of the study of the problem of adaptation of migrants). Nizhny Novgorod Educ. **3**, 176–182 (2013)
12. Strokov, A.A. (2020). Tsifrovizatsiya obrazovaniya: problemy i perspektivy (Digitalization of education: problems and prospects). Bull. Mininsky Univ. **8**(2). https://doi.org/10.26795/2307-1281-2020-8-2-15
13. Valeeva, N.G., Pavlova, E.B., Zakirova, Yu.L.: M-learning in teaching ESP: case study of ecology students. Eur. J. Contemp. Educ. **8**(4), 920–930 (2019). https://doi.org/10.13187/ejced.2019.4.920

Chapter 26
Toponymic Policy as a Prerequisite for Sustainable Socio-economic Regional Development

Dmitriy Y. Ilyin and Elena G. Sidorova

Abstract This study reveals the aspects of the present-day toponymic policy, providing cross-disciplinary insights. Its relevance is determined by the fact that said policy facilitates gaining scientific knowledge on the prevalence of symbolic signs over the public consciousness, as well as on the instrumental function of toponyms in terms of shaping and upkeep cultural and ethnic mentality. Special attention is paid to the increasing social importance of proper names assigned to urban elements in terms of various communicative aspects, indicating their inclusion in social and business activities and in the translation of meanings related to national and regional cultures. The principles of nominating geographic objects are of scientific interest. The total number of these names, understood as the hodonymic space, provides the focus of research on the toponymic policy notion and its theoretical and methodological reasoning. The study analyzes not only the historically established totality of names given to linear toponymic objects in three regional cities that reflects the diverse ethnic contacts of the Volgograd region (Russia) as a multiethnic territory but also the renaming of these objects as a result of government "intervention" into the hodonymic space, as toponymy is impacted by various extralinguistic factors, i.e., social, cultural, and historical circumstances.

26.1 Introduction

Toponymic policies of Russian federal subjects need to be institutionalized; in other words, they are to be coordinated by legally authorized regional structures that have to solve the challenging task of structuring, registering, and normalizing the hodonymic space of their region [1–3]. As O.V. Inshakov rightfully noted, the linguists should not only study the institutional theory "as is" but also develop it in multiple ways that include adapting innovative concepts [4, p. 47].

D. Y. Ilyin (✉) · E. G. Sidorova
Volgograd State University, Volgograd, Russia
e-mail: dilyin@volsu.ru

E. I. Inshakova and A. O. Inshakova (eds.), *New Technology for Inclusive and Sustainable Growth*, Smart Innovation, Systems and Technologies 287,
https://doi.org/10.1007/978-981-16-9804-0_26

Special attention is paid to the increasing social importance of proper names assigned to urban elements (streets, secures, alleys. etc.). We fully agree with M.V. Golomdiova, who believes that the changing demands of urban culture and communication, the necessity of cities as economic entities to successfully position themselves, coordinated activities of authorities in terms of urban nominations as part of planned and systemic urban development policy, and the need to establish reliable regulatory and legal grounds for toponymic nomination represent only some of the challenges the present-day toponymic policy needs to address when it comes to nominating urban objects [5, p. 37]. From the territorial development standpoint, it is worth noting that the modern onomastic space is actively expanding and becoming denser while keeping some of its permanent elements for centuries. However, the human being has always remained in its core as someone who invents, assigns, uses, criticizes, or changes the names [6, p. 102]. Artificially nominated toponymic objects represent the ethnic and cultural background, as well as the political discourse adopted by authorities in charge of naming and renaming geographic objects in the region.

26.2 Methodology

The materials chosen for study are hodonyms of streets, alleys, squares, avenues, and other address-forming elements of the street network, presented in the "Register of Volgograd and Volgograd Region Addresses" informational website (later referred to as the Register).

To analyze the hodonymic space specifies, we have selected three urban settlements: *Volgograd,* the regional capital and the largest city in the region, with a population of about 1 million people, *Kamyshin,* the town populated by approximately 100,000 people, and *Petrov Val,* the small town in the Kamyshin District of the region, with a population of about 10,000 people.

The hodonymic space is represented by approximately 2400 elements of the street network in Volgograd (in the descending order: streets, alleys, lanes, squares, avenues, embankments, boulevards, highways, cul-de-sacs, junctions) approximately 300 elements in Kamyshin (streets, alleys, lanes, squares) and 100 elements in Petrov Val (streets, alleys, avenues).

The hodonymic space of any settlement is diverse; the onomastic behavior of the name-givers and municipal authorities dictates the choice of names for various urban objects, their visual presentation in the urban environment, renovation or preservation of various onyms [6, p. 104]. Describing the elements of this network provides a comprehensive understanding of the structuring of urban nominations and helps specify the criteria for linguistic coding of linear objects in a settlement [7–9].

The research methodology includes general scientific methods of induction, generalization, analysis, synthesis, and description, as well as specialized linguistic methods of structural-semantic analysis.

26.3 Results

The study of names given to linear toponymic objects of the analyzed urban settlements has distinguished the linguistic coding principles applied for nominating said linear objects.

1. The most evident principle adopted by subjects authorized to regulate the toponymic policy is ***memorative***, i.e., an object is named after a person, event, place, or less commonly, an organization [10, p. 119]; thus, we can differentiate personal, group, or event memorative hodonyms. Let us study the examples of these types in the street networks of Volgograd, Kamyshin, and Petrov Val.

Personal memoratives, named after:

- outstanding national or global social and political personalities (these hodonyms are found in all three settlements):
- Volgograd: *Ulitsa imeni Yuriya Dolgorukova, Prospekt imeni V.I. Lenina, Bulvar imeni Engelsa* (Yuri Dolgorukiy Street, Lenin Avenue, Engels Boulevard);
  Kamyshin: *Ulitsa Lenina* (Lenin Street);
  Petrov Val: *Ulitsa Lenina, Ulitsa Karla Marksa, Ulitsa Fridrikha Engelsa* (Lenin Street, Karl Marx Street, Friedrich Engels Street);
- Soviet social and political personalities:
  Volgograd: *Ulitsa imeni Nadezhdy Krupskoy, Pereuok imeni Krasina, Ploshchad imeni Dzerzhinskogo (*Nadezhda Krupskaya Street, Krasin Alley, Dzerzhinsky Square);
  Kamyshin: *Ulitsa Kalinina, Ulitsa Kirova, Ulitsa Krupskoy* (Kalinin Street, Kirov Street, Krupskaya Street*)*;
  Petrov Val: *Ulitsa Kalinina, Pereulok Dzerzhinskogo* (Kalinin Street, Dzerzhinsky Alley);
- local social and political personalities:
  Volgograd: *Ulitsa imeni Grigoriya Zasekina* (Grigory Zasekin Street, named after the city founder);
  Kamyshin: *Ulitsa Sad Tkachecnko* (Tkachenko Garden, named after the town mayor);
- defenders of Tsaritsyn during the Civil War, including local residents: Volgograd: *Ulitsa imeni Budyonnogy, Ulitsa imeni Kikvidze* streets, *Pereulok imeni Rudneva* (Budyonny Street, Kikvidze Street, Rudnev Alley);
  Kamyshin: *Ulitsa Budyonnogo, Ulitsa Kosolapova* streets (Budyonny Street, Kosolapov Street, with the latter named after the commander of the 1st Kamyshin Division)*;*
- defenders of Stalingrad during WWII, including local residents:
  Volgograd: *Ulitsa imeni Serzhanta Voronina* street (Sargeant Voronin Street, named after the soldier who defended Pavlov's House in Stalingrad);
  Kamyshin: *Ulitsa Geroya Sovetskogo Soyuza A.P. Maresyeva* (A.P. Maresyev, Hero of the Soviet Union Street, named after the local resident and the Honorary citizen);

Petrov Val: *Ulitsa Matrosova* and *Pereulok Matrosova* (Matrosov Street and Matrosov Alley, named after the soldier of the 56th Rifle Division, the Hero of the Soviet Union)*;*

- modern-day heroes and local residents who died in the line of duty or a military conflict.

  Volgograd: *Ulista imeni Lyachina* (Lyachin Street, named after the commander of *Kursk* submarine, who died in 2000 and was posthumously awarded the Hero of the Russian Federation);

  Kamyshin: *Ploshchad imeni Geroya Rossii A.M. Kolgatina* (A.M. Kolgatin, the Hero of Russia Square, named after the senior lieutenant of the Russian Army, who died in 2000 during a military conflict in the Northern Caucasus region);

- cosmonauts:

  Volgograd: *Ulitsa imeni Gagarina, Ulitsa imeni Valentiny Tereshkovoy* (Gagarin Street, Valentina Tereshkova Street);

  Kamyshin: *Ulitsa Gagarina, Ulitsa Tereshkovoy* (Gagarin Street, Tereshkova Street), *Proezd Egorova, Proezd Feoktistova* (Egorov Lane, Feoktistov Lane), including local residents—*Ulitsa Malysheva* (Malyshev Street);

  Petrov Val: *Ulitsa Gagarina, Ulitsa Tereshkovoy* (Gagarin Street, Tereshkova Street);

- writers and literary critics:

  Volgograd: *Ulitsa Bunina, Ulitsa Pushkina* (Bunin Street, Pushkin Street), *Ploshchad Belinskogo* (Belinsky Square), including local residents—*Ulitsa imeni Lukonina, Ulitsa imeni Margarity Agashinoy* (Lukonin Street, Margarita Agashina Street);

  Kamyshin: *Ulitsa Gogolya, Ulitsa Pushkina* (Gogol Street, Pushkin Street);

  Petrov Val: *Ulitsa Gogolya, Ulitsa Pushkina, Ulitsa Turgeneva* (Gogol Street, Pushkin Street, Turgenev Street);

- composers and musicians:

  Volgograd: *Ulitsa imeni Chaykovskogo* (Tchaikovsky Street);

  Kamyshin: *Ulitsa Sviridova* (Sviridov Street);

- painters:

  Volgograd: *Ulitsa imeni Aivazovskogo* (Aivazovsky Street);

- architects and sculptors:

  Volgograd: *Ulitsa imeni Vuchecticha* (Vuchetich Street, named the author of the monument ensemble to the Battle of Stalingrad heroes);

- actors and directors:

  Volgograd: *Naberezhnaya imeni Vladimira Vysotskogo* (Vladimir Vysotsky Embankment), *Ulitsa imeni Ermolovoy* (Ermolova Street), including local theater actors - *Ulitsa imeni Ivana Lapikova* (Ivan Lapikov Street);

- scientists and researchers:

  Volgograd: *Ulitsa imeni Darvina (*Darwin Street), including local academicians—*Ulitsa imeni Professora Inshakova Ulitsa imeni Zagorulko* (Professor Inshakov Street, Zagorulko Street, both named after the rectors of Volgograd State University)*;*

Kamyshin: *Ulitsa Koroleva, Ulitsa Mendeleeva* (Korolev Street, Mendeleev Street);

Petrov Val: *Ulitsa Michurina* (Michurin Street).

Alongside personal memoratives, the hodonymic spaces of Volgograd and Kamyshin have group memoratives that are intended to preserve the memory of military units or other groups that took part in a military conflict: *Ulitsa 64-oy Armii* (64th Army Street) in Volgograd; *Ulitsa Voinov-Internatsionalistov* (Internationalist Warriors Street) in Kamyshin or emphasize the importance of labor in times of peace by nominating a linear object after a group of people united by occupation or professional activity: *Ulitsa Kosmonavtov* (Cosmonaut Street, *Prospekt Metallurgov* (Metallurgists Avenue) in Volgograd, *Ulitsa Neftyanikov* street (Oil Workers Street) in Kamyshin.

Event memoratives are the street network elements named after important dates and events in national history. In Volgograd, this linguistic code is represented in the following hodonyms: *50 let Oktyabrya; Stalingradskoy Pobedy* (50th anniversary of October Revolution; Stalingrad Battle Victory). The hodonymic space of Kamyshina and Petrov Val provides only a few examples of these nominations: *Ulitsa XXII Partsyezda* street (22nd CPSU Congress Street) in Kamyshin, *Ulitsa 30 let Pobedy* and *Ulitsa 40 let Pobedy* (30th anniversary of Victory Street and 40th anniversary of Victory Street) in Petrov Val, demonstrating that this approach is ineffective when it comes to nominating urban topographic objects in small towns.

The memorative linguistic code is quite productive: among a hundred street network elements in Petrov Val, 54 objects, i.e., more than a half of them, are named according to this principle, with Kamyshin and Volgograd having about 28% and 30% nominations of this type and personal memoratives comprising the overwhelming majority of them. The personalities commemorated in the names of linear urban objects vary a lot. The hodonymic spaces of all three analyzed settlements demonstrate personal memoratives related to significant national or global social and political persons, military conflict participants (including defenders of Stalingrad), scientists, cosmonauts, writers. Memoratives related to actors, architects, sculptors, and painters are presented in Volgograd only, as the largest city of the three. Memoratives related to local social and political personalities, defenders of Tsaritsyn during the Civil War, present-day heroes, local residents who died in the line of duty or a military conflict, composers, and musicians can be found in the hodonymic spaces of Volgograd and Kamyshin. The lack of such memoratives in Petrov Val is determined by its relatively short history (the town was founded in 1942) and small size, i.e., lower demand for nominating new elements of the street network.

2. Another highly demanded reason for nominating urban linear objects is the ***locative*** linguistic code, according to which the street network element is named after an already existing geographic object of another type. The analysis of the Register has revealed that the hodonymic spaces of Volgograd, Kamyshin, and Petrov Val have linear toponymic objects with the locative semantics, determined by:

- capitals of the former Soviet republics: Volgograd has streets named after 12 out of 15 capitals: *Ulitsa Vilnyusskaya, Ulitsa Erevanskaya* (Vilnius Street, Yerevan Street), etc., determined by the city status of the regional capital, while Kamyshin has seven streets of this type: *Ulitsa Bakinskaya, Ulitsa Kievskaya, Pereulok Moskovsky* (Baku Street, Kyiv Street, Moscow Lane), and no onyms of this code were found in Petrov Val;
- settlements of the former USSR and modern Russia: *Ulitsa Groznenskaya, Pereulok Pskovsky* (Grozny Street, Pskov Lane) in Volgograd; *Ulitsa Yaroslavskaya* (Yaroslavl Street) in Kamyshin; *Ulitsa Saratovskaya* (Saratov Street) in Petrov Val;
- settlements of the Volgograd Region: *Ulitsa Bykovskaya* (Bykovo Street, named after a local settlement), *Pereulok Kackalisnky* (Kachanlinskaya Lane, named after Kachalinskaya stanitsa) in Volgograd; *Ulitsa Volgogradskaya* (Volgograd Street) in Kamyshin; *Ulitsa Kamyshinskaya* (Kamyshin Street) in Petrov Val;
- foreign countries and their settlements were presented only in the hodonymic space of Volgograd: *Ulitsa Vengerskaya, Ulitsa Kitayskaya, Pereulok Varnensky* (Hunagey Street, China Street, Varna Lane), including twin cities of Volgograd: *Ulitsa Koventry, Ulitsa Port-Saida, Ulitsa Hirosimy* (Coventry Street, Port Said Street, Hiroshima Street);
- water objects: улица *Байкальская*, including those located in the Volgograd Region: *Ulitsa Donskaya, Ulitsa Nevskaya* (Don Street, Neva Street) in Volgograd; *Ulitsa Eltonskaya* (Elton Street) in Kamyshin; while *Ulitsa Lebyazhniskaya* (Lebyazhinskoye Street) is named after the lake located near Petrov Val;
- oronymic names: *Ulitsa Araratskaya* (Ararat Street), *Pereulok Alpiysky* (Alps Lane) in Volgograd; *Ulitsa Uralskaya* (Urals Street) in Kamyshin.

The analysis of the Register demonstrates that the locative coding is widely presented in the hodonymic spaces of settlements chosen for study, with the street network of the regional capital having most names of this type. For smaller cities and towns, "connection" to toponyms of their region or neighboring ones is more typical.

3. Many names of the street network elements are determined by their geographic position, landscape, size, and configuration, thus making it possible to define the ***topographic*** code. Examples of this code include *Ulitsa Bolshaya, Ulitsa Verkhnyaya* (Big Street, Upper Street) in Volgograd; *Ulitsa Zapadnaya, Ulitsa Ovrazhnaya* (Western Street, Gully Street), *Pereulok Gorny, Pereulok Yuzhny* (Mountain Lane, Southern Lane) in Kamyshin*; Ulitsa Rechnaya, Ulitsa Tsentralnaya* (River Street, Central Street) in Petrov Val.

4. The ***anthropogenic*** linguistic code that represents something man-made in the name of a geographic object is typical for settlements of any type. The examples of these nominations in the hodonymic space include *Ulitsa Bibliotechnaya, Ulitsa Vokzalnaya* (Library Street, Railway Station Street) *Pereulok Aeroportovsky, Pereulok Zavod\srkoy* (Airport Lane, Factory Lane), *Prospekt Universitetsky* (University Avenue), *Proezd Aptechny* (Apothecary Lane) in Volgograd*; Ulitsa*

Zavodskaya, Ulitsa Tokarnaya (Factory Street, Lathework Street) in Kamyshin; *Ulitsa Parkovaya, Ulitsa Sportivnaya* (Park Street, Sports Street), *Pereulok Tonnelny, Pereulok Shkolny* (Tunnel Lane, School Lane) in Petrov Val.

5. The ***symbolic*** code helps to preserve in the hodonymic space the names reflecting facts, values, and celebrations cultivated in the Soviet period. This coding is represented by such names of the street network elements as *Ulitsa Kolkhoznaya, Ulitsa Komsomolskaya* (Kolkhoz Street, Komsomol Street) in Volgograd; *Ulitsa Oktyabrskaya, Ulitsa Pionerskaya* (Octobrists Street, Pioneers Street) in Kamyshin; *Ulitsa Komsomolskaya, Ulitsa 1 Maya* (Komsomol Street, May 1st Street) *Prospekt Pionerov* (Pioneers Avenue) in Petrov Val. These nominations are frequently used even in Petrov Val, as this small town was founded in 1942 when such nominations were extremely popular.

For the hodonymic space of a big city, it is typical to have all linguistic codes of urban linear elements, with their memorative names related not only to influential national or regional individuals but also to worldwide famous social personalities, politicians, scientists, writers, painters, composers, actors, etc. The locative nominations of a big city represent not only geographical elements of its region or neighboring areas but also objects located in distant regions or even countries. The approach to establishing the hodonymic space shared by all analyzed settlements involves commemorating the names of local residents and people who largely contributed to the safety and development of this particular settlement.

The hodonymic space of the analyzed settlements demonstrates a high level of inclusion. 31 linear objects located in Petrov Val, i.e., 30% of all onyms of this type in the town have doubling elements in the street network of Volgograd; more similarities can be found among urban linear objects of Kamyshin and Vologorad: 196 nominations in Kamyshin, i.e., 65% of all town onyms have doubling street network elements in Volgograd. Most doubling nominations are memorative (*Dzerhzinskogo, Dostoyevskogo* in Petrov Val and Volgograd, *Budyonnogo, Timiryzeva* in Kamyshin and Volgograd); locative (*Altayskaya* (Altai), *Eltonskaya* in Kamyshin and Volgograd); topographic (*Malaya Zelyonaya* (Little Green) in Petrov Val and Kamyshin, *Rechnaya, Tsentralnaya* in Kamyshin and Volgograd); anthropogenic (*Zavodskaya, Sadovaya* in Kamyshin and Volgograd); and symbolic (*Kommunistechskaya* (Communist) in Petrov Val and Volgograd, *Pionerskaya, Revolutsionnaya* (Revolution) in Kamyshin and Volgograd). Finally, 33 nominations of street network elements are presented in the hodonymic space of all three settlements, with 20 (i.e., 60%) memorative (*Gagarina, Kalinina,* etc.), 6 symbolic (*Komsomolskaya, Sovetskaya,* etc.), 3 locative (*Kamyshinskaya, Saratovskaya*), 2 anthropogenic (*Sportivnaya, Shkolny)*), and 1 topographic (*Lugovaya*—Meadow) nomination.

26.4 Conclusion

Thus, the analysis of the hodonymic spaces of three urban settlements in the Volgograd region, with different statuses and population sizes, has found that the nomination of their street network elements is based on five dominating linguistic codes, with memorative, locative, anthropogenic, anthropogenic, and symbolic codes being the most demanded in all three analyzed settlements. Besides, the names of linear urban objects demonstrate the high rate of inclusion, which can be considered the evidence of onomastic universality.

Acknowledgements The study was supported by the Russian Foundation for Basic Research (project number 20-012-00217 "Linguistic codes of objects of artificial nomination in the toponymic policy of the region: problems, contradictions, and vectors of development").

References

1. Golomidova, M.: Municipal toponymic policy as a tool for translation of regional identity: a case of modern Kazan (Russia). J. Geogr. Polit. Soc. **10**(3), 42–54 (2020). https://doi.org/10.26881/jpgs.2020.3.05
2. Light, D., Young, C.: The politics of toponymic continuity: the limits of change and the ongoing lives of street names. In: Rose-Redwood, R., Alderman, D., Azaryahu, M. (eds.) The Political Life of Urban Streetscapes: Naming, Politics and Place, pp. 185–201. Routledge, New York (2018)
3. Wideman, T.J., Masuda, J.R.: Toponymic assemblages, resistance, and the politics of planning in Vancouver, Canada. Environ. Plan. C: Polit. Space **36**(3), 383–402 (2018). https://doi.org/10.1177/2399654417750624
4. Inshakov, O.V., Frolov, D.P.: Lingvistika Institutsionalnoy ekonomiki (Instutational Economy Lingusitics). Izdatelstvo VolGU, Volgograd (2010)
5. Golomidova, M.V.: Toponimicheskaya politika v sfere nominatsii vnutrigorodskikh obyektov: teoreticheskiye i prikladniye voprosy (Toponymic policy in terms urban nomination: theoretical and practical aspects). Voprosy onomastiki **15**(3), 36–57 (2018). https://doi.org/10.15826/vopr_onom/2018/15/3/028
6. Shmelyova, T.V.: Onomasticheskaya Kultura (Onomastic culture). Voprosy kultury rechi, pp. 101–106. Vinigradov institute of Russian, Moscow (2011)
7. Kostanski, L.: The Oxford Handbook of Names and Naming. Hough, C. (ed.). Oxford University Press (2016). https://doi.org/10.1093/oxfordhb/9780199656431.013.42
8. Quan, S., Kang, M., Weng, M.: Toponym mapping: a case for distribution of ethnic groups and landscape features in Guangdong. China. J. Maps **12**(1), 546–550 (2016). https://doi.org/10.1080/17445647.2016.1201017
9. Rose-Redwood, R.S., Alderman, D.H., Azaryahu, M.: Geographies of toponymic inscription: new directions in critical place-name studies. Prog. Hum. Geogr. **34**(4), 453–470 (2010). https://doi.org/10.1177/0309132509351042
10. Podolskaya, N.V.: Slovar russkoy onomaticheskoy terminologii (Vocabulary of Russian onomastic terms). Nauka, Moscow (1988)

Chapter 27
Practice of Infrastructural Transformations of Educational Environment as a Key Factor of Inclusive Growth of the Russian Economy

Elena A. Eltanskaya, Larisa M. Generalova, and Anastasia V. Arzhanovskaya

Abstract In the context of active international cooperation and intercultural dialogue affecting the Russian economy, there is a need for infrastructural changes in the educational environment, the incorporation of traditional and innovative teaching methods. The modern world has to face both the opportunities that the technologies provide and the challenges that they create in the local or global market as the conditions of the electronic environment are different from the traditional ones, therefore, it is important to not only understand the changing needs, the factors that influence digital interaction, but also to choose suitable strategies to satisfy these needs. The principle of inclusive growth implies overcoming structural restrictions and rigidity and introducing new technologies. Nowadays, the system of inclusive education is at its formation stage. The relevance of the research is that the practice of infrastructural transformations in the education sector is tasked with the inclusive growth of the Russian economy. The study substantiates the importance of such a structural modernization as one of the key factors for providing a perfect balance between society, state, and business. The contributors suggest considering crucial factors influencing the improvement of economic conditions through the integrating inclusive sector into the educational environment of Russia.

E. A. Eltanskaya · L. M. Generalova · A. V. Arzhanovskaya (✉)
Volgograd State University, Volgograd, Russia
e-mail: arjanovskaya@volsu.ru

E. A. Eltanskaya
e-mail: yeltanskaya@volsu.ru

L. M. Generalova
e-mail: l.m.generalowa@volsu.ru

E. I. Inshakova and A. O. Inshakova (eds.), *New Technology for Inclusive and Sustainable Growth*, Smart Innovation, Systems and Technologies 287,
https://doi.org/10.1007/978-981-16-9804-0_27

27.1 Introduction

A new buzzword has recently emerged in the economic policy debate: inclusive growth. This concept means creating favorable conditions for improving life quality and ensuring equal opportunities for all groups of the population. In other words, inclusive growth implies that the income from the economic rise should not belong to a limited group of people, but the largest number of people. This can be reflected by salaries, population employment, and extend to participation in public life in a broader sense. It also includes such aspects as family income, employment opportunities, security, and living standards. These aspects are indicators by which people judge the economic policy and political leadership of their country.

Strong, sustainable, balanced, inclusive growth expresses the desire to benefit from economic growth so that as many people as possible can benefit and not be left out. From an institutional economic perspective [1], the degree of growth inclusiveness crucially depends on whether all members of society or groups of the population are subjectively perceived as "suitable" for an inclusive economy and, consequently, for income generation.

Consequently, the consideration of inclusive economic growth is equally justified for both developing and developed countries. This phenomenon is explained by the following factors: in recent decades, many industrialized countries have severely neglected their political and institutional ecosystems, which has led to inequality in many societies due to technological changes, increased globalization, local deregulation, and increased immigration.

In turn, many developing countries have not been able to lay the foundation for successful economic development: industrialization and global integration have advanced, but the Government of these countries has missed the opportunity to involve the general population in the economic development process.

27.2 Methodology

The study was based on the analysis, comparison, and generalization of existing approaches in foreign and domestic theory and practice on the impact of the transforming educational environment on inclusive economic growth. The research is characterized by a multidisciplinary approach, as it integrates sociological, pedagogical, historical, and general scientific approaches and methods.

The convergent approach, which is actively used in politics, medicine, law, and sociology, serves as the basis for considering the transformation of the educational infrastructure in the context of widespread digitalization [2].

The competence-based approach [3, 4] is associated with the goals of education: the training of an all-around intellectually developed person who possesses basic and professional competencies and is able to effectively apply them within professional activities. Successful implementation of the competence-based approach will allow the specialist to reach a higher professional level in the current economic situation.

The system-activity approach [5, 6] allows considering the digital educational environment as an environment for constructing individual educational routes, an environment that develops and motivates students for active educational and cognitive activities. The methodology developed in the course of the research can be applied to the study of the impact of the educational environment transformation on sustainable economic development.

27.3 Results

More and more economic experts today agree that there is nothing wrong with the fundamental concept of inclusive growth. Such a "rethinking" of structural economic reform and its rebalancing concerning macroeconomic, economic, and trade policies is necessary for governments to respond more effectively to weak economic development and the resulting growing inequality [7].

Since the goal of any economy is to ensure the prosperity of the largest possible part of the population, the term "inclusive growth" clearly shows how economic growth is the main prerequisite for solving issues of social-economic policy, empowering people, strengthening the economy, and active participation in labour activity. "The priorities of the pro-inclusive economy should be reoriented to more effective countering insecurity and inequality that accompany external challenges and technological changes" [8].

Since "the inclusive model of any national economic development is based mainly on its internal factors", it can be assumed that in many respects this problem is connected "with the specifics of the industry of a particular country, and more specifically with the imbalance of its technological structure" [9]. In this regard, it is necessary to either replace technologically backward segments (branches, manufacturing) of industrial production with technologically advanced ones or compensate—in terms of reducing the level of scientific and technological equipment of production facilities [10]. Inclusive growth, according to scientists, should promote the development of the most vulnerable population groups in the present, as well as take into account the needs of future generations [11]. Practice shows that when it comes to the requirement of inclusion, it is often overlooked that autonomy and self-determination are linked to the personal responsibility of everyone in a competitive society.

Therefore, the measures recommended by many analytical experts are reasonable and serve this purpose. In particular, these are such measures as an investment in education and infrastructure, opening and promoting businesses, and a firm state budget policy.

Nowadays, economic thinking and activity affect every aspect of our lives. Thus since the beginning of the financial crisis, a neoclassical economic paradigm dominated, which was repeatedly criticized as a paradigm based on the practice of homo oeconomicus, which constantly produced winners and losers. Then today, first of all, it is necessary to understand how, from an economic perspective, education should be designed in order to achieve prosperity and success for everyone who participates in the economy.

Therefore, it is education that is the crucial key to the prospects in the labor market and the possibility of income generation in the market.

The pedagogical quest for inclusion is compensated by the tendencies of socio-economic isolation, which are reproduced through the education system and become visible in specialized educational institutions, as well as in various educational courses. Nowadays, the search for adequate concepts of economic education is of great importance for modern educational institutions.

An economic historical review and a critical analysis of institutional business ethics, as well as the on-going debate in the field of business education, show that inclusion is impossible without morally responsible actors. In this regard, today many researchers postulate the requirements for inclusive economic education, and already existing concepts of economic education are examined for their suitability within this context [12].

In pedagogical discourse, inclusion is understood as teaching children and young people with special educational needs at schools and universities to make them aware of their professional training in the primary labor market. However, the primary task of any professional education is to train a highly qualified specialist able to be employed in our economic system. It is the educational institutions that can encourage students to be responsible for the formation of a socially and economically developed society. The education system, in conjunction with the economic one, makes a significant contribution to the process of inclusion, providing opportunities and guaranteeing access to active participation in economic life. Therefore, the education system cannot be considered in isolation from the economy and society for which it trains professional personnel.

Vocational education in a broad socio-economic sense should mean full compliance with the United Nations Convention on the Rights of Persons with Disabilities—the full and effective inclusion and participation of all members of human society, as well as their equal opportunities in the economy of a particular state. In the economic sphere, this means the right to a dignified life and the opportunity to earn a living, protection from exploitation, and access to an educational environment. Achievement of economic inclusion will largely depend on access to the labor market and income prospects. It is not a coincidence that the paradox of inclusion in society is now particularly acute in vocational schools.

Until recently, in the Russian education system, there was some heterogeneity of the educational process associated with the division of groups and classes into general and special, for people with special needs. All efforts to overcome this division were called integration and had a positive connotation in the past.

Inclusion, on the other hand, means a new approach in which diversity of learning is the norm. Its qualitative feature is the adaptation of the system to the specific needs of interested persons, which until recently could not be taken for granted in a competitive economic system.

When it comes to the requirement of inclusion, it is often overlooked that autonomy and self-determination are linked to personal responsibility in a competitive society, which will allow students to find decent jobs and be able to earn a living.

The use of computer-based learning and studying has continued for many years and has been associated, typically, with the use of a portable computer, laptop, or video projector. Today, with the ubiquitous Internet and smartphones in the pockets of the students, computer-based learning is outdated. Digitization radically changes the modern education process. Digitalization is considered by both scientists and politicians as a key factor for enhancing domestic competitiveness. Research-oriented teaching and learning began with the Bologna Reform. And today, people are thinking more and more about how this can be implemented and supported with the help of digital media. Teaching staff creates a space for research orientation in training using digital media. Digital media can be used both to design individual processes or stages more efficiently (for research, data evaluation, or dissemination of results) and to incorporate processes that unlikely would be implemented without digital media (such as analyzing large amounts of data and compiling a database).

27.4 Conclusion

The convergence of digital technologies for the arts can go in several directions. In any case, a unique object with signs of copyright protection is created. Its circulation can be most effectively established by tokenization. The two functions of the blockchain network are most evident in the tokenization of intellectual property: the ability to authenticate the origin of an art object or intellectual right and verify it in transactions.

That confirms our belief that the development of blockchain technology and smart contracts enables artists and other creators to protect their works from misuse and expropriation. A robust system for the management of digital art objects is being established. In doing so, the goal of limited art token releases is usually not to impose stricter restrictions on use or a new form of digital rights management but to create new types of tradable digital assets.

Digitalization is a core force for economic growth and social development. In order to use the digitization potential for innovation, economic growth, and full employment, it is necessary to create the appropriate conditions and remove possible obstacles, expand infrastructure, improve the prospects for digital education and employment, develop and apply norms and standards, and be able to gain consumer confidence. In a sense, digital media can be considered as tools for managing certain situations, getting better in a research-oriented teaching–learning format, improving research methods, or generalizing heterogeneous data. Interactive tasks and assigned objectives facilitate research and communication processes.

Digital transformations provide new employment opportunities and new prospects for all segments of society, especially in developing countries. An inclusive economy will ensure expanded opportunity for more broadly shared prosperity. It will allow us to achieve a perfect balance between society, the state, and business.

References

1. Inshakov, O.: Social nature and specifics, typology and classification of economic institutions. Sotsiologicheskie Issledovaniya [Sociol. Res.] **9**, 42–51 (2003)
2. Arzhanovskaya, A.V., Eltanskaya, E.A., Generalova, L.M.: Convergence of technologies in education: new determinant of the society development. In: Popkova, E.G., Sergi, B.S. (eds.) "Smart Technologies" for Society, State and Economy. Lecture Notes in Networks and Systems, vol 155. Springer, Cham (2021). https://doi.org/10.1007/978-3-030-59126-7_69
3. Arzhanovskaya, A.V., Eltanskaya, E.A., Milovanova, M.V.: Use of convergent technologies in education as an essential prerequisite for neo-industrialization of Russia. In: Inshakova, A., Inshakova, E. (eds.) Competitive Russia: Foresight Model of Economic and Legal Development in the Digital Age. Lecture Notes in Networks and Systems, vol 110, pp. 395–399. Springer, Cham (2020). https://doi.org/10.1007/978-3-030-45913-0_45
4. Hutmacher, W.: Key competencies for Europe: Report of the Symposium Berne, Switzerland 27–30 March, 1996. Council for Cultural Co-operation (CDCC), Strasburg (1997)
5. Fayzullina, A.R., Saglam, F.A.: History and social sciences teacher's professional activity in the context of IT-development of education. J. Sustain. Dev. **8**(7) (2015). https://doi.org/10.5539/jsd.v8n7p107
6. Toisteva, O.S.: System-activity approach: its essential characteristics and ways of implementation. Pedagog. Educ. Russia **2**, 198–202 (2013)
7. Matytsin, D.E.: Neoindustrial tools for the circulation of uncertified securities: digital technologies for the implementation and protection of the rights of investors and issuers. Legal Concep. 19, 73–83 (2020). https://doi.org/10.15688/lc.jvolsu.2020.3.10
8. Novikov, A.I., Vitkina, M.K.: Inclusive economy and social responsibility in the regions of the world: dilemma or public consent. Reg. Econ. Manag. Electron. Sci. J. **2**(54) (2018). https://eee-region.ru/article/5401/ Last accessed 18 Sept 2021
9. Cherkasov, V.V., Zhdanov, D.A.: Inclusive development of Russia: industrial growth factors. Econ. Sci. **7**(164), 35–40 (2018)
10. Yaremenko, Y.V.: Theory and methodology of studying the multilevel economy. Nauka, Moscow (1997)
11. Meadows, D.H., Meadows, D.L., Randers, J., et al. The limits to growth. M.: MSU Publishing House (1991)
12. Petrova, E.A., Inshakova, A.O., Kalinina, V.V.: Information technology impact analysis on the structural changes dynamics in the regional economy. In: Popkova, E.G., Sergi, B.S. (eds.) "Smart Technologies" for Society, State and Economy. ISC 2020. Lecture Notes in Networks and Systems, vol 155. Springer, Cham (2021). https://doi.org/10.1007/978-3-030-59126-7_114

The manufacturer's authorised representative in the EU is Springer Nature Customer Service Centre GmbH, Europaplatz 3, 69115 Heidelberg, Germany. If you have any concerns regarding our products, please contact ProductSafety@springernature.com

Printed and bound by CPI Group (UK) Ltd, Croydon, CR0 4YY
30/09/2026
02225241-0006